Frana Do

March ...

Samir El-Gaby
Reinhard O. Greiling
(Eds.)

**The Pan-African Belt of
Northeast Africa and
Adjacent Areas**

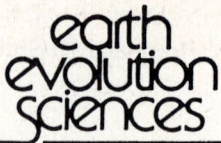

International Monograph Series
on Interdisciplinary
Earth Science and Applications

Editor
Andreas Vogel, Berlin

Samir El-Gaby
Reinhard O. Greiling
(Eds.)

The Pan-African Belt of Northeast Africa and Adjacent Areas

Tectonic Evolution and Economic Aspects of a Late Proterozoic Orogen

Friedr. Vieweg & Sohn   Braunschweig / Wiesbaden

A contribution to

project 215 Proterozoic Fold Belts

The Editors:
Prof. Dr. *Samir El-Gaby*, Geology Department, Assiut University, Assiut, Egypt.
Prof. Dr. *Reinhard O. Greiling*, Department of Geology, University College, P.O. Box 78, Cardiff
CF1 1XL, U.K.; present address: Geologisch-Palaeontologisches Institut, Universität Heidelberg,
Im Neuenheimer Feld 234, 6900 Heidelberg, FRG.

Vieweg is a subsidiary company of the Bertelsmann Publishing Group.

Produced by W. Langelüddecke, Braunschweig
Printed in Germany

ISBN   3-528-06325-4

Contents

Metallogenesis

Geophysics

Preface

It is now generally accepted that the Arabian-Nubian Shield, which forms large areas of the Middle East (mainly Saudi-Arabia, Egypt, Sudan, Yemen, Somalia and Ethiopia) is of Late Proterozoic, Pan-African age. The term Pan-African, first introduced by Kennedy in 1964, defines an "important and widespread tectonic and thermal event" which affected the African Continent during the Late Precambrian and Early Palaeozoic and led to its "structural differentiation into cratons and orogenic areas" some 500 ± 100 Ma ago. Subsequent studies on several Pan-African regions in the continent and the Arabian Shield have shown that the Pan-African is not restricted to the ~ 500 Ma episode, originally proposed by Kennedy, but it covered a longer lapse of time (Clifford, 1967; Cahen and Snelling, 1966; Gass, 1977; Al-Shanti, 1979; Kröner, 1979 and 1984). In the latter publication, Kröner suggested to assign the time span 950 - 450 Ma to the Pan-African.

The study of the crystalline rocks forming the Arabian-Nubian Shield started in Egypt at the beginning of this century and after the establishment of the Egyptian Geological Survey in 1896, by a number of eminent geologists who were fascinated by the excellent exposures and complexity of these rocks. Geological mapping was active by the Survey and the data accumulating were used in the compilation of the first geological map of the country published in 1928 at the scales of 1 : 1,000,000 and 1 : 2,000,000. Petrographic studies were carried out and attempts towards stratigraphic classifications (Hume, 1934) were made following the geosynclinal concept (El Shazly, 1964; El Ramly, 1972; Akaad and Noweir, 1980). In the last two decades special emphasis was put on the study of the geochemistry, genesis and tectonic setting of these rocks.

The recognition of ophiolitic assemblages among the basement rocks of the Eastern Desert by Garson and Shalaby (1976) and their presentation of a plate tectonic model for the evolution of these rocks involving arc accretion against the African craton with the elimination of oceanic crust through subduction, was accepted by a number of subsequent workers who supported the idea on the basis of geochemical and isotopic evidence (Engel et al., 1980; Dixon, 1981; Stern, 1981; Stern and Hedge, 1985; Shackleton et al., 1980; and Ries et al., 1983).

The serious study of the geology of the Arabian segment of the Shield began in 1950 by reconnaissance mapping and photo-interpretation and the resulting maps were published in 1963 at the scales of 1 : 500,000 and 1 : 2,000,000 (U.S.G.S. map I-270A, 1963). On these maps, the crystalline rocks are shown as petrologic units described mostly in field terms.

During the past two decades, more detailed mapping and laboratory studies by diverse individuals and groups helped to elucidate the structural complexities of these rocks, their chronologic episodes and sequences, sedimentary and volcanogenic facies changes, metallogenic epochs, geochemical characteristics and radiometric ages. As a result, the evolutionary model envolving crustal growth through accretion - obduction processes of island arcs or microcontinents and crustal thickening was proposed as a possible concept for the tectonic evolution of these rocks. It was in the Arabian segment of the Shield that this model was first applied (Brown, 1972; Greenwood et al., 1976; Fleck et al., 1978; and Gass, 1981).

In the Red Sea Hills of the Sudan, regional studies on the basement rocks started in the fifties and early sixties (Gass, 1955; Ruxton, 1956; Gabert et al., 1960 and Kabesh and Lotfi, 1967) and the similarity of the rock types encountered and their tectonic setting with those in Saudi-Arabia and the Eastern Desert of Egypt, were recognized by Neary et al. (1976) who also supported the evolutionary model developed in Arabia for these rocks. But the state of knowledge on the Sudan and Ethiopia segments of the Nubian Shield is still far behind that of the better known terrains in Saudi-Arabia and Egypt.

This brief review of the history of geological studies on the Pan-African Arabian-Nubian Shield is intended to show the voluminous wealth of knowledge, which accumulated during the last fifteen years, on the different aspects of the geology of this Shield. But these research works are dispersed among a large number of publications and periodicals, a fact which urged the two editors of this book, Samir El Gabi and Reinhard Greiling, to bring together in one publication those who have been contributing for a long time to the geology of the "Pan-African belt of northeast Africa and adjacent areas", to express their latest views on the subject.

Now that this book has been published, I am sure that it will stimulate further research, in the same manner as the volumes dealing with the "Evolution and Mineralization of the Arabian-Nubian Shield" did in the late seventies, which recorded the proceedings of a symposium held in Jeddah, Saudi Arabia, in 1978 (published by the Institute of Applied Geology).

M.F. El-Ramly, Cairo, Egypt

(For references see 'INTRODUCTION')

Chapter 1

Introduction

S. El-Gaby / R. O. Greiling

THE PAN-AFRICAN EVENT: HISTORY AND EVOLUTION

The Precambrian of Africa consists of three major and several minor Archaean cratons surrounded and separated by Upper Precambrian terrains (Clifford 1970; Fig. 1). Kennedy (1964) noted the dominance of 650 - 450 Ma K-Ar ages in the Upper Precambrian terrains but did not recognize indications of classic orogenesis in the non-Archaean terrains, except in the Damara-Katanga belt, and hence used the term "Pan-African thermo-tectonic event" to identify this major episode in the African geological evolution. He believed the Pan-African episode had led to the structural differentiation of virtually the entire continent into cratons and reactivated circum-cratonic areas. He emphasized that the evidence did not support the concept of continental accretion around cratonic nuclei, but rather the disruption of an existing craton and its partial regeneration into mobile zones. Since then, the term 'Pan-African' has been used to describe much of the non-Archaean African basement and the time span has been expanded by some authors (e.g. Gass 1981) to 1200 - 450 Ma. Others (e.g. Kröner 1984) proposed a more limited time span of 950 - 450 Ma.

It is apparent from the detailed geological reviews on the Pan-African belt system (Clifford 1970, Kröner 1979 a, b) that it is not uniformly built. Three categories can broadly be defined (Fig. 1):

1. ensimatic terrains of volcanic arcs and associated ophiolitic mélanges, characteristic of the Arabian-Nubian Shield,

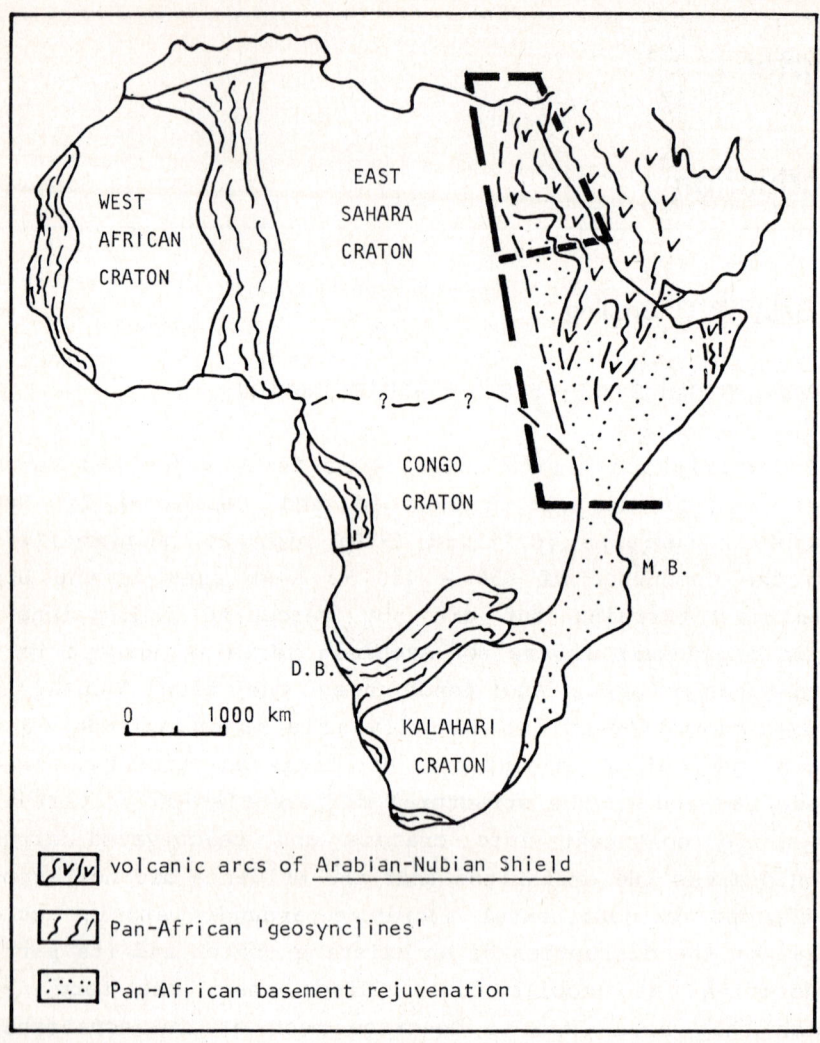

FIGURE 1.
    Pan-African domains in Africa enclosing older cratons,
    compiled by Pohl (1984). Arabia is re-rotated into a pre-
    Red Sea position. The eastern and southern limits of the
    East Sahara craton are not yet well established. The
    quadrilateral enclosed by the thick line covers Egypt and
    northern Sudan, parts of which are treated in chapters 2,
    3, 4, 5, 6 and 12.
    Chapters 7, 8, 9, 10 and 11 cover parts of a wider area to
    the east of the thick, broken line, including the
    quadrilateral.
    D.B. - Damara belt, M.B. - Mozambique belt.

2. orogenically deformed Upper Proterozoic sediments and igneous rocks deposited in geosynclines or aulacogenes (e.g. Damara belt),

3. belts of deformed and rejuvenated older basement rocks (e.g. Mozambique belt).

In the case of ensimatic orogenic belts, for example in the Arabian-Nubian Shield, higher Pan-African ages up to about 1000 Ma may be expected from ophiolites predating island arc magmatic rocks. Island arc rocks are represented by low-potash volcanic and plutonic rocks comprising tholeiite-andesite-dacite and their plutonic counterparts (Schmidt and Brown 1983, El-Gaby et al. 1987). In the following Cordilleran or culminant-orogeny stage, calc-alkaline igneous rocks with moderate potash content predominate, namely rhyodacite-rhyolite and granodiorite-granite (Schmidt and Brown 1983, El-Gaby et al. 1987). The culminant orogeny took place about 680 - 580 Ma ago in Arabia (Schmidt and Brown 1983, Stoeser and Stacey 1987) and 655 - 570 Ma ago in Egypt (El-Gaby et al. 1987). Subduction related Pan-African granites intrude both the Pan-African belt and the stable foreland and induce, together with post-tectonic granites, the observed thermal effects on adjoining stable zones. Ages for this late-tectonic episode range mainly from 550 - 500 Ma ago (e.g. Cahen et al. 1984, Stern 1985).

In West Africa, Black (1966) showed that Late Precambrian orogeny took place at Hoggar, Nigeria and Dahomey, followed in the Early Palaeozoic by several post-tectonic granitic intrusions. Recently, Black et al. (1979) and Caby et al. (1981) produced convincing evidence of a Wilson cycle in the Pharusian and Dahomian belts ending with the collision of the West African craton and a (Sahara) craton in the east.

The Pan-African belt in East Africa is constituted of two contrasting units: the Arabian-Nubian Shield in the north and the Mozambique belt in the south. The Mozambique belt (Holmes

1951) comprises a N-S trending structural belt extending from south of the Zambesi to the extreme north of Kenya and Uganda. It is composed of gneisses and migmatites with infolded schists, marbles and amphibolites. Kennedy (1964) refrained from using the term orogeny to describe the Pan-African events in the Mozambique belt itself as it 'involved only basement rocks, which recorded the event mainly by granitization or as a thermal effect resulting in low absolute ages impressed on the crystalline rocks'. The ideas of Cahen and Snelling (1966), Clifford (1970) and Shackleton (1977) that the Mozambique belt represents a Late Precambrian orogenic belt is highly disputed. There appears to be a general consensus that the Mozambique belt is a rejuvenated pre-Pan-African cratonic basement. However, its relation to the Late Proterozoic rocks of the Arabian-Nubian Shield is still unsettled.

The Arabian-Nubian Shield is characterized by extensive thick volcanic and volcaniclastic successions enclosing several opiolite complexes (Bakor et al., 1976, Shackleton et al., 1980). Metamorphism is of greenschist grade, leading to the term 'Greenschist assemblage' of Vail (1983). The Precambrian of Arabia, largely formed of Greenschist assemblage, is intruded by calc-alkaline granitoids and believed to represent cratonized, late Proterozoic ensimatic island arcs without much contribution of older continental crust (Greenwood et al., 1976; Gass, 1977). The non-existence of older continental crust in Arabia is challenged by Stoeser and Stacey (1987). The Greenschist assemblage extends from the Eastern Desert of Egypt to Lake Turkana (Rudolf) in N Kenya. The boundary between the Greenschist assemblage and the basement rocks of the Mozambique belt is irregular. Kazmin et al. (1978, 1979) and Kröner (1979) assumed the Upper Complex (= Greenschist assemblage) of Ethiopia to have originated in rift zones narrowing southwards within the ensialic Mozambique belt. Conversely, Hepworth (1979) interpreted the interdigitated boundary between the gneissose basement and the Greenschist assemblage as defining an intensely folded surface with the basement plunging north-eastwards on a regional scale beneath a greenschist cover. The contact between the Greenschist assemblage in the Eastern Desert of Egypt and in the Sudan and the underlying remobilized older continental crust, i.e. the northern continuation of the Mozambique belt, is always occupied by thrust faults that have been folded subsequently (El-Gaby et al. 1987, Vail 1987). It seems most probable that the northern continuation of the Mozambique belt in Egypt, Sudan and western Ethiopia represents an active continental margin towards the Late Proterozoic orogenic belt of the Arabian-Nubian Shield. Likewise, the Mozambique belt from Ethiopia to Mozambique might represent the active continental margin(s) of colliding East and West Gondwana cratons whereby the oceanic island-arc material exposed in the Arabian-Nubian Shield was destroyed by subduction farther south rather than by erosion from above (Kröner, 1979; McWilliams, 1981; Vail, 1983; Warden and Daniels, 1983).

THE PRESENT VOLUME

The Pan-African belt of Northeast Africa, containing the oldest ophiolites known as yet, displays spectacular examples of 'modern-type' plate tectonic features in the Proterozoic, for example sea-floor spreading, island-arc formation, arc-arc and arc-continent collision, possible exotic terrains and large-scale low angle thrusting. This has been documented during the last decade and we feel it is time now to put together different pieces of evidence to give both a review of the 'state of the art' and provide a case example for the study of a (Late) Proterozoic fold belt. In particular the tectonic evolution and related metallogenesis will be of interest as a reference to other orogens, Phanerozoic or Precambrian.

The present volume is primarily concentrated on the Pan-African belt of Northeast Africa and Arabia (Fig. 1), since important and extensive geological knowledge has accumulated recently on this area. In particular the IGCP project 164 'Pan-African crustal evolution in the Arabian-Nubian Shield', led by A.M.

Al-Shanti (Jeddah, Saudi Arabia) has to be acknowledged.
However, information is widely scattered and we therefore tried
to bring together all those interested in the geology of Northeast Africa/Arabia to compile the known facts and find out
where a sound knowledge of the geological evolution exists and
where further research is needed. This question is not only of
general scientific interest for the improvement of our
knowledge on the evolution of fold belts but is also vital for
the exploration of mineral resources.

We divided the contributions into four groups. The first group
of papers deals with the Pan-African evolution in Egypt and
northern Sudan from the margin of an earlier craton in the west
to the mainly ensimatic parts of the orogen east of the Nile.
Ophiolites and the evolution of the Arabian-Nubian Shield are
treated in the second group of papers, covering a wider region
(Fig. 1) and the more general aspects of Late Proterozoic
orogeny. Subsequent chapters review the metallogeny of (Pan-
African) Northeast Africa and geophysical aspects of crustal
evolution in Egypt.

Unfortunately, the time limits set for the submission of
manuscripts turned out too narrow for some potential
contributors, including most of our Egyptian colleagues, from
whom we expected a wealth of geological and geophysical data
accumulated at the universities and, in particular, at the
Egyptian Geological Survey. We hope, these data will be
included in a future volume on the progress of geological
knowledge in Northeast Africa.

The present volume is a contribution to IGCP project 215
'Proterozoic fold belts', led by R. Caby (Montpellier, France).

Editing was supported by the following colleagues, who reviewed
one or more of the contributions: R.W.H. Butler (Durham,
England), W.R. Church (London, Ontario), M. Eyal (Beer Sheva,
Israel), W. Frisch (Tübingen, W.Germany), J.W. Gabelman
(Danville, California), N.B.W. Harris (Milton Keynes, England),

W. Jacoby (Mainz, W.Germany), B. Kober (Heidelberg, W.Germany),
A. Kröner (Mainz, W.Germany), P.N. Mosley (Nottingham,
England), W. Pohl (Braunschweig, W.Germany), T. Reischmann
(Mainz, W.Germany), H. Schandelmeier (Berlin, W.Germany),
R.J. Stern (Dallas, Texas), W. Todt (Mainz, W.Germany),
J.R. Vail (Portsmouth, England) and A. Vogel (Berlin,
W.Germany). Their cooperation is gratefully acknowledged. We
also thank the series editor, A. Vogel (Berlin) and his
assistants K. Hoffmann and G. Wagenhäuser for their technical
support. Last but not least, A.A. Weis, representative of
Vieweg Publishers, is thanked for his cooperation.

Transcription from Arabic to English did apparently not improve
very much since the somewhat whimsical approach by
T.E. Lawrence in the 1920s (Lawrence 1935) and we do not know
of any generally accepted rules. Hence, spelling of place names
may vary slightly from contribution to contribution.

REFERENCES (including those for the Preface)

Akaad, M.K. and A M. Noweir, 1980. Geology and
    lithostratigraphy of the Arabian Desert orogenic belt of
    Egypt between Lat. 25°35′ and 26°30′. - Inst. Appl. Geol.,
    Jeddah, Bull. 4, v. 3, 127-134.
Al-Shanti, A.M., 1979. The aims, objectives and scope of IGCP
    project no. 164 ′Pan-African crustal evolution in the
    Arabian-Nubian Shield′. - Newsletter ′Pan-African crustal
    evolution in the Arabian-Nubian Shield′ 2, 9-13.
Bakor, A.R., I.R. Gass and C.R. Neary, 1976. Jabal El-Wask: An
    Eocambrian, back-arc ophiolite in NW Saudi Arabia. - Earth
    Planet. Sci. Lett., 30, 1-9.
Black, R., 1966. Sur l′existence d′une orogenie riphéenne en
    Afrique de l′Ouest. - C.R.Ac.Sc., 262, ser.D, 1046-1049.
Black, R., R. Caby and others, 1979. Evidence for Late
    Precambrian plate tectonics in West Africa. - Nature, 278,
    223-227.
Brown, G.F., 1972. Explanatory text, tectonic map of the
    Arabian Peninsula. - DGMR, Jeddah.
Caby, R., J.M.L. Bertrand and R. Black, 1981. Pan-African ocean
    closure and continental collision in the Hoggar-Iforas
    segment, Central Sahara.- In: A. Kröner (ed.), Precambrian
    plate tectonics, Elsevier, Amsterdam, 407-434.
Cahen, L. and N.J. Snelling, 1966. The geochronology of
    Equatorial Africa. - North-Holland, Amsterdam, 195 pp.
Clifford, T.N., 1967. The Damaran episode in the Upper
    Proterozoic-Lower Palaeozoic structural history of
    southern Africa. - Geol. Soc. Amer., Spec. Paper 92.

Clifford, T.N., 1970. The structural framework of Africa. In:
    T.N. Clifford and I.G. Gass (eds.), African magmatism and
    tectonics. Oliver and Boyd, Edinburgh, 1-19.
Dixon, T.H., 1981. Gabal Dahanib, Egypt: A Late Precambrian
    layered sill of komatiitic composition. - Contrib.
    Mineral. Petrol., 76, 42-52.
El-Ramly, M.F., 1972. A new geological map for the basement
    rocks in the Eastern and Southwestern Deserts of Egypt. -
    Annals Geol. Surv. Egypt, 2, 1-18.
El-Shazly, E.M., 1964. On the classification of the Precambrian
    and other rocks of magmatic affiliation in Egypt, U.A.R. -
    24th Int. Geol. Congr. India, 10, 88-101.
Engel, A.E.J., T.H. Dixon and R.J. Stern, 1980. Late
    Precambrian evolution of Afro-Arabian crust from ocean arc
    to craton. Geol. Soc. Am. Bull., 91, 699-706.
Fleck, R.J., W.R. Greenwood, D.G. Hadley, R.E. Anderson and
    D.L. Schmidt, 1978. Age and evolution of the southern part
    of the Arabian Shield. - Precambrian Res. 6, A21-22.
Gabert, G., B.P. Ruxton and H. Venzlaff, 1960. Über
    Untersuchungen im Kristallin der nördlichen Red Sea Hills
    im Sudan. - Geol. Jb. 77, 241-270.
Garson, M.S. and I.M. Shalaby, 1976. Precambrian-Lower
    Paleozoic plate tectonics and metallogenesis in the Red
    Sea region. - Geol. Assoc. Canada, Sp. paper No. 14,
    573-596.
Gass, I.G., 1955. Geology of Dungunab, Sudan. - Thesis, Univ.
    of Leeds.
Gass, I.G., 1977. The evolution of the Pan-African crystalline
    basement in NE Africa and Arabia. J. Geol. Soc. London,
    134, 129-138.
Gass, I.G., 1981. Pan-African (Upper Proterozoic) plate
    tectonics of the Arabian-Nubian shield. In: A. Kröner
    (ed.), Precambrian plate tectonics, Elsevier, Amsterdam,
    388-405.
Greenwood, W.R., D.G. Hadley, R.W. Anderson, R.J. Fleck and
    D.L. Schmidt, 1976. Late Proterozoic cratonization in
    south-western Saudi Arabia. - R. Soc., Phil. Trans. Ser.
    A, 280, 517-527.
Hepworth, J.V., 1979 Does the Mozambique belt continue into
    Saudi Arabia? - Inst. Appl. Geol., Jeddah, Bull. 3,
    Vol. 1, 39-51.
Holmes, A., 1951. The sequence of Precambrian orogenic belts in
    South and Central Africa. - 18th Int. Geol. Congr. London,
    14, 254-269.
Hume, W.F., 1934. Geology of Egypt, Vol. 2, The fundamental
    Precambrian rocks of Egypt and the Sudan; Part 1: The
    metamorphic rocks. - Geol. Surv. Egypt, 300 pp.
Kabesh, M.L. and M.A. Lotfi, 1967. On the basement complex of
    the Red Sea Hills, Sudan. - Bull. Inst. Desert Egypt 12,
    1.
Kazmin, V., A. Schiferaw and T. Balcha, 1978. The Ethiopian
    basement: stratigraphy and possible manner of evolution. -
    Geol. Rdsch., 67, 531-546.
Kazmin, V., M. Teferra, S.M. Behre and S. Chowaka, 1979.
    Precambrian structure of Western Ethiopia. - Annals. Geol.
    Surv. Egypt, 9, 1-8.

Kennedy, W.Q., 1964. The structural differentiation of Africa
    in the Pan-African tectonic episode. - Ann. Rept. Res.
    Inst. African Geol., Univ. Leeds, 8, 48-49.
Kröner, A., 1979. Pan-African plate tectonics and its
    repercussions on the crust of northeast Africa. - Geol.
    Rdsch., 68, 565-583.
Kröner, A., 1984. Late Precambrian plate tectonics and orogeny:
    a need to redefine the term 'Pan-African'. - In:
    J. Klerkx and J. Michot (eds.), African Geology, Musée r.
    Afr. centr., Tervuren, 23-28.
Lawrence, T.E., 1935. Seven pillars of wisdom - a triumph. -
    Cape, London.
McWilliams, M.O., 1981. Palaeomagnetism and Precambrian
    tectonic evolution of Gondwana. - In: A. Kröner (ed.),
    Precambrian plate tectonics, Elsevier, Amsterdam, 649-687.
Neary, C.R., I.G. Gass and B.J. Cavanagh, 1976. Granitic
    association of northeastern Sudan. - Geol. Soc. Amer.
    Bull. 87, 1501-1512.
Pohl, W., 1984. Large scale metallogenetic features of the Pan-
    African in East Africa, Nubia and Arabia. - Bull. Fac.
    Earth Sci., King Abdulaziz Univ. 6, 591-601.
Ries, A.C., R.M. Shackleton, R.H. Graham and W.R. Fitches,
    1983. Pan-African structures, ophiolites and mélanges in
    the E.D. of Egypt: a traverse at 26°N. - J. geol. Soc.
    London, 140, 75-95.
Ruxton, B.P., 1956. The major rock groups of the northern Red
    Sea Hills, Sudan. - Geol. Mag. 93, 314-330.
Schmidt, D.L. and G.F. Brown, 1983. Major-element chemical
    evolution of the Late Proterozoic shield of Saudi Arabia.
    - Bull. Fac. Earth Sci., King Abdulaziz Univ., 6, 1-21.
Schürmann, H.M.E., 1964. Rejuvenation of Precambrian Rocks
    under epirogenetical conditions during old Palaeozoic
    times in Africa. - Geol. Mijnbouw, 43, 196-200.
Shackleton, R.M., 1977. Possible Late Precambrian ophiolites in
    Africa and Brazil. - Ann. Rept. Res. Inst. African Geol.,
    Leeds Univ., 20, 3-7.
Shackleton, R.M., A.C. Ries, R.H. Graham and W.R. Fitches,
    1980. Late Precambrian ophiolitic mélange in the eastern
    Desert of Egypt. - Nature 285, 472-474.
Stern, R.J., 1981. Petrogenesis and tectonic setting of Late
    Precambrian ensimatic volcanic rocks, CED of Egypt. -
    Precambrian Res., 16, 195-230.
Stern, R.J. and C.E. Hedge, 1985. Geochronologic and isotopic
    constraints on Late Precambrian crustal evolution in the
    E.D. of Egypt. - Am. J. Sci., 285, 97-127.
Vail, J.R., 1983. Pan-African crustal accretion in north-east
    Africa. - J. African Earth Sci., 1, 285-294.
Warden, A.J. and J.L. Daniels, 1983. Evolution of the
    Precambrian of Northern Somalia. - Bull. Fac. Earth Sci.,
    King Abdulaziz Univ., 6, 145-164.

The Pan-African of Egypt and Northern Sudan

The Precambrian of Egypt and northern Sudan is treated in the following contributions by El-Gaby et al. (Chapter 2), Schandelmeier et al. (3), Kröner et al. (4), Greiling et al. (5) and Stern et al. (6). It is also included in the metallogenetic study by Pohl (11). Geophysical aspects of the crustal structure are discussed by Makris et al. (12).

El-Gaby et al. believe that the Mid-Proterozoic continental crust extends into the Eastern Desert of Egypt beneath obducted Pan-African ophiolites and island arc volcanics and volcani-clastics. Schandelmaier et al. demonstrate the Pan-African thermal and structural effects on primarily Mid-Proterozoic continental crust in southwestern Egypt. In the Eastern Desert of Egypt Kröner et al. assume a crust formed mainly in the Late Proterozoic. Greiling et al. show that potential 'infrastruc-tural' (sensu El-Gaby et al.) gneisses at Hafafit are allochthonous parts of a fold and thrust belt and do not represent a basement. Stern et al. address the problem of the late-tectonic granites, which are formed after Pan-African crustal shortening in an extensional environment.

The Preface by M.F. El-Ramly gives a short introduction into the Pan-African basement rocks of Egypt and their history of exploration.

Chapter 2

Geology, Evolution and Metallogenesis of the Pan-African Belt in Egypt

Samir El-Gaby[1] / Franz K. List[2] / Resa Tehrani[2]
[1] Geology Department, Assiut University, Assiut, Egypt.
[2] Institute of Applied Geology, F. U. Berlin, Malteserstraße 74–100, D-1000 Berlin 46

Keywords: Proterozoic, Pan-African, Egypt, NE Africa, Arabian-Nubian Shield, Structure, Metamorphism, Metallogenesis

Abstract: Geologic investigations on the basement of the Eastern Desert and Sinai, Egypt, has revealed that it constitutes the fold and thrust belt of a late Proterozoic (Pan-African) continental margin orogen. Late Proterozoic ophiolites and island arc volcanics and volcaniclastics occur as thrust sheets, the suprastructure, thrust from an easterly direction over an early Proterozoic continental crust, the infrastructure. The rocks of the infrastructure are largely mylonitized at shallow depths along a décollement surface, or remobilized at greater depths. The northern Eastern Desert and Sinai were upheaved during the Pan-African orogeny along a major right-lateral shear zone trending nearly NE-SW, whereby the Pan-African suprastructural rocks were almost entirely eroded away. In the central and southern Eastern Desert, the deformed and largely mylonitized rocks of the infrastructure crop out in gneiss domes disposed along the axes of two geanticlines trending NNW-SSE which also functioned as magmatic

arcs. Molasse sediments essentially derived from the erupting
continental margin calc-alkaline volcanics accumulated in the
intermontane basins between the raised geanticlines.

Wadi Allaqi area, southern Eastern Desert, is believed to once
have been a foreland basin, in which a great thickness of
molasse sediments accumulated above shelf sediments
intercalated with abundant limestone layers; the shelf
sediments are considered of comparable age to the island arc
rocks. Synorogenic granodiorite diapirs induced wide-spread
low-pressure metamorphism when intruded into molasse sediments,
a case particularly pronounced at Wadi Allaqi area.

In the infrastructure, Nb-Ta and emerald mineralizations are
locally found in gneissose leucocratic granite and adjoining
mica schists, respectively. Boundinaged chromite lenses occur
in ophiolitic serpentinites while banded iron ores and base
metal sulphides are found in island arc volcaniclastics.
Hydrothermal Au and Cu mineralizations are associated with
I-type granites and related volcanics, while Cu-Ni sulphides
occur in some ultrabasic intrusions. Sn, W, Mo and a younger
generation of Nb-Ta are associated with S- and A-type granites.

INTRODUCTION

Precambrian rocks occupy about one tenth of the land surface of
Egypt. They form mountainous terrains in southern Sinai and the
Eastern Desert as well as small, low-lying inliers in the
southern parts of the Western Desert. The basement of Sinai,
the Eastern Desert, Sudan and Ethiopia constitutes the Nubian
Shield that had formed a contiguous part with the Arabian
Shield in the Arabian Peninsula before the opening of the Red
Sea less than 30 Ma ago. It is generally accepted that the
basement of the Arabian-Nubian Shield exposed on both sides of
the Red Sea has been cratonized during the Pan-African orogeny
around 600 Ma ago. The occurrence of pre-Pan-African rocks in
the Arabian-Nubian Shield is highly disputed. The basement in
the Eastern Desert, NE Sudan and Arabia is chracterized by the
abundance of volcano-sedimentary rock sequences metamorphosed
in the greenschist facies.

Hume (1934, 1935) established a four-fold subdivision of the
Precambrian rocks in Egypt into Protarchean, Metarchean,
Eparchean and Gattarian (Table 1), a pattern considered
analogous to that of the rest of Africa at that time. The
oldest unit, the "fundamental gneiss" represented the
substratum of all other rocks present. The existence of this
unit is maintained in the classifications of Schürmann (1953,
1966) and El-Ramly (1972) but denied by most recent authors. An
unconformity between the "fundamental gneiss" and the overlying
late Proterozoic volcano-sedimentary rocks has never been found
and the contact is always structural. Said (1962) raised doubt
about the validity of a lithostratigraphic separation of an
older fundamental gneiss which appears to be based primarily on
the degree of metamorphism. This doubt has been aggravated to
the extent that the "fundamental gneiss" has been recently
considered as a high-grade geosynclinal filling or shelf
sediments of an age comparable to the structurally overlying
low-grade volcano-sedimentary rocks (Ries et al., 1983;
El-Ramly et al., 1984; Bentor, 1985). These authors believe
that no old sialic crust underlies the Arabian-Nubian Shield.

Table 1. Classification of the Precambrian rocks of Egypt.

	HUME (1935)	EL-RAMLY (1972)	Western Desert	PRESENT Wadi Allaqi	STUDY Eastern Desert & Sinai	Isotopic age in Ma
Gattarian (Late pre-Cambrian)	Alk. granite (?) Felsite & dolerite dykes	Alk. granite & syenite			Subalk. to peralkaline silicic igneous rocks	570
	Red granite Bi- & Hb-granites Diorite Peridotite (?)	Younger Granites Gabbro (?) Post-Hammamat Felsites			CORDILLERAN STAGE Calc-alk. granites (G-1 & G-2) and their volcanic brethren, the Dokhan Volcanics, penecontemporaneous molasse Hammaat sediments and lherzolite-gabbro-diorite.	
Eparchean (Upper Proterozoic)	Hammamat or Upper Paraschists Dokhan Volcanics	Hammamat Group Dokhan Volcanics		(Molasse sediments & Dokhan Volcanics)		655
Metarchean (Lower Proterozoic)	Shadli or Middle Paraschists	Metagabbro-diorite Serpentinites Geosynclinal Shadli Metavolcanics Geosynclinal Meta-sediments	Metasedimentary Formation	FORELAND ASSOC. (Shelf sediments)	ISLAND ARC STAGE Island Arc Assemblage: Intermed. metavolcanics, tuffs & volcanogenic greywackes. Ophiolites: Metabasalts, metadolerite, metagabbro and serpentinites.	720
Protarchean	Shait Hb-granite Acid gneisses	Migif-Hafafit Paragneiss and migmatites	Anatexite Formation	?	Granites (incl. Shait granite) and high-grade schist and gneiss, largely mylonitized or remobilized during the Pan-African orogeny.	1800
	Fundamental Gneiss		Granoblastite Formation			2900

They rely on the absence of Rb-Sr dates older than c. 1000 Ma
and the generally low Sr_i in the Arabian-Nubian Shield.
Moreover, the abundance of volcaniclastic sequences, the
ubiquitous presence of calc-alkaline magmatic rocks and the
identification of several ophiolite occurences in linear zones
of mafic-ultramafic complexes led these authors, as well as
Gass (1979) and Shackleton (1979), to envisage the geological
history of the shield to have started only in the late
Proterozoic when one or several ensimatic island arcs were
swept together and subsequently cratonized. However, recent
lead-isotope data and reliable zircon ages of 1.6 - 2.0 Ga from
the southern Arabian Shield (Stacey et al., 1984; Stacey and
Hedge, 1984; Stoeser and Stacey, 1987) persuaded Kröner (1985)
and Kröner et al., (1987) to view the shield as an
agglomeration of old continental fragments (exotic terrains)
and juvenile arcs that were swept together during the accretion
process.

On the other hand, granites, migmatites and high-grade gneisses
and schists containing almandine, staurolite and kyanite have
been found in the Meatiq area that have suffered mylonitization
and diaphthoresis concomitant with the overthrusting of Pan-
African ophiolites and associated volcano-sedimentary rocks
(El-Gaby and El-Nady, 1983; Habib et al., 1985). A similar
situation exists on the eastern flank of the Hafafit swell,
particularly well displayed around Wadi Sikait, where
garnetiferous mica schists and gneisses have been deformed by
the overthrusting of the overriding ophiolite sheet (Hegazy,
1984). Moreover, Abdelmonem and Hurley (1979) obtained a
$^{207}Pb/^{206}Pb$ zircon age of 1770 Ma from the gneissose granites
occupying the core of a small anticline at Wadi Sikait which
they considered, however, as "psammitic gneiss" as described by
Hassan (1973). The granitic character of this mass was already
recognized by Hume (1934) as well as by Basta and Zaki (1961)
and Hegazy (1984) and mapped as such by Ries et al. (1983,
Fig. 8) and by Elbayoumi and Greiling (1984, Fig. 1). In
addition to that, the Aswan granite gave a high Sr_i (Hashad et
al., 1972; Cahen et al., 1984) and has a high radiogenic lead

content (Abdel-Monem and Hurley, 1978; Gillespie and Dixon, 1983; Stacey and Stoeser, 1983) suggesting the presence of older continental crust. Moreover, the rather low positive $_{Nd}$ of + 2.5 reported from the Aswan granite (Harris et al., 1984) as compared to the strong positive $_{Nd}$ values of + 6.6 and + 7.6 obtained from the Gabal El-Wask and Gabal Ess ophiolites in the Arabian Shield (Claesson et al., 1984) may signify contamination by old continental crust (Kröner, 1985). We suggest, therefore, that the Archaean to Early Proterozoic nucleus of the Gabal Uweinat inlier (Klerkx and Deutsch, 1977) is fringed by mid-Proterozoic crust (Harris et al., 1984) which extends eastwards past Aswan into the Eastern Desert as far as Wadi Sikait. Furthermore, the presence of old continental crust underneath the ophiolite and volcano-sedimentary associations in the Eastern Desert, i.e. a Cordilleran structure, can be indirectly inferred from the presence of molasse sediments and by the abundance of granite (s.s.) intrusions. Molasse sediments characterize continental margin orogenic belts (Windley, 1977) and granite magmas (s.s.) do not form by partial melting of subducted oceanic crust or of upper mantle material, but originate through interaction of subduction-related magmas with continental crust at depth (Wyllie, 1983 a,b).

GEOLOGICAL OVERVIEW OF THE EASTERN DESERT AND SINAI

Crystalline basement rocks occupy the southern part of the Sinai Peninsula and a triangular area in the Eastern Desert that widens southwards to cover the whole width of the Eastern Desert at about the latitude of Aswan. The Eastern Desert is divided by Qena-Safaga and Aswan-Ras Banas lines into the northern Eastern Desert (NED), central Eastern Desert (CED) and southern Eastern Desert (SED).

The main classifications of the Precambrian rocks in Egypt are shown in Table 1. The proposed classification is based on our geologic investigations in the Egyptian basement within the

frame of producing a new geologic map of Egypt, scale
1 : 500 000 (Klitzsch et al., 1986). The identified rock units
are classified into four main groups according to their space
and time relationships:
- Pre-Pan-African granites, gneisses and schists and their
  mylonitized and remobilized equivalents,
- Pan-African ophiolites and island arc assemblage,
- Dokhan volcanics and molasse-type Hammamat sediments,
- Foreland assemblage of Wadi Allaqi.

The distribution of these groups, excluding the intrusive
rocks, is shown in Figure 1, where it is demonstrated that the
basement in Sinai and the NED differs from that exposed in the
CED and SED in the absence of ophiolites and almost all of the
island arc assemblage, as well as in the prevalence of granites
(not shown on the sketch map) and gneisses. The remobilized
infrastructural rocks in the NED and Sinai are extensively
intruded by late Proterozoic and younger granites so that the
gneisses and migmatites occupy relatively small areas, e.g. the
Feiran gneisses in SW Sinai and the Barud gneisses in southern
NED. The Qena-Safaga line represents the southern boundary of a
major shear zone, more than 50 km wide, along which the
northern block has been pushed north-eastwards and upheaved
(cf. Stern et al., 1984). The upheaval favoured complete
erosion of the obducted ophiolites and most of the island arc
volcano-sedimentary rocks (El-Gaby, 1983). A left-lateral shear
zone trending NW-SE along the Kom Ombo basin is suspected to be
present and to have moved the foreland basin of the Wadi Allaqi
area into the Eastern Desert.

PRE-PAN-AFRICAN ROCKS

Archaean to Early Proterozoic rocks crop out at Gabal Uweinat
in the southwestern corner of Egypt (Klerkx and Deutsch, 1977;
Klerkx, 1980; Harris et al., 1984). The Precambrian rocks in
the southwestern Desert, which are highly affected by the Pan-
African thermal event, are treated and described by

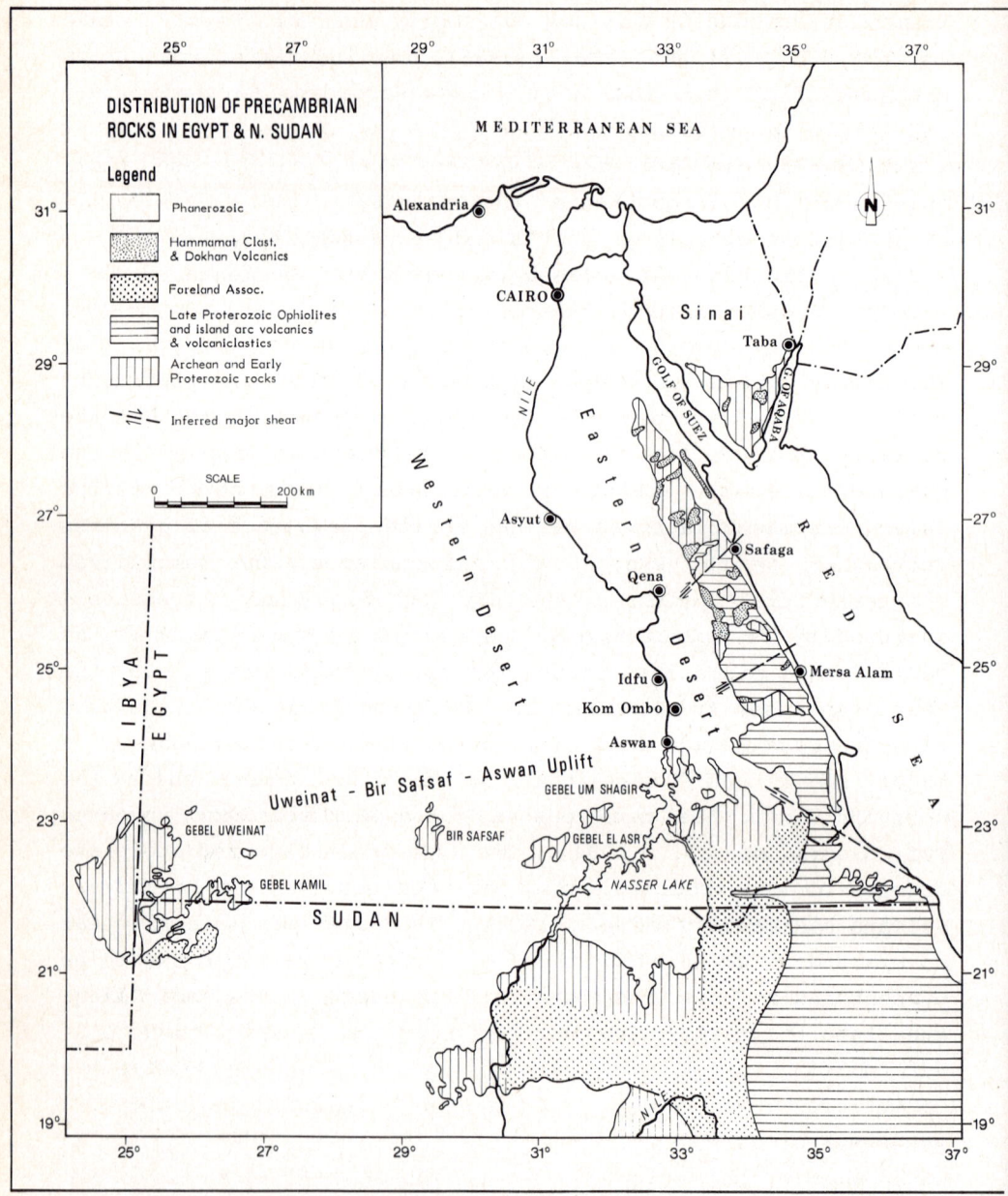

FIGURE 1.
   Geologic sketch map showing the distribution of major Precambrian rock units in Egypt and northern Sudan; compiled
   and modified after El-Gaby (1983), Vail (1983), Schandelmeier et al. (1983) and Kröner et al. (1987). Note, the
   remobilized basement of Sinai and NED is extensively
   intruded by Pan-African and later granites leaving only
   limited gneiss outcrops.

Schandelmeier et al. (1983, 1987), and Richter and
Schandelmeier (1987). We believe, as elaborated in the
Introduction, that the early Proterozoic continental crust
extends eastwards into the Eastern Desert underneath the Pan-
African ophiolite and volcano-sedimentary cover. It crops out
in the cores of large swells or gneiss domes in the CED and SED
and occupies considerable areas in the NED and Sinai. It is
clearly affected by the Pan-African orogeny. The effect of this
orogeny upon the rocks of the early Proterozoic continental
crust varied according to the prevailing temperatures. Granites
and granite gneisses were subjected to plastic deformation at
relatively shallow levels of moderate temperature leading to
formation of quartzo-feldspathic blastomylonites, the "acid
gneisses" of Hume (1934), at Meatiq, Sibai and Shalul gneiss
domes in the CED (Fig. 4). The pre-Pan-African rocks were
remobilized at high-temperature deeper levels and occur largely
at present as migmatites and granite-gneisses, commonly
associated with autochthonous to parautochthonous synkinematic,
calc-alkaline granites of tonalitic, granodioritic and locally
monzogranitic composition. Remobilized infrastructural rocks
are well displayed at Wadi Feiran, SW Sinai (El-Gaby and Ahmed,
1980), along the Qena-Safaga road, Eastern Desert (Sabet et
al., 1972; Akaad et al., 1973) and around Aswan (Gindy, 1954).
The large Hafafit culmination or swell (El-Ramly et al., 1984;
Elbayoumi and Greiling, 1984) offers a rather complete cross-
section of the infrastructure. On the eastern flank of this
culmination, garnetiferous mica schists are deformed and
diaphthorized along many small, listric thrust faults below a
Pan-African serpentinite-metagabbro thrust sheet. Variably
deformed gneissose granites of calc-alkaline and alkaline
composition occur at Wadi Sikait and within the "psammitic
gneisses" of El-Ramly and Akaad (1960) at the upper reaches of
Wadi Nugrus. At lower or deeper levels, the fine-grained
hornblende gneiss exposed along the western banks of Wadi
Nugrus grades downward into medium to coarse-grained banded and
homogenized hornblende gneiss which is intruded by
parautochthonous granitoids occupying the core of the
culmination. On the western flank, the granite of Wadi Shait is

deformed and even mylonitized along several shear zones
trending nearly NW-SE (El-Gaby and El-Aref, 1977).

The pre-Pan-African rocks comprise, as stated above, medium to
high-grade metamorphic rocks and granites that have been
deformed and diaphthorized as well as old rocks that have been
migmatized during the Pan-African orogeny. On the other hand,
low-pressure metamorphism at Wadi Kid area, SE Sinai (Shimron
and Zwart, 1970; Reymer et al., 1984), and at Wadi Um Had area,
CED (Kamal El-Din, 1986), developed biotite, almandine and
staurolite contact schists and hornfelses that grade through a
relatively narrow anatectic zone, carrying the critical mineral
assemblage cordierite + andalusite/sillimanite + potash
feldspar + garnet + biotite, into granite-gneiss that forms
part of a granodiorite diapir. In both areas, low pressure
metamorphism and anatexis affected primarily molasse-type
Hammamat sediments and associated andesitic and dacitic members
of the Dokhan Volcanics. It seems that geologic conditions
pertinent to the Hammamat sediments, or rather to molasse
basins, are favourable to low-pressure metamorphism. Great
thicknesses of terrigenous molasse sediments (and Dokhan
Volcanics) accumulating in molasse basins, coupled with later
folding and imbrication by thrusting, would furnish the
necessary pressures of 3 - 4 kb, values obtained by Reymer et
al. (1984). In the foreland molasse basin believed to have once
occupied the greater Wadi Allaqi area, a thicker pile of
sediments must have accumulated and thus accounting for the
higher pressures required for the development of kyanite in the
thermal aureole at Nasb Aloba (Hussein and Rasmy, 1976) and
together with andalusite and sillimanite at Wadi Dif area. The
gneisses and schists spatially associated with this low-
pressure metamorphism at Wadi Allaqi area incorporate most
likely late Proterozoic sediments.

PAN-AFRICAN ROCK ASSOCIATIONS

1. Ophiolites and Island Arc Association

Ophiolites and island arc association constitute a highly tectonized sequence of low-grade, regionally metamorphosed serpentintes, gabbros, volcanics and volcaniclastics. They correspond to the "ophiolitic mélange" of Shackleton et al. (1980) and to the "greenschist assemblage" of Vail (1983). They cover wide areas in the CED and are preserved in a synform between the Khuda swell and the East Aswan swell in the SED. Ophiolites are entirely absent in the NED and Sinai whereas the island arc association occupies narrow stretches in the blockfaulted southern part of the NED. The whole succession of the ophiolites and island arc association is dissected by many listric thrust faults causing repetitions and tectonic mixing. Nevertheless, it can be subdivided into a tectonically lower ophiolite sequence, always forming the lower part of major thrust sheets, and a tectonically upper island arc association.

1.1 Ophiolites

Ophiolites always occur as allochthonous and commonly dismembered ultrabasic to basic bodies. More or less complete sections are described from Wadi Ghadir (El-Sharkawi and Elbayoumi, 1979) and near Fawakhir gold mine (Garson and Shalaby, 1976; Nasseef et al., 1980). A compiled section is composed as follows, starting from the base:

a)   Serpentinites, essentially formed after harzburgite and to a lesser extent after dunite and lherzolite, are frequently transformed into talc-carbonates particularly along thrust faults and shear zones. They sometimes enclose boundinaged chromite lenses and may contain enstatite, diopside and rarely olivine relics. They are characterized by an almost constant, high Mg/Fe ratio.

b) Metagabbros of tholeiitic composition comprise lower cumulate pyroxenites and gabbros followed upward by isotropic gabbro and hornblende gabbro, which frequently enclose small dioritic bodies or appinite veins. Olivine, fresh or altered, has never been recorded so that cumulate dunites appear to be missing. The contact between metagabbros and the underlying serpentinites is always structural. The thickness of the metagabbro layer is about 1 km at Fawakhir, and 2 km at Gabal El-Rubshi.

c) Massive diabase and sheeted dyke complexes, 100 - 200 m thick, are also of tholeiitic composition. The best occurrence is described from Wadi Ghadir (El-Sharkawy and Elbayoumi, 1979).

d) Low-potash tholeiitic basalts, several hundreds of meters thick, are sometimes pillowed and pillows are always right way up (Stern, 1981). Alkali olivine basalts are rare.

e) Thinly bedded metasediments of deep water facies, 100-200 m thick at Wadi Ghadir, are quite rare and are essentially formed of pelites intercalated with thin chert and calacareous bands.

The ophiolites are commonly dismembered and even the large serpentinite masses are frequently intercalated by thin, highly foliated pelitic layers (Shackleton et al., 1980; Khudeir, 1983) marking minor thrust faults. The ophiolites are believed to have developed in back-arc basins (Stern, 1981) and were later obducted from the east over the old continental margin (El-Gaby et al., 1984; Kröner, 1985).

The proponents of an arc accretion evolutionary model for the Arabian-Nubian Shield consider the ophiolites as remnants of oceanic crust obducted along destructive plate boundaries during the closure of small ocean basins, thus delineating, at least tentatively, sutures or arc margins (Bakor et al., 1976;

Gass, 1979, 1982; Shackleton, 1979). Church (1979) raised
doubts against this assumption and believes that the ophiolites
in the Eastern Desert occupy a certain stratigraphic horizon,
and that their distribution on the geologic map of Egypt is due
to the presence of first-order, NW-trending anticlinal
structures which have been eroded sufficiently deep to expose
stratigraphic levels beneath the ultramafic-bearing units of
the geosynclinal sequence. El-Gaby (1983) and El-Gaby et al.
(1984) have shown that the ophiolites are commonly located
along the outcrops of major thrust faults and particularly
along the décollement surface between the overthrusted Pan-
African island arc rocks (suprastructure) and the underlying
old continental margin (infrastructure). This would explain why
ophiolite outcrops are concentrated around gneiss domes or
swells in the Eastern Desert. The distribution of ophiolites
might also help in unravelling the gross structure of the
Eastern Desert. For example, in the Qift-Quseir area one can
recognize two major thrust sheets: a lower Atalla thrust sheet
verging to the West along Wadi Atalla, and an upper Rubshi
thrust sheet verging also to the West along Gabal El-Rubshi and
Gabal Um Selimat, to the north and south of the Meatiq gneiss
dome, respectively.

1.2 Island Arc Assemblage

The ophiolites are structurally overlain and partly imbricated
with a series of weakly metamorphosed calc-alkaline
intermediate volcanics essentially formed of andesites,
dacites, and volcaniclastics of comparable composition; basalts
and ryhodacites are subordinate while true rhyolites are almost
entirely absent. Rhyolites frequently mentioned in literature
(e.g. Shukri and Mansour, 1980; Hafez and Shalaby, 1983) are
actually high-silica rhyodacites with potash contents of less
than 1.8 wt %. The tuffs and volcanogenic greywackes are often
banded and graded, indicating subaqueous deposition. They are
frequently intercalated with banded iron ores, particularly in
the northern half of the CED, and locally associated farther to
the south with stratabound base-metal sulphides (Garson and

Shalaby 1976; El-Aref et al., 1985; Fig. 4). The tuffs and volcanogenic greywackes are commonly mapped as metasediments. This volcanic association is referred to as "Younger Metavolcanics (YMV)" to differentiate it from the "Older Metavolcanics (OMV)" of the ophiolite association (Stern, 1981). The metavolcanics around Sheik Shadli, the type locality of the geosynclinal Shadli metavolcanics (El-Ramly, 1972) are formed of island arc volcanics and volcaniclastics, i.e. they belong to the YMV. Kröner (1985) reported a preliminary Rb-Sr age of 714 Ma from these metavolcanics at the Um Samiuki area.

Plutonic rocks belonging to the island arc association are least recognized and identified. They might include diorites and tonalites, the plutonic equivalents of island arc andesites and dacites, as well as mantle-derived gabbros. The small, sheared tonalite mass of Gabal El-Mayit (Akaad and El-Ramly, 1963) could be the only known representative of island arc granites (s.l.) in the Eastern Desert. Metagabbros and metadiorites occur intruded into the YMV to the north of Sheikh Shadli. Similar rocks occupy wide areas between Idfu-Mersa Alam road and Wadi Mubarak (Hassan and Essawy, 1977); some of these gabbroic bodies are so large and so little tectonized that we believe they were intruded after the overthrusting of the YMV and are probably genetically related to the Idfu-Mersa Alam shear zone.

2. Cordilleran Stage Associations

Emplacement of the ultramafic and mafic rocks (i.e. ophiolites) and the overthrusting of the island arc rock assemblage marks the culminant orogeny stage, or tectogenesis, in the Eastern Desert. This stage is characterized by the onset of an active phase of calc-alkaline magmatic activity. The calc-alkaline silicic rocks possess moderate potash content and comprise granites (s.s.) among plutonic rocks, and rhyolites among volcanic rocks - a feature that heralds participation of sialic crust in magma generation or modification (Wyllie, 1983 a,b). In other words, the Eastern Desert and Sinai acquired a

Cordilleran structure where the ophiolites and island arc rock association were thrust over the old continental margin. Moreover, the volcanic activity during this stage was mainly subaerial and ignimbrites were quite abundant. The associated sediments are essentially coarse terrigenous clastics with abundant conglomerates generally deposited in non-marine fluvial and fanglomerate environments, i.e. molasse sediments. The lithologies of magmatic and sedimentary rocks corroborate the end of the oceanic/island arc stage and the passage into a Cordilleran stage. The Cordilleran-stage associations comprise penecontemporaneous (a) subduction-related, Andean-type, calc-alkaline magmatic rocks, (b) Hammamat molasse sediments, and (c) intrusive mantle-derived, commonly fresh peridotites, gabbros and diorites.

2.1 Calc-alkaline Magmatic Rocks

About 680 to 570 Ma ago, the Eastern Desert and Sinai were sites of pronounced calc-alkaline magmatic activity in which the silicic members prevail. This magmatic activity is manifested in the emplacement of syn- to late-orogenic calc-alkaline granites of tonalitic to ideal granitic composition, and in the eruption of their volcanic and subvolcanic analogues, the Dokhan Volcanics and Post-Hammamat Felsites.

2.1.1 Calc-alkaline Granites

Hume (1935) and Schürmann (1953) consider the Egyptian granites (apart from the deformed Wadi Shait granite) of Gattarian (Late Precambrian) age and younger than the Dokhan Volcanics and Hammamat series. El-Ramly and Akaad (1960) differentiated the Egyptian granites into (a) "Older Granites" of tonalitic to granodioritic composition and unfailing gray colour, and (b) "Younger Granites" of pink and red colour and of granitic to alaskitic composition. The Older and Younger Granites are separated stratigraphically by the Dokhan Volcanics and Hammamat sediments. El-Ramly (1972) introduced the post-orogenic alkaline granites to incorporate the riebeckitebearing granites. El-Gaby (1975) showed that the granites of Egypt constitute one continuous series in which the granitic rocks become with time more silicic and richer in potash, and that the hypersolvus alkaline granites represent a side-branch developed during the late phases of differentiation, probably under subvolcanic conditions and through gas transfer. El-Gaby and Ahmed (unpublished work) found in the Wadi Feiran area, SW Sinai, that highly differentiated calc-alkaline granite intrusions were followed by the less differentiated Goza-Banat quartz syenite and then by the peralkaline ring intrusion of Gabal Serbal. They suggested, as adopted in El-Gaby and Habib (1982), to classify the Egyptian granites into:

a)   An older, syn- to late-orogenic calc-alkaline granite series comprising the "Old Grey Granite", the porphyritic granite of Aswan, and the two-feldspar "Younger Granites". This series corresponds to Gr. A grantoids of El-Shatoury et al. (1984) and encompasses both G-1 and G-2 granites of Hussein et al. (1982).

b)   A younger, post-tectonic alkaline to peralkaline granite series comprising quartz syenite, alaskite and aegirine- or riebeckitebearing leucocratic granites. This series correspond to Gr. B granitoids of El-Shatoury et al. (1984) and to G-3 granite of Hussein et al. (1982).

The calc-alkaline granites are rather arbitrarily separated into G-1 and G-2 according to their mineralogical composition which is reflected by colour and susceptibility to weathering and erosion, i.e. relief. G-1 granites occur as autochthonous, parautochthonous and intrusive bodies elongated parallel to the regional setting of the enclosing country rocks. They form the largest plutonic masses commonly possessing marginal zones with well-developed planar foliation. The early members are quartz dioritic, mafic-rich and better foliated, while the later members are of granodioritic to monzogranitic composition, lighter in colour and commonly structureless. Enclosed mafic endogenic xenoliths progressively decrease in size and abundance in the later phases.

The G-2 granites, on the other hand, always have intrusive contacts and commonly occur in the form of smaller intrusions that are also conformable to the surrounding regional setting. Marginal foliation is normally absent and the mafic endogenic xenoliths are small and few in number. Hornblende, which is very common in G-1 granites, is typically absent in G-2 granites. The Um Had granite pluton is the most typical - "Younger Granite" representative according to El-Ramly and Akaad (1960) since it is intruded into the Hammamat sediments. It possesses, however, a monzogranitic margin containing hornblende beside biotite; the Um Effein "Younger Granite" intrusion, just to the north of Um Had, shows the same mineralogical features.

The calc-alkaline granites are of the I-type; however, the G-1 granites are affiliated to the Cordilleran type while the G-2 granites resemble the Caledonian type, according to the scheme of Pitcher (1984). The distinction between Cordilleran and Caledonian type granites might reflect merely the depth of erosion (i.e. old or young orogenic belts), or the situation within the main magmatic arc or the back-arc of a continental margin orogenic belt.

The isotopic ages of the calc-alkaline granites range between 680 and 570 Ma but the maximum age can be reduced to 655 Ma which represents the mean value obtained from the parautochthonous granites from Mons Claudianus and Hafafit and the gneisses of Sinai (Hashad, 1980; Stern and Hedge, 1985; Bentor, 1985).

2.1.2 Dokhan Volcanics

Dokhan Volcanics are the surface and near-surface manifestations of the plutonic calc-alkaline magmatic activity comprising G-1 and G-2 granites. Discrimination of the Dokhan

Volcanics from the older, island arc YMV on the basis of metamorphism, as proposed by El-Ramly and Akaad (1960) and adopted by later authors, is inadequate. It has caused serious misidentifications, for example in the cases of the Kid and Sa'al volcanics in Sinai (Shimron, 1980; Bentor, 1985) although Bentor (1985) noticed that the high proportion of acidic volcanics in the Sa'al volcanics is remarkable for the island arc stage.

The Dokhan Volcanics are mostly subaerial, distinctly silicic and intimately associated and intercalated with Hammamat molasse sediments. Rhyolites and rhyodacites predominate while andesites are far less abundant; extensive welded-tuff units are particularly conspicuous. The common intercalation with the Hammamat molasse sediments and the ubiquitous presence of ignimbrites and other features of subaerial volcanism indicate the dominantly terrestrial setting of this volcanic activity (Gass, 1982). Geochemical investigations indicate that the Dokhan Volcanics are calc-alkaline and of Andean-type, erupted along an active continental margin (Basta et al., 1980; Gass, 1982; Furnes et al., 1985). Stern et al. (1984) believe, however, that the common bimodality of the Dokhan Volcanics expressed by andesites (56-62 % Si_2) in association with rhyolitic ignimbrite and related felsic rocks (72-76 % SiO_2) indicates that the crust was actively extending and undergoing rifting.

Available radiometric ages from the Dokhan Volcanics range between 639 Ma and 581 Ma (Dixon, 1979; Stern and Hedge, 1985; Bielski, 1982 in Bentor, 1985); they are comparable to the coeval calc-alkaline G-1 and G-2 granites. It is worth mentioning that the metamorphosed andesitic Dokhan Volcanics (recognized on the basis of their association with molasse sediments and/or with abundant silicic volcanics) commonly give apparent older Rb-Sr ages. The Abu Swayel rhyodacites yielded 655-654 Ma (El-Shazly et al., 1973) and 768 Ma (Stern and Hedge, 1985), while the Nuqarah andesites gave 686 Ma, and Wadi Massar - Wadi Arak volcanics gave 632 Ma, or rather 690 Ma for

the andesites (Stern and Hedge, 1985). A Rb-Sr age of 734 Ma
was obtained from the metamorphosed volcanics of Wadi Sa'al,
while granite-gneiss boulders from the associated Sa'al
conglomerates (=Hammamat conglomerates) yielded only 641 Ma
(Bielski, 1982 in Bentor, 1985). The weakly metamorphosed
volcanics from Wadi El-Mehdaf, unfortunately one of the type
localities of the YMV according to Stern (1981), gave 622 Ma
(Stern and Hedge, 1985) and thus must belong to the Dokhan
Volcanics; they also include proper rhyolites containing up to
2.84 % K_2O (Stern, 1981) which are common in the Dokhan
Volcanics.

2.2 Hammamat Molasse Sediments

The Hammamat sediments are largely composed of poorly sorted
clastic sediments intercalated with minor impure calcareous
layers. Akaad and Noweir (1980) described the Hammamat
sediments as formed at the base of an alternation of
predominately brick-red and green siltstones, the Igla
Formation, which pass upwards into sub-greywackes and volcanic
arenites frequently enclosing thick conglomerate banks. The
Igla Formation was deposited in fresh-water basins (Samuel,
1977). Grothaus et al., (1979) produced several models for the
deposition of the Hammamat sediments in intermontane, alluvial
and fresh-water basins, whereby variation in grain-size of
clastic sediments is attributed to lateral facies variation and
not to tectonic events.

During the Cordilleran stage, the basement was thrown into a
series of large, open anticlinal folds and swells disposed
along two geanticlines trending NNW-SSE (Fig. 5). These
geanticlines acted also as Andean-type magmatic arcs along
which calc-alkaline magmas erupted. Erosional products of the
Dokhan Volcanics extruded along the raised magmatic arcs
accumulated as Hammamat molasse sediments in intermontane and
foreland basins. Consequently, the Dokhan Volcanics and the
Hammamat sediments are characterized by their intimate temporal
and spatial association.

The Hammamat sediments display their main development in the northern part of the CED but occupy smaller and smaller areas farther south due to the northward slope of the basement. They are well developed in the NED and Sinai suggesting that the upheaval of the northern block along the Qena-Safaga line took place during the Cordilleran stage while eruption of the Dokhan Volcanics and deposition of the Hammamat sediments are still in progress (El-Gaby, 1983). The model proposed by Stern et al., (1983) for the NED suggesting deposition of the Hammamat sediments in downfaulted grabens trending nearly NE-SW cannot be substantiated. We noticed that the Hammamat sediments occur on the dip-slopes of a series of faulted blocks bounded from the south by reverse (?) faults that obviously have been rejuvenated during the Tertiary. In the SED, recognized Dokhan Volcanics and Hammamt sediments are very much reduced on the Geological Map of Egypt (El-Ramly, 1972); this point will be further treated together with the foreland association.

2.3 Younger Ultrabasic to Basic Intrusions

A suite of intrusive ultrabasic and basic rocks comprising spinel lherzolite, clinopyroxenite, troctolite, olivine gabbro and meladiorite occur as small, frequently layered intrusions and sills. They were intruded contemporaneously with the Andean-type calc-alkaline magmatic rocks. A layered gabbro intrusion is intruded by granodiorite near the mouth of Wadi Mubarak, CED (Kabesh et al., 1967), while the granodiorite intrusion of Wadi El-Sheikh, SW Sinai, is dissected by a meladiorite dyke-like body which is, in turn, truncated by the more differentiated Ma'in granite. The younger ultrabasic and basic rocks are commonly not metamorphosed, but at Abu Swayel they have clearly undergone, together with the enclosing country rocks, low-pressure metamorphism that affected the Wadi Allaqi region.

The younger peridotites and gabbros differ from their ophiolitic and island arc counterparts in the following aspects:

a)   fresh olivines are commonly present with marked variation in their Fo-content indicating crystallization from a differentiating gabbroic magma, and

b)   clinopyroxenes, Mg and/or Cr-spinels and sometimes Cu-Ni sulphides are typically present in the peridotites.

In this context, we believe, until conclusive radiometric dates have been obtained, that the peridotites of St. John's (Zabargad) Island in the Red Sea also belong to this suite, though Bonatti et al. (1981, 1983) insinuate that the peridotites represent an uplifted fragment of sub-Red Sea lithosphere (upper mantle) connected with the early stages of the rift development. Our interpretation is based on the following facts:

a)   Bulk composition and mineral paragenesis of the Zabargad peridotites are very much similar to spinel lherzolites described from Gabal Dahanib (Dixon, 1981) and from Gabal El-Motaghayerat (Abu El-Ela, 1985) from the northern part of the Khuda swell, SED. Spinel lherzolites also occur in SE Sinai (Shimron, 1984) and at Wadi Allaqi.

b)   Cu-Ni sulphide mineralization associated with the Zabargad peridotites also occur in some ultrabasic bodies in the SED, for example at Abu Swayel (El-Goresy, 1964), at Gabbro-Akarem (Bugov and Shalaby, 1973), and at El-Geneina West (Kamel et al., 1980).

c)   The old metamorphic country rocks of Zabargad Island are composed of coarsely banded hornblende-plagioclase gneisses and amphibolites similar to the gneiss exposures widespread in the Eastern Desert (Bonatti et al., 1981, 1983), and particularly to the nearby Khuda gneisses. This signifies that the island constitutes a detached block of

the basement of the Eastern Desert, as also acknowledged
by Bonatti et al. (1981, 1983), which is more probable
than to assume vertical upheaval of the order of 30 km,
particularly if we take into consideration that the
peridotites are in tectonic contact with Upper Cretaceous
sediments (Bonatti et al., 1983) and unconformably
overlain by basal Miocene sediments (Emad Philobbos,
oral communication).

3. Foreland Association

Hunting (1967) noticed that the metasediments of the greater
Wadi Allaqi region differ markedly from the geosynclinal
metasediments described by El-Ramly and Akaad (1960) from the
CED, and they therefore assumed the presence of another
geosynclinal basin. The sedimentary rocks are characterized by
the frequent occurence of limestone (metamorphosed into marble)
in the lower part of the succession, and by poorly sorted
arenites and siltstones in the upper part; intercalations of
fine conglomerates are also not infrequent in the upper part.
The facies of the clastic sediments is very similar to that of
the Hammamat molasse sediments. Rhyodacites and rhyolites,
though commonly metamorphosed, are quite abundant among the
associated volcanics, so that they should be considered
pertaining to the Dokhan Volcanics as already recognized by
El-Shazly et al. (1973) and reported by El-Ramly (1972).

At least the lower part of the sedimentary succession of the
greater Wadi Allaqi region is equivalent to the metasedimentary unit of Vail (1983), laid down as a sedimentary
prism on a rifted or attenuated pre-Pan-African continental
margin. During the Cordilleran stage, the greater Wadi Allaqi
region was transformed into a foreland basin in which a thick
pile of molasse sediments accumulated above the shelf sediments
containing limestone layers. The foreland basin extends farther
south across the border into northern Sudan (Fig. 1).

Available isotopic ages from the Wadi-Allaqi region are
questionable or inconclusive. El-Shazly et al. (1973) reported
a Rb-Sr isochron age of 1195 Ma from a garnetiferous mica
schist from the Abu Swayel area. However, schists in general
and metamorphosed clastic sediments in particular are generally
difficult to date (Jäger, 1977; Hashad, 1980). The model Nd age
of c. 1500 Ma from the metamorphosed sediment at Wadi Allaqi
(Harris et al., 1984) could simply signify mixed clastic
sediments from early Proterozoic sources. El-Shazly et al.
(1973) reported Rb-Sr ages of 890 - 980 Ma from intrusive
quartz diorites which are based, however, on only two whole-
rock data points, or even on one point and an assumed initial
ratio.

PHANEROZOIC ALKALINE ROCKS

From the end of the Pan-African orogeny, about 570 Ma ago,
until the Tertiary the basement was intermittently intruded by
a number of subalkaline to peralkaline A-type granite bodies,
mostly as shallow cauldron complexes or as small discordant
intrusions having almost circular outlines; dyke-like
intrusions, chimneys or volcanic necks are not uncommon. These
post-kinematic silicic rocks, generally have alkaline
tendencies and are lumped together as G-3 granites, which
correspond to the "Younger Granite Complexes" of Almond (1979),
and to the "Katherina Province" of Bentor (1985). Available
radiometric ages show an overlap in the time of intrusion with
G-2 and G-3 granites. Elsewhere in the world S-type granites
are commonly intruded in the back-arc, concurrently with the
emplacement of calc-alkaline granites in the mobile belt
(Pitcher, 1983); they are not confidently identified in the
Eastern Desert and Sinai (cf. Hussein et al., 1982). The G-3
group comprises a vast array of rocks, including two-mica
granite, alkaline granite and syenite, and was intruded over a
very long time span; it needs to be separated into several
subgroups. The earlier phases of G-3 granites are clearly
potassic and are unrelated to gabbroic rocks. They comprise

subalkaline to peraluminous leucocratic biotite granites and two-mica granites which are difficult to distinguish petrographically from G-2 granites. All known Mo and most of the Sn and W mineralizations are associated with these leucocratic two mica granites. The peraluminous character seems of secondary origin due to pneumatolysis and greisenization; andalusite or cordierite are never reported but garnet may occur locally. Quartz syenites and riebeckite-bearing alkaline granite, that commonly form ring complexes, also occur among the early phases. Associated volcanics, namely the Katherina Volcanics, are rare and are only described from Sinai as a remnant of a caldera surrounded and intruded by a large ring-dyke (Shimron, 1980). The volcanic rocks are essentially formed of comendites and their pyroclastics (Bentor, 1985). It still remains controversial, however, whether the late Precambrian/Cambrian G-3 granites evolved from G-2 granites by differentiation and modification through gas transfer, or they had an independent magmatic source. We support the latter alternative since we observed, particularly in Sinai, that highly differentiated G-2 granites were followed by less differentiated quartz syenite and post-kinematic granite.

Mesozoic alkaline silicic igneous rocks are always sodic and are frequently associated with minor gabbroic rocks and rarely carbonatites (El-Ramly et al., 1971). They are essentially composed of syenites which sometimes carry nepheline. The associated volcanic rocks are well represented by the Wadi Natash Volcanics which occur in the lower part of the "Nubia" sandstones of Cretaceous age. Similar volcanics occur also at Gabal Abraq, SED. The Meozoic ring complexes are apparently confined to the SED. On the basis of our preliminary discrimination characteristics, namely soda/potash ratio and association with gabbroic rocks, the albite-granites commonly known as "apo-granites" (Sabet et al., 1973) which are commonly associated with Nb-Ta, Sn and F mineralizations, and the El-Bakreya granite intrusion, all in the CED, might belong to the Mesozoic magmatic association.

STRUCTURAL AND METAMORPHIC HISTORY

1. Metamorphism

The metamorphic history of the basement rocks in the Eastern Desert and Sinai is a topic rarely treated. Geologic investigations revealed that parts of the basement underwent several phases of metamorphism, and that several types of metamorphism are also present. The oldest phase of metamorphism is detected in pre-Pan-African relics preserved within sheared and diaphthorized early Proterozoic rocks of the infrastructure. El-Gaby and El-Nady (1983) and Habib et al. (1985) reported from Meatiq area the occurrence of migmatites and metamorphites containing kyanite, staurolite and almandine that have been extensively mylonitized and diaphthorized concomitant with the overthrusting of Pan-African supracrustal rocks. The described mineral assemblages and the composition of analyzed almandine garnets which range from Alm $_{69.2}$ Pyr $_{21.8}$ Gr $_{8.7}$ Sp $_{0.4}$ (associated with staurolite) to Alm $_{71.6}$ Pyr $_{15.9}$ Gr $_{11.8}$ Sp $_{0.6}$ (El-Gaby, unpublished data) indicate a medium pressure of Barrovian type regional metamorphism. Mica schists containing almandine crystals, 1 - 2 cm across, occur among the infracrustal rocks exposed in the cores of minor ESE-plunging anticlines, on the northern side of Wadi El-Gemal.

During the Pan-African orogeny, the rocks of the early Proterozoic continental margin incorporated within the orogenic belt were subjected to mylonitization and diaphthoresis or remobilization depending on the temperature prevailing at the respective sites. Dynamic metamorphism is prominent at low-temperature shallow levels along the dècollement surface between Pan-African supracrustal rocks and the infrastructure, leading to the formation of mylonites and diaphthorites. Coarse-grained quartzo-feldspathic rocks like granites and migmatites were transformed into flaggy muscovite + quartz + feldspars + biotite + garnet schists, the "acid gneiss" of Hume (1934), which are widespread at Meatiq, Sibai and Shalul gneiss domes. These blastomylonites are commonly intercalated in their upper parts with actinolite schists and foliated serpentinites that represent detached fragments from the overlying Pan-African ophiolite thrust sheets. Sturchio et al. (1983) recognized the mylonitic nature of the quartzo-feldspathic schists forming a carapace around a granite protolith still preserved in the core of the Meatiq gneiss dome; they overlooked, however, the presence of diaphthorized relictic metamorphic rocks as well as the complex nature of the granite-gneiss core which comprises variably deformed calc-alkaline and peralkaline granites. The obtained Rb-Sr isochron age of about 620 Ma, assumed to be the age of the granitic protolith (Sturchio et al., 1983), must be considered as the age of mylonitization.

Farther to the south, mylonite occurrences are developed along thrust faults on the eastern flank of Hafafit swell. On the western side of the swell, deformed tonalites and granodiorites traversed by several mylonite zones trending nearly NW-SE (El-Gaby and Al-Aref, 1977) occupy wide areas at Wadi Shait. In other words, the Shait granite is considered once more a very old granite (i.e. pre-Pan-African) as previously recognized by Hume (1934) and Schürmann (1953).

Pre-Pan-African infracrustal rocks underwent remobilization at higher-temperature deeper levels, and the newly formed gneisses and migmatites display only Pan-African tectonic elements. Remobilized infracrustal rocks are exposed in the NED and Sinai where the Pan-African supracrustal cover is largely eroded away, and also in the cores of large swells in the CED and SED such as Hafafit and Khuda swells. The Feiran-Solaf gneiss belt, SW Sinai, is largely composed of variably migmatized and homogenized hornblende and hornblende-biotite gneiss, biotite gneiss and locally cordierite-sillimanite-biotite-gneiss (El-Gaby and Ahmed, 1980). The gneisses are folded into three huge, doubly plunging asymmetrical anticlines trending NNW-SSE and verging to the SW. The gneiss belt is intruded from the east by the Wadi El-Sheikh granodiorite along a conformable contact. The Barud gneisses exposed along the Qena-Safaga road are composed largely of granite-gneisses enclosing fine-grained gneiss skialiths and migmatites. They grade imperceptibly into autochthonous G-1 granites of tonalitic to granodioritic and locally porphyritic monzogranitic composition (Sabet et al., 1972; Akaad et al., 1973), Radiometric ages from the Feiran and Fjord gneisses in Sinai and from the parautochthonous G-1 granites from Mons Claudianus, NED, and the Migif dome in the Hafafit swell range between 682 Ma and 641 Ma, with a mean value of about 655 Ma (Stern and Hedge, 1985; Bentor, 1985).

Obducted Pan-African ophiolites and island arc volcanics and volcaniclastics show a complicated metamorphic history. Ophiolitic gabbros and basalts were subjected to non-uniform submarine metamorphism; uralitized gabbros and weakly metamorphosed basalts and dolerites are not infrequent. Ultramafic harzburgite and dunite are entirely transformed into serpentinites; relict pyroxenes, and rarely olivine, may occur. The island arc rock assemblages were subjected to burial metamorphism up to the greenschist facies. The ophiolites and island arc assemblage suffered later diaphthoresis during thrusting; obvious is the transformation of serpentinites into talc-carbonate, which took place preferably along shear zones, while the metavolcanics were transformed into actinolite schist.

Low-pressure or Buchan type (regional) metamorphism has just started to gain recognition as an important event in the evolution of the Egyptian basement. Hume (1935) and Schürmann (1966) reported the presence of andalusite and cordierite hornfelses from the Wadi Kid area, SE Sinai, and Wadi Dib, NED, which they attributed to contact effects of granite intrusions. The Wadi Kid area has been recently studied in detail by Shimron and Zwart (1979) and Reymer et al. (1984) who ascribed the development of successive biotite, garnet, staurolite, cordierite and andalusite/sillimanite mineral zones to thermal domes. Low-pressure metamorphic rocks containing identical mineral assemblages are described from Wadi Um Had area (Kamal

El-Din, 1986) and from Wadi Ara, a tributary of Wadi El-Miyah
(Hafez and El-Amin, 1983), both in the CED. All the authors
cited noticed that the metamorphic mineral zones are not
related to the outcropping granite intrusions; thermally
metamorphosed rocks in the contact aureoles of granite
intrusions may contain cordierite and/or biotite but never
garnet, staurolite or andalusite. Recently, Kamal El-Din (1986)
found that the isogrades of garnet, staurolite and
andalusite/sillimanite (the latter in the migmatized zone) do
not conform with the Um Had granite pluton; the proper contact
aureole of this pluton developed only chlorite and biotite
zones.

It is worth to mention that anatectic rocks occur in the
highest grade zones at Wadi Kid, Wadi Ataqa (Beyth et al.,
1985), Wadi Zaghra and Wadi Um Had areas. Another important
feature of this low-pressure metamorphism is its occurrence in
the terrigenous clastic sediments associated with the Dokhan
Volcanics, i.e. the Hammamat molasse sediments. At both Wadi
Kid and Wadi Um Had areas, andesites and dacites representing
the early members of the Dokhan Volcanics are metamorphosed,
while the later acidic members, viz. rhyolites and granophyres,
are not affected. Even at Wadi Ara, Hafez and El-Amin (1983)
described the metamorphosed (meta-)sediments as terrigenous,
ill-sorted and enclosing conglomerate bands - these are the
typical lithological features of the Hammamat sediments, which
differ from the volcanogenic island arc greywackes and tuffs
commonly described as eugeosynclinal metasediments (El-Ramly,
1972; Akaad and Noweir, 1980). It might be concluded that
geologic conditions pertinent to the Hammamat sediments, or
rather to molasse basins, are favourable to low-pressure
metamorphism. Reymer et al. (1984) obtained for the Wadi Kid
area pressure estimates of 3.2 kb and temperature estimates of
565 $^{\circ}$C and 620 $^{\circ}$C, corresponding to geothermal gradients of
47 $^{\circ}$C/km and 50 $^{\circ}$C/km, for staurolite-andalusite zone
assemblages and anatectic rocks, respectively. Great
thicknesses of sediments (and Dokhan Volcanics) accumulating in
molasse basins, coupled with later folding and imbrication by

thrusting, would furnish the necessary operative pressures of
3-4 kb. In the foreland basin once occupying the greater Wadi
Allaqi region, SED, still greater thicknesses of sediments must
have accumulated so that resulting higher operative pressures
might account for the development of kyanite in the thermal
aureoles at Nasb Aloba (Hussein and Rasmy, 1976) and together
with andalusite and sillimanite at Wadi Dif area. Compressive
stresses were also operative in the Wadi Allaqi region.
However, Atherton et al. (1975) believe that higher pressures
are not a prerequisite for the development of kyanite in
thermal aureoles. A multitude of large G-1 granodiorite diapirs
in the Wadi Allaqi region might account for the widespread and
profound metamorphism oberserved. It is of academic interest to
find out whether low-pressure metamorphism was induced directly
by the G-1 granodiorite diapirs or by the thermal domes
preceding the uprising diapirs. In most low-pressure regional
terrains, synkinematic granitic masses appear to have been
emplaced late during or slightly after the end of regional
metamorphism (Miyashiro, 1973). It remains, however, to examine
which parts of the gneisses and schists at the greater Wadi
Allaqi region belong to the infrastructure and which are
derived from the Pan-African foreland sediments and overlying
molasse sediments.

2. Structure

The Eastern Desert lies within the fold and thrust belt of the
Pan-African continental margin orogen (El-Gaby, 1983). It
consists of relatively thin and imbricated thrust sheets
overlying an attenuated early Proterozoic continental margin.
Two tectonic trends, namely NNW-SSE and ENE-WSW, prevail in the
Eastern Desert and Sinai. There is no consensus whether these
two trends are coeval or which trend is the primary one.
El-Gaby (1983), El-Ramly et al. (1984), Kröner (1985) and
Elbayoumi and Greiling (1984) believe that the Pan-African belt
was created by compression from an easterly direction, while
Shackleton et al. (1980), Ries et al. (1983) and Habib et al.

(1985) consider the direction of tectonic transport is towards the NNW.

The distribution of the major rock units (Fig. 4) suggests that the basement in the Eastern Desert and Sinai is composed of large fragments separated by major zones of tectonic dislocation. Sinai and the NED are stripped off their Pan-African supracrustal cover suggesting that they were raised along the right-lateral shear zone running along the Qena-Safaga line and dipping steeply northwards (El-Gaby, 1983). The shear zone is more than 50 km wide. The southern part of the NED is occupied by a major domal structure (Fig. 5) elongated ENE-WSW. The faulted blocks are tilted to the north and bounded on the south by reverse faults; the Dokhan Volcanics and the Hammamat sediments occupy the dipslopes. Rejuvenation along some shear faults during the Tertiary is believed to have been responsible for the creation of the Qena bend in the course of the River Nile.

In SW Sinai, the NNW-trend is well displayed in the Feiran gneiss belt. The regional setting of SE Sinai is not quite clear: the Dokhan Volcanics and Hammamat sediments are laid into open folds trending ENE-WSW at Wadi Kid, whereas the Sa'al volcanics and associated Sa'al conglomerates at Wadi Zaghra as well as the Fjord gneisses display the NNW-trend.

The CED is largely covered by ophiolites and island arc assemblage that are locally interrupted by outcrops of infrastructural rocks in gneiss domes. The gneiss domes are disposed along the axes of two geanticlines having the same regional trend (Fig. 5), namely NNW-SSE. Minor folds show southwesterly vergence and the trust faults verge to the west. However, the regional stretching along the Idfu-Mersa Alam road displays the ENE-trend and is considered as another, but deeper, right-lateral shear zone running nearly parallel to the Qena-Safaga shear zone. Greiling and El-Ramly (1985) consider this region as a lateral ramp, though the clockwise deflection of the strike around Mersa Alam does not conform with the

assumed direction of movement. Further to the south, the area
between the Hafafit swell and the Ras Banas line is slightly
different from the central and northern parts of the CED, where
it is occupied exclusively by island arc assemblages which
reveal almost E-W trends; a structural explanation cannot be
given at present.

The SED is composed of the large Khuda gneiss dome to the east
and the huge composite East Aswan swell to the west, separated
by a syncline largely occupied by ophiolites. The East Aswan
swell is composed to a large extent of shelf and foreland
molasse sediments that are pierced by large granodiorite
diapirs which induced, directly or indirectly, regional low-
pressure metamorphism. It is truncated on the north by the
Phanerozoic Kom Ombo basin forming at present a composite
graben structure. This graben might occupy an old weak zone, or
rather a shear zone, that truncated and terminated the western
geanticline and moved the foreland molasse basin to the SE into
the SED. This points to the possible presence of a left-lateral
shear zone trending NW-SE similar to the Najd shears in the
Arabian Peninsula.

Consistent with a principal compressional stress from the ENE,
and probably due to continental collision after the closure of
the oceanic tract, the first-order left-lateral Najd fault-
system of Arabia (Schmidt et al., 1979) and its complimentary
right-lateral Qena-Safaga and Idfu-Mersa Alam shear zones
(El-Gaby, 1983) were created. The Najd trend is subdued in the
Eastern Desert, but it is probably represented by the left
lateral wrench-faults along the Kom Ombo basin and the upper
course of Wadi Sikait. Wrench-fault tectonics (Moody and Hill,
1964) suggest that a primary compressional stress acting from
the east would create second-order shears, some of which are
well represented in the Eastern Desert (Fig. 2):

a) left-lateral, NNW-SSE Shear, or Gabal Atalla trend, along
    which the Atalla felsite was intruded and later sheared,
    and

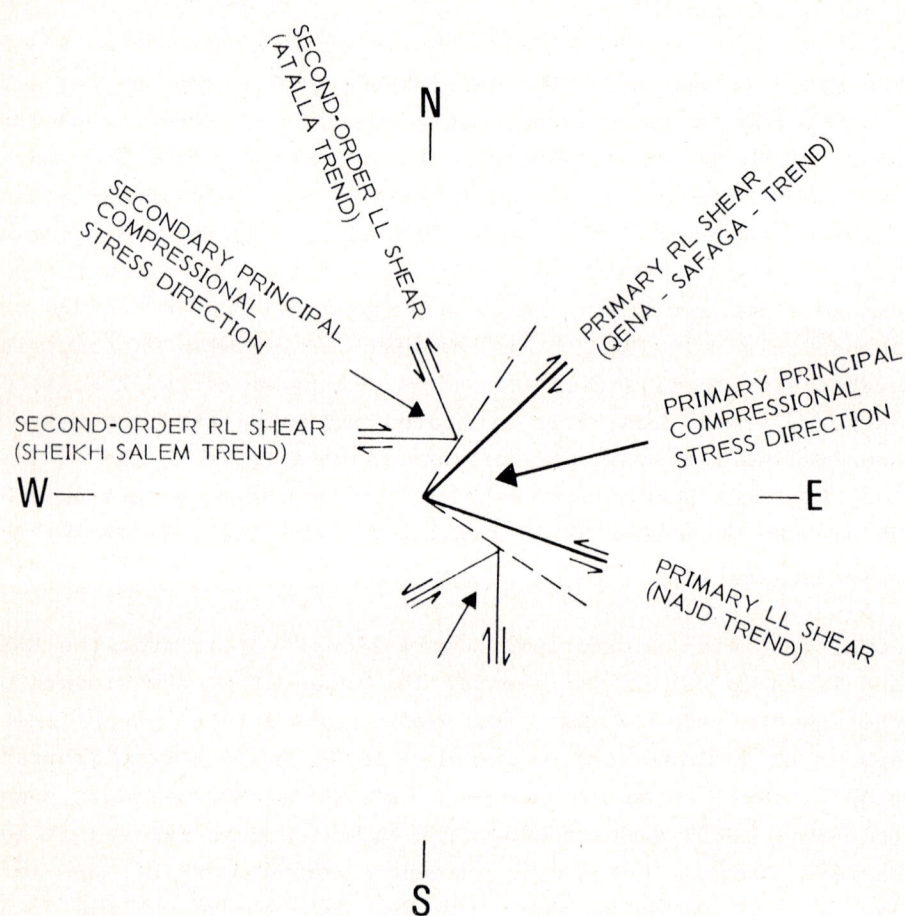

FIGURE 2.
   Major shear trends in the Eastern Desert of Egypt.

b)   right-lateral,   E-W   shear   faults,   or   Sheikh   Salem
     trend, which occur along the Idfu-Mersa Alam road near
     Sheik Salem and in the area of Wadi Zeidoun.

3. Evolution of the Pan-African Belt

The evolution of the basement complex in the Eastern Desert and
Sinai is shown diagrammatically in Figure 3. An Archaean
nucleus located in the southwestern corner of Egypt was fringed
by an early Proterozoic continental mass (Harris et al., 1984)
that extended eastward past Aswan into the Eastern Desert, at
least as far as Wadi Sikait on the eastern flank of the Hafafit
swell. This assumption is based on the high initial Sr ratio
and high radiogenic lead content in the Aswan granite (Hashad
et al., 1972; Stacey and Stoeser, 1983), and on the 1770 Ma
zircon age (Abdel-Monem and Hurley, 1979) from the deformed
granites exposed between Wadi Sikait and Wadi Abu Rusheid. The
old continent was fringed by an island arc about 800-700 Ma
ago; reliable isotopic dates from the Egyptian ophiolites are
not available. Kröner (1985) reported a preliminary Rb-Sr age
of 715 Ma from the typical island arc volcanics of the
Um Samiuki area.

Somewhat later, probably around 700 Ma ago, the island arc was
swept against the old continent, thereby thrusting back-arc
ophiolites and the island arc volcanics and volcaniclastics
onto the margin of the old continent (Fig. 3b, c). The Pan-
African belt then acquired a Cordilleran character. Intrusion
of   subduction-related   and   mantle-derived   magmas   induced
softening   and   remobilization   of   the   early   Proterozoic
continental crust or "infrastructure" around 655 Ma ago; this
age is the mean value of radiometric ages ranging between 682
Ma and 641 Ma obtained from the parautochthonous G-1 granites
at Mons Claudianus and at Hafafit in the Eastern Desert, and
from the Feiran and Fjord gneisses in Sinai (Stern and Hedge,
1985, Bentor, 1985). According to Ramberg (1981), the presence
of density differences between the overthrusted sheets of
ophiolites and island arc assemblages, and the underlying

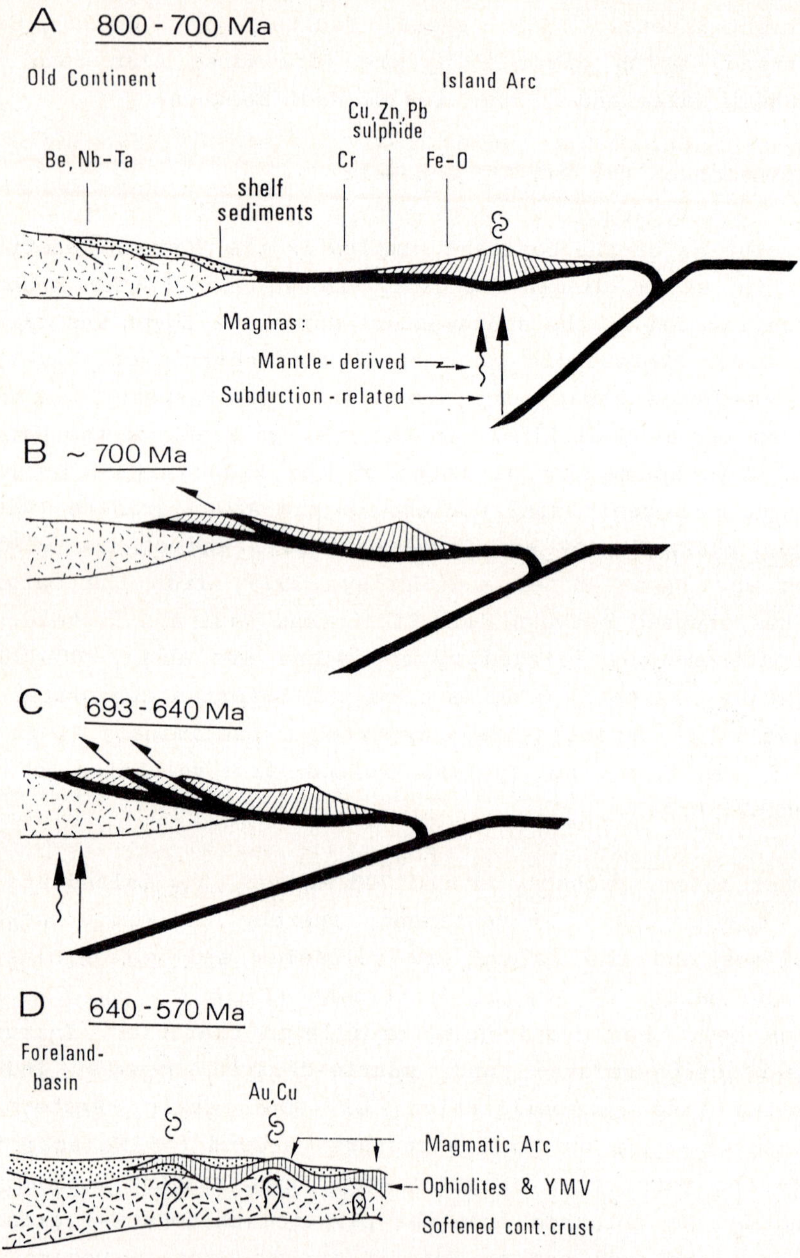

A   800 - 700 Ma

Old Continent   Island Arc

                                              Cu, Zn, Pb
                                               sulphide
Be, Nb-Ta   Cr   Fe-O

                       shelf
                       sediments

          Magmas:

              Mantle - derived

              Subduction - related

B ~ 700 Ma

C   693 - 640 Ma

D   640 - 570 Ma

Foreland-
basin
                           Au, Cu

                                         Magmatic Arc
                                         Ophiolites & YMV
                                         Softened cont. crust

FIGURE 3.
Cartoon illustrating the tectonic evolution of the Eastern
Desert of Egypt: (a) island arc stage, (b) overthrusting
of back-arc ophiolites and island arc volcanics and
volcaniclastics   over   the   old   continental   margin,
(c) Cordilleran stage, and (d) a sketch profile along the
Qift-Quseir road during the Cordilleran stage.

remobilized granitoids of the infrastructure causes the formation of large scale undulations (Fig. 3d) that are reflected on the surface as geanticlines by the buoyant migration of granitoid material. Two geanticlines are recognized in the CED (El-Gaby, 1983) that were sites of pronounced magmatic activity. Calc-alkaline magmas erupted along the geanticlines, i.e. they functioned as magmatic arcs. This magmatic activity was manifested in the calc-alkaline granites (G-1 and G-2) and their volcanic brethren, the Dokhan Volcanics; granite porphyries and felsites are their subvolcanic equivalents. Virtually undeformed calc-alkaline granites intruded 640-570 Ma ago (Hashad, 1980, Stern and Hedge, 1985; Bentor, 1985). Terrigenous Hammamat sediments were deposited in intermontane and foreland basins. Naturally, a substantial part of the clastic ingredients were derived from the Dokhan Volcanics erupting along the raised magmatic arcs.

With regard to tectonic movements, thrusting of the ophiolites and island arc assemblages took place in a westerly direction. Thrusting preceded and continued during the upswelling of the geanticlines (El-Gaby and El-Nady, 1983). The Rb-Sr isochron age of 626 ± 2 Ma obtained from the quartzo-feldspathic mylonites and deformed granite-gneisses (Sturchio et al., 1983), defines the age of the last main thrust movement and the recrystallization of the blastomylonites at the Meatiq mantled gneiss dome. The blastomylonites were intruded along the southern border of the Meatiq dome by a virtually undeformed, conformable sheet-like tonalite-granodiorite mass 614 ± 8 Ma ago (Stern and Hedge, 1985). Moreover, in the Wadi Hammamat - Quseir region, the Hammamat sediments and associated Dokhan Volcanics are not tectonically deformed, whereas they are affected by thrust faults to the east of the Fawakhir Gold Mine and along Wadi Atalla to the west (El-Gaby et al., 1984). Besides, gravel analysis of the Hammamat conglomerates at Wadi Kareim area indicated eastward deposition suggesting the existence of the raised eastern geanticline (Atalla, 1979). On the other hand, the Hammamat sediments at Wadi Hammamat do not show any preferred direction of deposition (Grothaus et al.,

1979) suggesting that the western geanticline did not function as a water-divide at that time, i.e. the geanticline developed later. Accordingly, we might conclude that tectonic movements and upswelling become progressively younger westward.

Mantle-derived spinel lherzolite, olivine gabbro and meladiorite also intruded during the Cordilleran stage.

The cratonized basement was intermittently intruded by sub-alkaline to peralkaline silicic rocks since the end of the Pan-African orogeny, around 570 Ma ago, and most probably until the Tertiary.

METALLOGENESIS

The pioneering work of Hume (1937) and the classification of the mineral deposits of Egypt by El-Shazly (1957) still influence local geologic thinking. However, in the last decade several publications appeared on metallogenesis and on the relation between plate tectonics and mineral occurrences in the basement of Egypt; reference should be made to Volumes 3 and 6 of the Annals of the Geological Survey of Egypt (1973, 1976), to Garson and Shalaby (1976) and to Al-Shanti and Roobol (1979). Tectonics control magmatic activity and sedimentation which have profound bearing on ore formation. In the following, the ore deposits of the Eastern Desert and Sinai are classified with respect to the tectonic environment prevailing during their development. The ore mineral occurrences discussed here are obtained mainly from the Mineral Map of Egypt (Afia and Imam, 1979) and the Metallogenic Map of Aswan Quadrangle (Shalaby and Rossman, 1983).

1. Pre-Pan-African Mineral Deposits

Pre-Pan-African rocks cropping out in southern Sinai, NED and in gneiss domes in the CED and SED are in their majority either remobilized or mylonitized, so that eventual mineral deposits

were apt to be destroyed. Well-preserved infrastructural rocks
occur on the eastern flank of Hafafit swell, e.g. at Wadi
Nugrus and its tributaries Wadi Sikait and Wadi Abu Rusheid.
This area is characterized by a beryl (emerald) mineralization,
extending for about 45 km from Gabal Zabara to Gabal Um Kabu
(Hassan and El-Shatoury, 1976), which is confined to the mica
schist underneath an obducted ophiolite thrust sheet. Beryl
occurs in the mica schist along the contacts with deformed
leucocratic alkaline granites. The deformed alkaline granite of
Wadi Sikait - Wadi Abu Rusheid, described as columbite-bearing
metasomatized "psammitic gneiss" (Hassan, 1973), contains
thorite, cassiterite and fluorite in addition to columbite.
Abdel Monem and Hurley (1979) obtained a U-Pb zircon age of
1770 Ma from this mass.

2. Pan-African Ore Deposits

Ore deposits formed during the Pan-African Cycle are treated in
connection with the two successive stages of its evolutionary
history, namely
2.1 the island arc stage, and
2.2 the Cordilleran stage.

2.1 Island Arc Stage

Chromite deposits
Chromite deposits occur as boundinaged lenses, several metres
long, in serpentinites at different localities extending from
Gabal El-Rubshi (near Meatiq) in the north till Wadi Allaqi in
the south. The ophiolites, in general, are believed to
represent back-arc oceanic crust. Sulphides are not recorded in
the associated metabasalts.

Banded iron ores and base metal sulphide
Banded iron ores and base metal sulphides occur exclusively in
the island arc assemblage which consists of weakly
metamorphosed andesite-dacite volcanics, subaqueous tuffs and
volcanogenic greywackes. Iron oxides occur in the northern

part, whereas base metals are found in the southern part of the
CED (Fig. 4). This might be attributed to variation in the
depositional environment, where iron oxides formed in an
aerated near-shore facies, while the sulphides were deposited
in a deeper water euxinic environment. The two southernmost
iron ore occurences at Gabal El-Hadid and Um Nar contain
pyrite, chalcophyrite and siderite beside iron oxide minerals
(Sabet et al., 1976); these occurrences may represent
transitional conditions.

Iron ores occur as thin beds, 1-3 m thick, commonly thinly
banded and interbedded with tuff layers that are commonly
described as metasediments, and were therefore considered of
purely sedimentary origin (El-Shazly, 1957). Bishara and Habib
(1973) noticed the intimate association between the iron ore
layers and the enclosing metavolcanics at Semna and considered
it a special type. A volcanogenic origin is generally accepted
at present (Ivanov et al., 1973; Garson and Shalaby, 1976).

Syngenetic, stratabound to stratiform Fe-Zn-Cu-Pb sulphide
deposits are described only from the Um Samiuki area (Garson
and Shalaby, 1976). Pyrite, chalcopyrite, sphalerite and
locally galena frequently occur as stratabound disseminations,
lenses and streaks in banded, weakly metamorphosed andesitic
and dacitic crystal tuffs; they occur in the interstitial
spaces between the feldspar and quartz crystals and crystal
fragments, and were deposited before final lithification of the
tuffs and deposition of cementing colloidal silica and sparry
calcite (El-Aref et al., 1985). The old mines are dug into a
shear zone where massive sulphides occur; these may represent
mobilizations from nearby primary stratabound sulphides. Talc
is developed at higher levels along the shear zone and
metavolcanics are strongly propylitized. Base metal sulphides
are also present at the deeper levels of Darhib and Atshan talc
mines (Shukri and Basta, 1959) which are situated along shear
zones in island arc rock assemblage; they may also represent
mobilizations of stratabound sulphides occurring at depth.

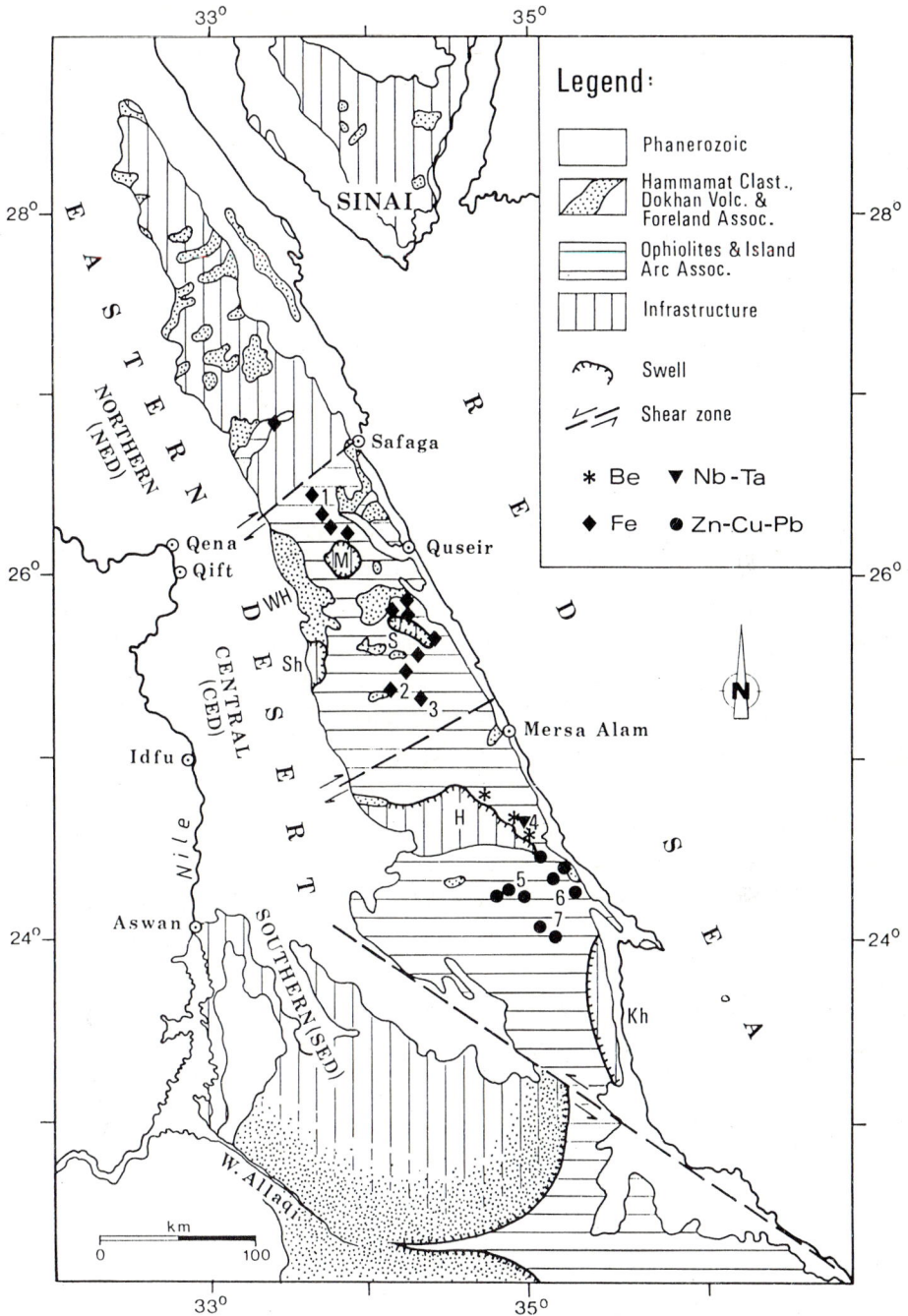

FIGURE 4.
Distribution of pre-Pan-African Be and Nb-Ta mineralization, and island arc banded iron ores and Zn-Cu-Pb
sulphide ocurrences. M = Meatiq, WH = Wadi Hammamat,
S = Sibai, Sh = Shalul, H = Hafafit, Kh = Kuda. (1) Semna,
(2) Gabal El-Hadid, (3) Um Nar, (4) Wadi Abu Rusheid,
(5) Um Samiuki, (6) Atshan, (7) Darhib.

2.2 Cordilleran Stage

During this stage, ore deposits and mineralizations are
directly related to magmatic rocks that can be divided into:
a)   subduction-related calc-alkaline granites and Dokhan
      Volcanics, and
b)   mantle-derived lherzolites and gabbros.

Gold and copper mineralization

All known Au and Au-Cu mineralizations in the Eastern Desert
are of the hydrothermal quartz-vein type; disseminations in the
wall-rocks are not infrequent. Copper mineralization in SE
Sinai and in the Eastern Desert as well as the argentiferous
galena occurrence at Siwikat El-Soda, Barramiya (Sabet and
Saleeb-Roufaiel, 1968), seem to belong to this group. Gold
mineralization is attributed to hydrothermal activity
accompanying diorites of Metarchaean age (Hume, 1937) or
Gattarian granites (El-Shazly, 1957). El-Gaby (1983) noted that
the two gold belts recognized by El-Ramly et al. (1969) in the
CED coincide with the two geanticlines or magmatic arcs present
(Fig. 5). In other words, Au, Au-Cu and Cu mineralizations are
genetically connected to the subduction-related calc-alkaline
magmatic activity, which is manifested in calc-alkaline
granites (G-1 and G-2), subvolcanic granite porphyries and
felsites, and Dokhan Volcanics. This assumption may explain the
spatial relationship between this type of mineralization and
the Dokhan Volcanics as well as the occurrence of Au mineralization in felsite and quartz porphyry dykes and stocks at Um
Mongul Fatiri and Abu Mereiwat (Sabet et al., 1976) and at
Hammash and Um Garayat (Garson and Shalaby, 1976). This would
also suggest that the Au and Au-Cu mineralizations are
epithermal and were formed at relatively shallow depths. The
commonly stressed intimate spatial relationship between the
Dokhan Volcanics and the Hammamat sediments, on the one hand,
and the gold mineralization, on the other hand, would explain
the prolific gold occurrences along Wadi Allaqi, where Hammamat
sediments and Dokhan Volcanics (in the upper part of the
foreland association) form a skirt around the East Aswan swell;

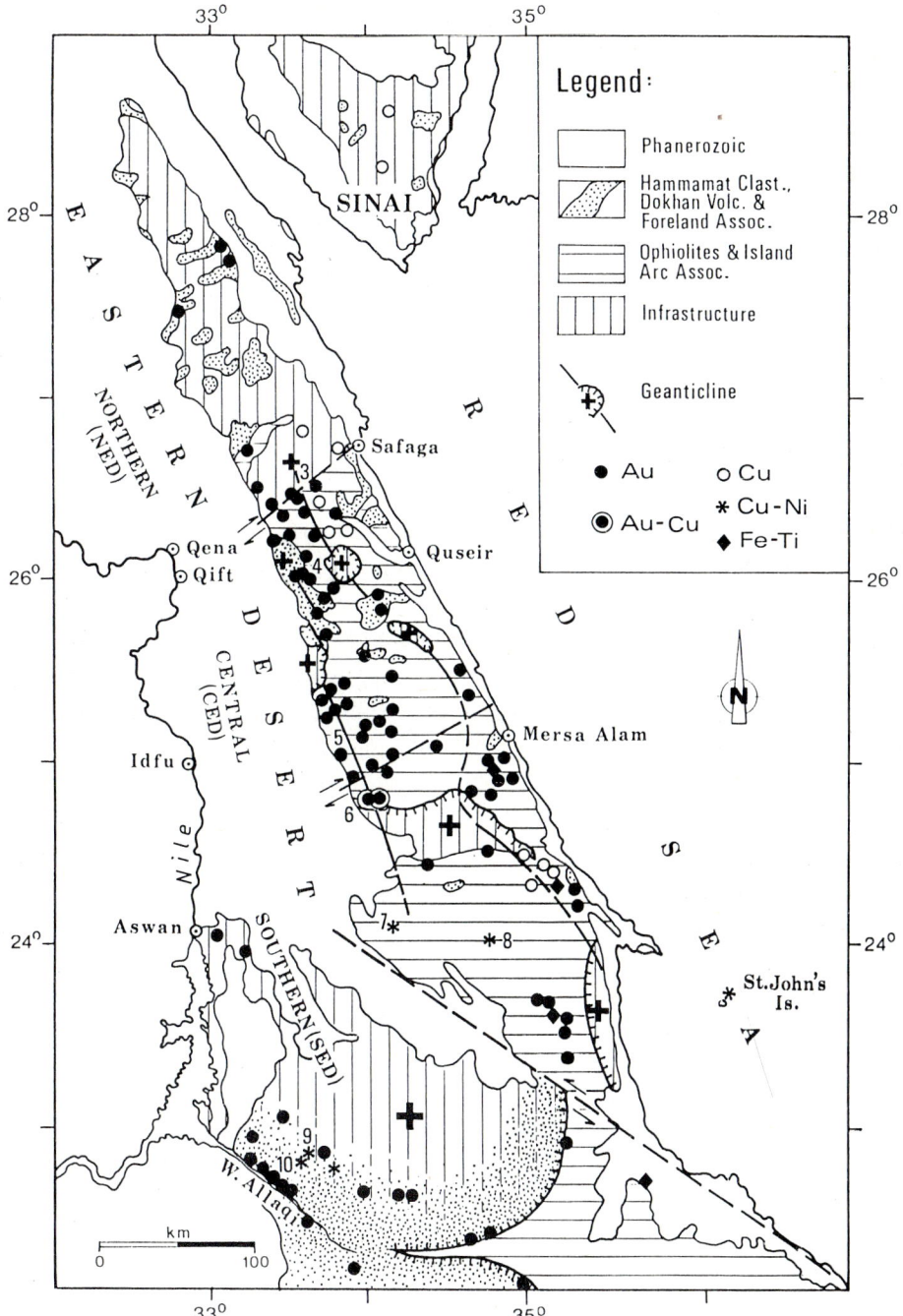

FIGURE 5.
Distribution of ore mineral occurrences formed during the Cordilleran stage. (1) Um Mongul, (2) Fatiri, (3) Abu Mereiwat, (4) Fawakhir, (5) Baramiya, (6) Hamash, (7) Gabbros Akarem, (8) El-Genena West, (9) Abu Swayel, (10) Um Garayat.

the mineralized cover rocks are eroded away from the raised
central part of the swell. Similarly, the gold potential of the
eastern gold belt is less than that of the western belt, since
in the former the mineralized cover rocks are largely removed
by erosion and only the roots of the auriferous quartz veins in
calc-alkaline granites are left over.

Gold mineralization is also associated with G-3 granites, since
auriferous quartz veins occur in alkali-feldspar granite at
Atalla gold mine.

Copper-nickel sulphide deposits

The ancient Egyptians exploited the oxidized top part of the
Abu Swayel Cu-Ni deposit for copper (Hume, 1937). The ore is
essentially composed of pyrrhotite, chalcopyrite and
pentlandite. It occurs in amphibolite (El-Shazly et al., 1969),
which is a thermally metamorphosed peridotite band containing
chromite grains. The deposit is believed to have formed by
magmatic segregation (El-Goresy, 1964). Beside St. John's
(Zabargad) Island, similar Cu-Ni mineralizations have been
discovered in the peridotites of virtually unmetamorphosed
layered basic-ultrabasic intrusions at Gabbro-Akarem (Bugrov
and Shalaby, 1973) and El-Genena West (Kamel et al., 1980). The
oxidized sulphide ores in the peridotites of Zabargad Island
contain appreciable amounts of Ag, Au and Pt as in the case of
Abu Swayel, though not as much (Sabet et al., 1973).

Ilmenite deposits

Ilmenite forms large lenses and bands in intrusive gabbros at
Um Effein, Abu Ghalaga, Abu Dahr and Hamra Dome (Abdel-Tawab,
1978). The gabbros represent the more differentiated phase of
mantle-derived ultrabasic-basic magma, and the ilmenite and
titanomagnetite have crystallized out of the magma under
increasing oxygen fugacity (Basta and Takla, 1974) in the
course of magma differentiation.

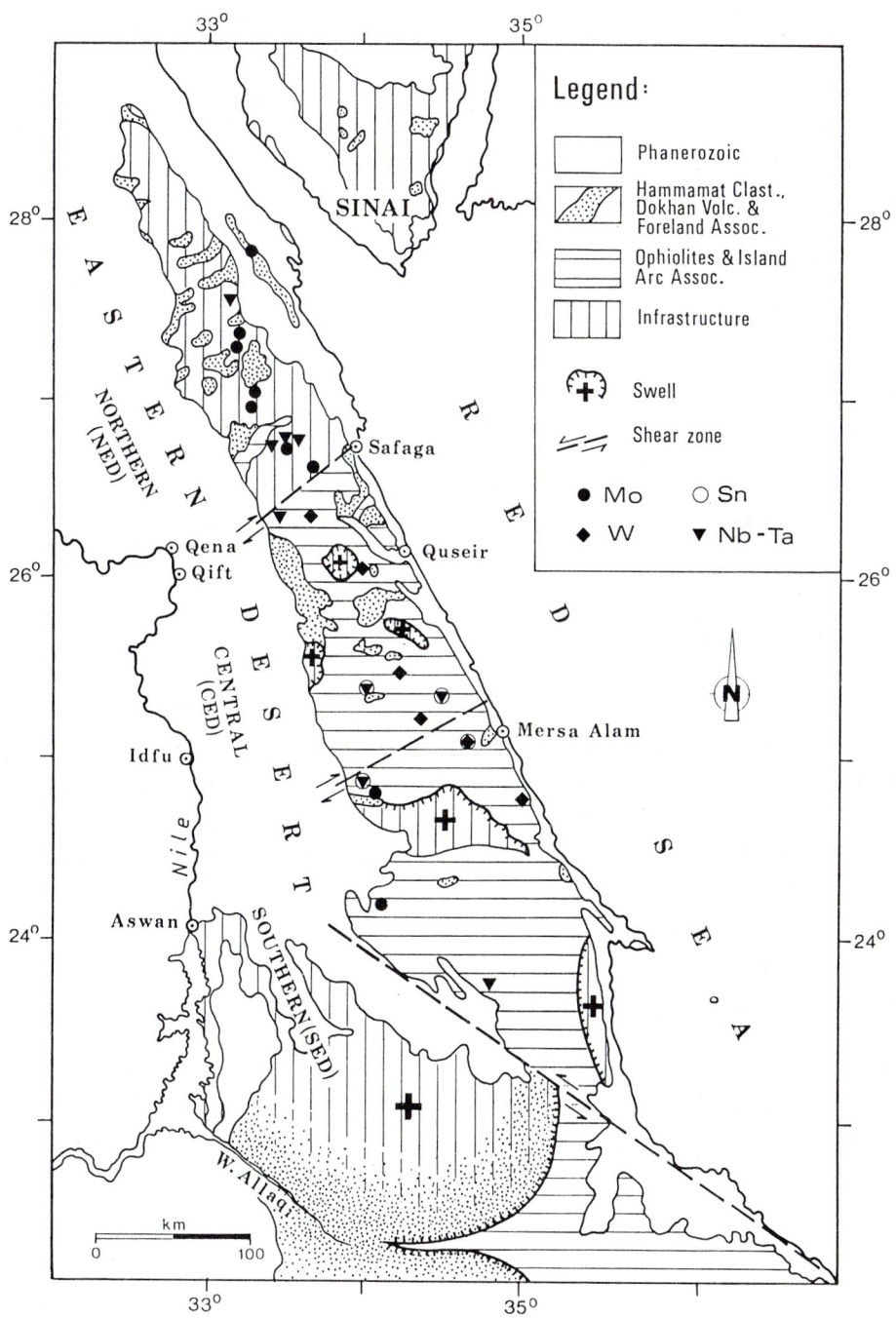

FIGURE 6.
   Distribution of ore mineral occurrences associated with
   G-3 granites.

3. Late Pan-African and Phanerozoic Ore Deposits

About the end of the Pan-African orogeny and during the Phanerozoic, the back-arc region and the cratonized basement were intruded by several anorogenic subalkaline to peralkaline granite plutons, which sometimes contain Mo, Sn, W, Nb-Ta and U mineralizations in the form of disseminations, stockworks or in quartz veins within the plutons or in their close vicinity. A metasomatic origin is ascribed to the Nb-Ta and Sn mineralization in the CED and the Russian term "apogranite" is thus commonly used (Sabet et al., 1973). However, the sequence of crystallization of the rock-forming minerals and accessories in conjunction with the petrochemistry of the enclosing rocks favours direct crystallization from an albite-granite magma rich in fluorine for the so-called "apogranites" of Nuweibi and Abu Dabbab (Asran, 1985).

SUMMARY AND CONCLUSIONS

An Archaen nucleus in the southwestern corner of Egypt is fringed by early Proterozoic continental crust which extends eastwards deep into the Eastern Desert. Late Proterozoic ophiolites and island arc volcanics and volcaniclastics were thrusted from the east over the old continental margin. The older rocks were largely mylonitized at shallow depths, or remobilized at greater depths. The NED and Sinai were upheaved along a right-lateral shear zone trending nearly NE-SW, whereby the Pan-African supracrustal rocks were almost entirely removed by erosion. In the CED, largely mylonitized infrastructural rocks crop out in gneiss domes. The gneiss domes or swells are disposed along the axes of two geanticlines that also functioned as magmatic arcs. Molasse sediments accumulated in the intermontane basins between the raised geanticlines; they are constituted to a large extent from the debris of the erupting Andean-type calc-alkaline volcanics.

The greater Wadi Allaqi region, SED, is believed to have been once a foreland basin in which a great thickness of molasse sediments accumulated above a sedimentary shelf sequence intercalated with abundant limestone layers.

Granodiorite and monzogranite diapirs induced wide-spread low-pressure metamorphism when intruded in molasse basins; anatectic rocks are locally developed.

Emerald and Nb-Ta mineralizations occur in association with gneissose leucocratic granites, about 1800 Ma old, on the eastern flank of Hafafit swell.

Boudinaged chromite lenses occur in ophiolitic serpentinites while banded iron ores and base metal sulphides are associated with Pan-African island arc volcaniclastics.

During the Cordilleran stage, hydrothermal Au and Cu mineralizations were formed in intimate association with calc-alkaline granites and related volcanics.

Post-tectonic subalkaline to peralkaline silicic rocks were emplaced about the end of the Pan-African orogeny and through the Phanerozoic, and probably till the Tertiary. They brought about Mo, Sn, W, U and Nb-Ta mineralizations.

ACKNOWLEDGEMENT

The authors wish to express their gratitude to Mr. Coy Squyres, President of Conoco Coral Inc., Egypt, for logistic and financial support within the frame of the CONOCO-EGPC Map Project. Thanks are also due to Prof. A. Kröner and Prof. W. Pohl for reading the manuscripts and for their valuable suggestions.

REFERENCES

Abdel-Monem, A.A. and P.M. Hurley, 1978. Age of Aswan monumental granite, Egypt, by U-Pb dating of zircon.- Precambrian Res., 6, A4.

Abdel-Monem, A.A. and P.M. Hurley, 1979. U-Pb dating of zircons from psammitic gneisses, Wadi Abu Rusheid-Wadi Sikait area, Egypt, - Inst. Appl. Geol., Jeddah, Bull.3,v.2, 165-170.

Abdel Tawab, M.M. 1978. A comparative study on the metal-logenic provinces in the ophiolite belts and ring complexes in both Egypt and Saudi Arabia between Lat. 22° - 24^ON. - In H. Zapfe (ed.) Scientific results of the Austrian Projects of the IGCP until 1976. - Österr. Adad. d. Wissensch., Schriftr. d. Erdwissenschaftl. Komm., Band 3, 45-55.

Abu El-Ela, E.M., 1984. Mineralogical and geochemical studies on Gebel El-Motaghairat area, SW Berenice, Egypt. - M. Sc. thesis, Assiut Univ., 115 p.

Afia, M.S. and I. Imam, 1979. Mineral map of Egypt, scale 1 : 2.000.000, with explanatory notes and lists. - Geol. Surv. Egypt., 44 p.

Akaad, M.K., S. El-Gaby and M.S. Habib, 1973. The Barud Gneisses and the origin of the grey granite. - Fac. Sci Bull., Assiut Univ., 2, 55-69.

Akaad, M.K. and M.F. El-Ramly, 1963. The cataclastic-mylonitic gneisses north of Gabal El-Maiyit and the origin of the granite of Shaitian type. - Geol. Surv. Egypt, paper No. 26.

Akaad, M.K. and A.M. Noweir, 1980. Geology and lithostratigraphy of the Arabian Desert orogenic belt of Egypt between Lat. $25^O35'$ and $26^O30'$. - Inst. Appl. Geol., Jeddah, Bull. 4, v. 3., 127-134.

Almond, D.C., 1979. Younger granite complexes of Sudan. - Inst.Appl. Geol., Jeddah, Bull. 3, v. 1, 151-164.

Al-Shanti, A.M.S. and M.J. Roobol, 1979. Some thoughts on metal-logenesis and evolution of the Arabian-Nubian Shield. Inst. Appl. Geol., Jeddah, Bull. 3, v. 1, 87-96.

Asran, M.H.A., 1985. Geology, petrography and geochemistry of the apogranites at Nuweibei and Abu Dabbab areas, E.D., Egypt. - M. Sc. thesis, Sohag, Assiut Univ., 149 p.

Atalla, R.F., 1978. Geology and petrography of the Hammamat Conglomerate of Wadi Kareim. - M.Sc. thesis, Assiut Univ., 106 p.

Bakor, A.R., I.G. Gass and C.R. Neary, 1976. Jabal El-Wask, Northwest Saudi Arabia: An Eocambrian back-arc ophiolite. Earth Planet. Sci. Lett., 30, 1-9.

Basta, E.Z., H. Kotb and M.F. Awadallah, 1980. Petrochemical and geochemical characteristics of the Dokhan Formation at the type locality, Jabal Dokhan, E. D., Egypt. - Inst. Appl. Geol., Jeddah, Bull. 3, v. 3, 121-140.

Basta, E.Z. and M.A. Takla, 1974. Distribution of opaque minerals and the origin of the gabbroic rocks of Egypt. - Fac. Sci. Bull., Cairo Univ., 47, 347-363.

Basta, E.Z. and M. Zaki, 1961. Geology and mineralization of Wadi Sikait area, Southeastern Desert. - J. Geol. U.A.R., Cairo, 5, 1-37.

Bentor, Y.K., 1985. The crustal evolution of the Arabia-Nubian massif with special reference to the Sinai Peninsula. Precambrian Res., 28, 1-74.

Beyth, M., G. Grunhagen, A. Itamar and A. Zilberfarb, 1985. A thermal dome in the Ataqa metamorphic wedge in SE Sinai. Sixth Intern. Conf. Basement Tect., Santa Fe, N.M., USA.

Bishara, W.W. and M.E. Habib, 1973. The Precambrian banded iron ore of Semna, E. D., Egypt, - N. Jb. Mineral. Abh., 120, 108-118.

Bonatti, E., P. Hamlyn and G. Ottonello, 1981. Upper mantle beneath a young oceanic rift: Peridotites from the island of Zabargad (Red Sea). - Geology, 9, 474-479.

Bonatti, E., R. Clocchiatti, P. Colantoni, R. Glemini, G. Marinelli, G. Ottonello, R. Santacroce, M. Taviani, A.A. Abdel-Meguid, H.S. Assaf and M.A. El-Tahir, 1983. Zabargad (St. John's) Island an uplifted fragment of sub-Red Sea lithosphere. - J. geol. Soc. London, 140, 677-690.

Bugrov, V.A. and I.M. Shalaby, 1973. First discovery of Cu-Ni sulphides mineralization in gabbro-peridotitic rocks in the Eastern Desert of Egypt. - Annals. Geol. Surv. Egypt., 3, 177-183.

Cahen, L., N.J. Snelling, J. Delhal and J.R. Vail, 1984. The geochronology and evolution of Africa. - Clarendon Press, Oxford, 512 p.

Church, W.R., 1976. Late Proterozoic ophiolites. - Coll. intern. du C.N.R.S. No. 272, Assoc. mafiques-ultramafiques orogenes, 105-117.

Claesson, S., J.S. Pallister and M. Tatsumoto, 1984. Samarium-Neodymium data on two Late Proterozoic ophiolites of Saudi-Arabia and implications for crustal and mantle evolution. - Contrib. Mineral. Petrol., 85, 244-252.

Dixon, T.H., 1981. Gabal Dahanib, Egypt: A Late Precambrian layered sill of komatiitic composition. - Contrib. Mineral. Petrol., 76, 42-52.

El-Aref, M., A. Khudeir and G Hamed, 1985. On the geometry of strata-bound Fe, Cu, Zn and Pb sulphides in the metapyroclastics at Um Samiuki area, E. D., Egypt. - Second Jord. Geol. Conf., Amman, (abstract).

Elbayoumi, R.M.A. and R. Greiling, 1984. Tectonic evolution of a Pan-African plate margin in Southeastern Egypt - A suture zone overprinted by low angle thrusting? In: J. Klerkx and J. Michot (eds). - African Geology, Tervuren, 47-56.

El-Gaby, S., 1975. Petrochemistry and geochemistry of some granites from Egypt. - N. Jb. Mineral. Abh., 124, 147-189.

El-Gaby, S., 1983. Architecture of the Egyptian basement complex. - Proc. Fifth Intern. Conf. Basement Tectonics, Cairo (in press).

El-Gaby, S. and A.A. Ahmed, 1980. The Feiran-Solaf gneiss belt, SW Sinai, Egypt. - Inst. Appl. Geol., Jeddah, Bull. 3, v. 4, 95-105.

El-Gaby, S. and M. El-Aref, 1977. Geological, petrochemical and geochemical studies on the Shait granite at Wadi Shait, E.D., Egypt. - Fac. Sci. Bull., Assiut Univ., 6, 307-239.

El-Gaby, S. and O.M. El-Nady, 1983. Meatiq mantled gneiss dome. - Proc. Fifth Intern. conf. Basement Tectonics, Cairo (in press).

El-Gaby, S., O.M. El-Nady and A.A. Khudeir, 1984. Tectonic evolution of the basement complex in the CED of Egypt. - Geol. Rdsch., 73, 1019-1036.

El-Gaby, S. and M.S. Habib, 1982. Geology of the area SW of Port Safaga, with special emphasis on the granitic rocks, E. D., Egypt. - Annals Geol. Surv. Egypt, 12, 47-71.

El-Goresy, A., 1964. Neue Beobachtungen an der Nickel-Magnet-kies-Lagerstätte von Abu Swayel, Ägypten. - N. Jb. Mineral. Abh., 102, 107-113.

El-Ramly, M.F., 1972. A new geological map for the basement rocks in the Eastern and Southwestern Deserts of Egypt. - Annals Geol. Surv. Egypt, 2, 1-18.

El-Ramly, M.F. and M.K. Akaad, 1960. The basement complex in the CED of Egypt between lat. $24^{\circ}30'$ and $25^{\circ}40'$. - Geol. Surv. Egypt, paper No. 8, 33 pp.

El-Ramly, M.F., S.S. Ivanov and G.G. Kochin, 1969. The occurrence of gold in the Eastern Desert of Egypt. In: O.M. Mahgoub and M.F. El-Ramly (eds), Studies on some mineral deposits of Egypt. - Geol. Surv. Egypt.

El-Ramly, M.F., R. Greiling, A. Kröner and A.A. Rashwan, 1984. On the tectonic evolution of the Wadi Hafafit area and environs, Eastern Desert of Egypt. - Bull. Fac. Sci., King Abdulaziz Univ., 6, 113-126.

El-Sharkawy, M.A. and R.M. Elbayoumi, 1979. The ophiolites of Wadi Ghadir, E.D., Egypt. - Annals Geol. Surv. Egypt, 9, 125-135.

El-Shatoury, H.M., M.E. Mostafa and F.E. Nasr, 1984. Granites and granitoid rocks - a statistical approach of classification. - Chem. Erde, 43, 229-246.

El-Shazly, E.M., 1957. Classification of Egyptian mineral deposits. - Egypt. J. Geol., 1, 1-22.

El-Shazly, E.M., I.A. Farag and F.A. Bassyuni, 1969. Contributions to the geology and mineralization at Abu Swayel area, E.D., Part II: Abu Swayel copper-nickel deposit - Egypt. J. Geol. 13, 1-14.

El-Shazly, E.M., A.H. Hashad, Y.A. Sayyah and F.A. Bassyuni, 1973. Geochronology of Abu Swayel area, SED, Egypt. Egypt. J. Geol., 17, 1-18.

Furnes, H., A.E. Shimron and D. Roberts, 1985. Geochemistry of Pan-African volcanic are sequences in SE Sinai Peninsula and plate tectonic implications. - Precambrian Res., 29, 359-382.

Garson, M.S. and I.M. Shalaby, 1976. Precambrian-Lower Paleozoic plate tectonics and metallogenesis in the Red Sea region. - Geol. Assoc. Canada, Sp. paper No. 14, 573-596.

Gass, I.G., 1979. Evolutionary model for the Pan-African crystalline basement. - Inst. Appl. Geol., Jeddah, Bull. 3, v. 1, 11-20.

Gass, I.G., 1982. Upper Proterozoic (Pan-African) calc-alkaline magmatism in north-eastern Africa and Arabia. In: R.S. Thorpe (ed), Andesites - Orogenic andesites and related rocks, 591-609, Wiley, Chichester.

Gillespie, J.G. and T.H. Dixon, 1983. Lead isotope systematics of some igneous rocks from the Egyptian shield. - Precambrian Res., 20, 63-77.

Gindy, A.R., 1954. The plutonic history of Aswan area, Egypt.-Geol. Mag., 91, 484-497.

Greiling, R.O. and M.F. El-Ramly, 1985. Thrust tectonics in the Pan-African basement of SE Egypt.- CIFEG, Publ. Occ. 1985/3, 73-73

Grothaus, B., D. Eppler ans R. Ehrlich, 1979. Depositional environments and structural implications of the Hammamat Formation, Egypt. - Annals Geol. Surv. Egypt., 9, 70-80.

Habib, M.S, A.A. Ahmed and O.M. El-Nady, 1985. Two orogenies in the Meatiq area of the CED, Egypt. - Precambrian Res., 30, 83-111.

Hafez, A.M.A. and H. El-Amin, 1983. Structure and metamorphism of a Precambrian sequence in Wadi Al-Miah, Barramiya area, E. D., Egypt. - Proc. Fifth Intern. Conf. Basement Tectonics, Cairo (in press).

Hafez, A. and I.M. Shalaby, 1983. On the geochemical characteristics of the volcanic rocks at Um Samiuki, Eastern Desert, Egypt. Egypt. J. Geol., 27, 73-92.

Harris, N.B.W., C.J. Hawkesworth and A.C. Ries, 1984. Crustal evolution in northeast Africa from model Nd ages. - Nature, 309, 773-776.

Hashad, A.H., 1980. Present status of geochronological data on the Egyptian basement complex. - Inst. Appl. Geol., Jeddah, Bull. 3, v. 4. 31-46.

Hashad, A.H., T.A. Sayyah, S.B. El-Kholy and A. Yousef, 1972. Rb-Sr isotopic age determinations of some basement Egyptian granites, Egypt. - Egypt. J. Geol., 16, 269-281.

Hassan, M.A., 1973. Geology and geochemistry of radioactive columbite-bearing psammitic gneiss of Wadi Abu Rusheid, SED, Egypt. - Annals. Geol. Surv. Egypt, 3, 207-225.

Hassan, M.A. and H.M. El-Shatoury, 1976. Beryl occurrences in Egypt. - Mining Geol., 26, 253-262.

Hassan, M.A. and M.A. Essawy, 1977. Petrography of the metagabbro-diorite complex of Wadi Mubarak-Gabal Atud area, E. D., Egypt. - J. Univ. Kuwait (Sci.), 4, 203-213.

Hegazy, H.A.M., 1984. Geology of Wadi El-Gemal area, E.D., Egypt. - Ph. D. Diss., Assiut Univ., Egypt, 238 pp.

Hume, W.F., 1934. Geology of Egypt, Vol. 2, The fundamental Precambrian rocks of Egypt and the Sudan; Part 1: The metamorphic rocks. - Geol. Surv. Egypt, 300 pp.

Hume, W.F., 1935. ditto, Vol. 2, Part 2: The later plutonic and intrusive rocks, 301-688 pp.

Hume, W.F., 1937. ditto, Vol. 2, Part 3: The minerals of economic value associated with the intrusive Precambrian igneous rocks, 689-990 pp.

Hunting Geology and Geophysics Ltd., 1967. Assessment of the mineral potential of the Aswan region, U.A.R.: Photogeological survey; U.N. Dev. Progr.; U.A.R. Regional planning of Aswan, 138 p.

Hussein, A.A.A., M.M. Ali and M.F. El-Ramly, 1982. A proposed new classification of the granites of Egypt. - J. Volc. Geoth. Res., 14, 187-198.

Hussein, A.A.A. and A.H. Rasmy, 1976. Kyanite in a contact
    aureole at Nasb Aluba, SED. - 14th Ann. Meet. Geol. Soc.
    Egypt (abstract).
Jäger, E., 1977. The evolution of the central and west European
    continent. In: La Chaìne Varisque d'Europe Moyenne et
    Occidentale. - Coll. Intern. CNRS Rennes No. 243, 227-239.
Kabesh, M.L., M.E. Hilmy and A.M. Bishady, 1967. Geology of the
    basement rocks in the area around Um Rus Gold mines,
    E.D. - J. Geol. Egypt, 11, 59-85.
Kamal El-Din, G.M., 1986. Geology of Wadi Um Had area, E.D.,
    Egypt. - M. Sc. thesis, Fac. Sci., Qena, Assiut Univ.
    Egypt.
Kamel, O.A., I.M. Shalaby and M.M. El-Mahallawy, 1980. Petrolo-
    gical study of basic-ultrabasic suite at El-Genena
    El-Garbia, SED, Egypt. - Annals Geol. Surv. Egypt, 10,
    725-749.
Khudeir, A.A., 1983. Geology of the ophiolite suite of
    El-Rubshi area, E. D., Egypt. - Ph. D. diss., Assiut
    Univ., Egypt.
Klerkx, J. and S. Deutsch, 1977. Resultats preliminaires
    obtenus par la methode Rb/Sr sur làge des formations
    Précambriennes de la région d'Uweinat (Libye). - Musée
    Royal d'Afrique Centrale, Tervuren (Belgium). Dept. Geol.
    Min., Rap. Ann., 83-94.
Klitzsch, E., F.K. List, G Pöhlmann, R. Handley, M. Hermina and
    B. Meissner (Eds.), 1986 ff. Geological map of
    Egypt 1 : 500.000, 20 sheets. - Cairo (E.G.P.C.).
Kröner, A., 1985. Ophiolites and the evolution of tectonic
    boundaries in the Late Proterozoic Arabian-Nubian shield
    of Northeast Africa and Arabia. - Precambrian Res., 27,
    277-300.
Kröner, A., R. Greiling, T. Reischmann, I.M. Hussein,
    R.J. Stern, J. Kruger, S. Durr and M. Zimmer, 1987. Pan-
    African crustal evolution in the Nubian segment of
    northeast Africa.-Am. Geophys. Union, Spec. Publ. 17 (in
    press).
Moody, J.D. and M. J. Hill, 1956. Wrench-fault tectonics. -
    Geol. Soc. Am. Bull., 67, 1207-1246.
Nasseef, A.O., A.R. Bakor and A.H. Hashad, 1980. Petrography of
    possible ophiolitic rocks along Qift-Quseir road, E.D.,
    Egypt. - Inst. Appl. Geol., Jeddah, Bull. 3, v. 4, 77-82.
Pitcher, W.S., 1983. Granite: Topology, geological environment
    and melting relationshiphs. In: M.P. Atherton and
    C.D. Gripple (eds.), Migmatites, melting and metamorphism.
    - Shiva Pub. Ltd., Cheshire, UK, 277-285.
Ramberg, H., 1981. The role of gravity in orogenic belts. In:
    K.R. McClay and N.J. Price (eds.), Thrust and nappe
    tectonics. - Geol. Soc. London, Sp. Publ. No. 9, 125-140.
Reymer, A.P.S., A. Mathews and O. Navon, 1984. Pressure-
    temperature conditions in the Wadi Kid metamorphic
    complex: implications for the Pan-African event in SE
    Sinai. - Contrib. Mineral. Petrol. 85, 336-345.
Richter, A. and H. Schandelmeier, 1987. The Precambrian
    basement inliers of the Western Desert - geology,
    petrology and structural evolution. - In: R. Said (ed.),
    The Geology of Egypt, (in press).

Ries, A.C., R.M. Shackleton, R.H. Graham and W.R. Fitches, 1983. Pan-African structures, ophiolites and mélanges in the E. D. of Egypt: a traverse at 26°N. - J. geol. Soc. London, 140, 75-95.

Sabet, A.H., V. Chabanenco and V. Tsogeov, 1973. Tin-tungsten and rare-metal mineralisation in the CED of Egypt. - Annals Geol. Surv. Egypt, 3, 75-86.

Sabet, A.H., S. El-Gaby and A.A. Zalata, 1972. Geology of the basement rocks in the northern parts of El-Shayib and Safaga sheets. - Annals Geol. Surv. Egypt, 2, 111-128.

Said, R., 1962. The geology of Egypt. - Elsevier, 377 p.

Samuel, M.D., 1977. Lithological and sedimentological studies on the Red Beds of Wadi Igla, E. D., Egypt.- N.R.C. Bull., 2, 287-297.

Schandelmeier H., A. Richter and G. Franz, 1983. Outline of the geology of magmatic and metamorphic units from Gebel Uweinat to Bir Safsaf (SW Egypt / NW Sudan).- J. Afr. Earth Sci, 1, 275-283.

Schandelmeier, H., D.P.F. Daryshire, U. Harms and A. Richter, 1987. The East Sahara Craton: Evidence for pre-Pan-African crust in NE Africa wet of the Nile. - In S. El-Gaby and R.O. Greiling (eds.), The Pan-African belt of Northeast Africa and adjacent areas, Earth Evol. Sci., (in press).

Schmidt, D.L., D.G. Hadley and D. Stoeser, 1979. Late Protero-zoic crustal history of the Arabian shield. - Inst. Appl Geol., Jeddah, Bull. 3, v. 1, 41-58.

Schürmann, H.M.E., 1953. The Precambrian of the Gulf of Suez area. - 19th Intern. Geol. Cong., Algiers, C.R., 1, 115-135.

Schürmann, H.M.E., 1966. The Precambrian along the Gulf of Suez and the northern part of the Red Sea. - Brill. Leiden, 404 p.

Shalaby, I.M. and D.L. Rossmann, 1983. Metallogenic map of the Aswan Quadrangle, Egypt, scale 1 : 500 000. - Geol. Surv. Egypt.

Shackleton, R.M, 1979. Precambrian tectonics in north-east Africa. - Inst. Appl. Geol., Jeddah, Bulol. 3, v. 2, 1-6.

Shackleton, R.M., A.C. Ries, R.H. Graham and W.R. Fitches, 1980. Late Precambrian ophiolitic mélange in the E.D. of Egypt. - Nature, 285, 472-474.

Shimron, A.E., 1980. Proterozoic island arc volcanism and sedimentation in Sinai. - Precambrian Res., 12, 437-458.

Shimron, A.E. and H.J. Zwart, 1970. The occurrence of low pressure metamorphism in the Precambrian of the Middle-East and North East Africa. - Geol. Mijnb., 49, 369-374.

Shukri, N.M. and E.Z. Basta, 1959. A note on the occurence of a polysulphide deposit at Gabal Derhib, SED, Egypt. - Egypt. J. Geol., 3, 167-173.

Stacey, J.S. and C.E. Hedge, 1984. Geochronologic and isotopic evidence for early Proterozoic continental crust in the eastern Arabian Shield. - Geology 12, 310-313.

Stacey, J.S. and D. Stoeser, 1983. Distribution of oceanic and continental leads in the Arabian-Nubian shield.- Contrib. Mineral. Petrol., 84, 91-105.

Stacey, J.S., D.B. Stoeser, W.R. Greenwood and L.B. Fischer, 1984. U-Pb zircon geochronology and geological evolution of the Halaban - Al Amar region of the eastern Arabian shield, Kingdom of Saudi Arabia. - J. geol. Soc. London, 141, 1043-1055.

Stern, R.J., 1981. Petrogenesis and tectonic setting of Late Precambrian ensimatic volcanic rocks, CED of Egypt. - Precambrian Res., 16, 195-230.

Stern, R.J., D. Gottfried and C.E. Hedge, 1984. Late Precambrian rifting and crustal evolution in the NED of Egypt. - Geology, 12, 168-172.

Stern, R.J. and C.E. Hedge, 1985. Geochronologic and isotopic constraints on Late Precambrian crustal evolution in the E. D. of Egypt. - Am. J. Sci., 285, 97-127.

Stoeser, D.B. and J.S. Stacey, 1987. Evolution, U-Pb geochronology and isotope geology of the Pan-African Nabitah orogenic belt of the Saudi Arabian Shield. - In: Pan-African Belt of NE Africa and adjacent areas, S. El-Gaby, and R.O. Greiling (eds.). Earth evol. sci. (Vieweg).

Sturchio, N.C., M. Sultan and R. Batiza, 1983. Geology and origin of Meatiq dome, Egypt: a Precambrian metamorphic core complex? - Geology, 11, 72-76.

Vail, J.R., 1983. Pan-African crustal accretion in north-east Africa. - J. Afr. Earth Sci., 1, 285-294.

Windley, B.F., 1977. The evolving continents. - J. Wiley and Sons Ltd., New York, 385 p.

Wyllie, P.J., 1983a. Experimental studies on biotite- and muscovite-granites and some crustal magmatic sources. In: M.P. Atherton and C.D. Gribble (eds.), Migmatites, melting and metamorphism. - Shiva Pub. Ltd., Cheshire, UK, 12-26.

Wyllie, P.J., 1983b. Experimental and thermal constraints on the deep-seated parentage of some granitoid Magmas in subduction zones. In: M.P. Atherton and C.D. Gribble (eds.), Migmatites, melting and metamorphism. - Shiva Pub. Ltd., Cheshire, UK, 37-51.

Chapter 3

The East Saharan Craton: Evidence For Pre-Pan-African Crust in NE Africa West of the Nile

H. Schandelmeier[1] / D. P. F. Darbyshire[2] / U. Harms[1] / A. Richter[1]
[1] Special Research Project Arid Areas, TU Berlin, Ackerstraße 71, 1000 Berlin 65, FRG.
[2] British Geological Survey, 64 Gray's Inn Road, London WC1X 8NG, UK)

Keywords: Archean, Proterozoic, Pan-African, Egypt, Sudan, NE Africa, East Sahara Craton, Arabian-Nubian Shield, Geochronology, Rb/Sr, Sm/Nd

Abstract: A Precambrian basement west of the Nile (East Saharan Craton) in southern Egypt and northern Sudan comprises rock assemblages like granulites, migmatites, migmatitic gneisses, calcsilicates, marbles and other metasedimentary rocks which are generally considered to be typical for cratonic areas. They are significantly different in their lithology, structure and metamorphism from the volcano-sedimentary-ophiolitic sequences which are exposed east of the Nile in the Late Proterozoic Arabian-Nubian Shield. Although these rocks are generally believed to be of Pre-Pan African age, no radiometric age data are available to prove this view, except in the Gebel Uweinat basement inlier, where undoubtedly Archean/Early Proterozoic and Mid-Proterozoic rocks are exposed. However, new Sr and Nd istopic data suggest a major crustal formation event in the Mid-Proterozoic. Nd Model ages derived from gneisses and from granitoid rocks point to an Early- to Mid-Proterozoic

protolith. Another series of calc-alkaline granitoids yielded
Nd model ages between 1400 Ma and 900 Ma. However, the
intrusive rocks have all Rb/Sr whole rock ages between 680 Ma
and 560 Ma, reflecting intensive alkaline to peralkaline
within-plate magmatism in the adjacent Arabian-Nubian Shield.

Pan African accretional and collisional tectonics in the
Arabian-Nubian Shield had its impact on the foreland of the
cratonic continental plate, expressed as intensive overprinting
of older rocks, resetting of isotope systems, generation of
calc-alkaline granitoid bodies and widespread intraplate
fracturing and mylonitization.

INTRODUCTION

The Arabian-Nubian Shield of NE Africa developed through a
process of horizontal crustal accretion during the Pan African
period (Gass, 1981; Vail, 1983; Kröner, 1985) from about 950 Ma
to 550 Ma. The western edge of this Shield with its extensive
volcano-sedimentary-ophiolitic   and   granitoid   assemblages
changes west of the Nile into a series of high grade migmatitic
gneisses, migmatites and partly granulites with intercalated
high grade supracrustal rocks like marbles, calcsilicates and
amphibolites. Although no discrete Pre-Pan African ages from
these rocks in southern Egypt and northern Sudan are yet
available, there is considerable evidence that these rocks
belong to an older sialic continental plate (Abdel-Monem and
Hurley, 1979; Harris et al., 1984; Bernau et al., 1987) which
was named Nile Craton (Rocci, 1965; referring to the Uweinat
inlier) or East Sahara Craton (Kröner, 1979; referring to the
area between Hoggar and Arabian-Nubian Shield). Since there is
enough evidence now, that the area between Tibesti and Arabian-
Nubian Shield was underlain by a continuous Pre-Pan African
crust, we decided to apply for this area the term "East Saharan
Craton".

Rb/Sr whole rock ages (Klerkx and Deutsch, 1977) and Nd model
ages (Harris et al., 1984) which were obtained from granulitic
rocks and anatectic gneisses from the Uweinat area in SE Libya
prove a Late Archean age, as well as a Mid-Proterozoic event
which had affected these rocks.

The area between Gebel Uweinat in the west and the Arabian-
Nubian Shield in the east (Fig. 1) is part of the eastern
foreland of a Pre-Pan African continent in NE Africa and it can
be shown that the impact of Pan African tectonic and magmatic
events (high grade metamorphism, low grade overprint,
mylonitization and emplacement of granitic bodies) within this
zone increases considerably from west to east (Richter, 1986;
Bernau et al., in press). The deformational, metamorphic and
magmatic history of the easternmost margin of this plate

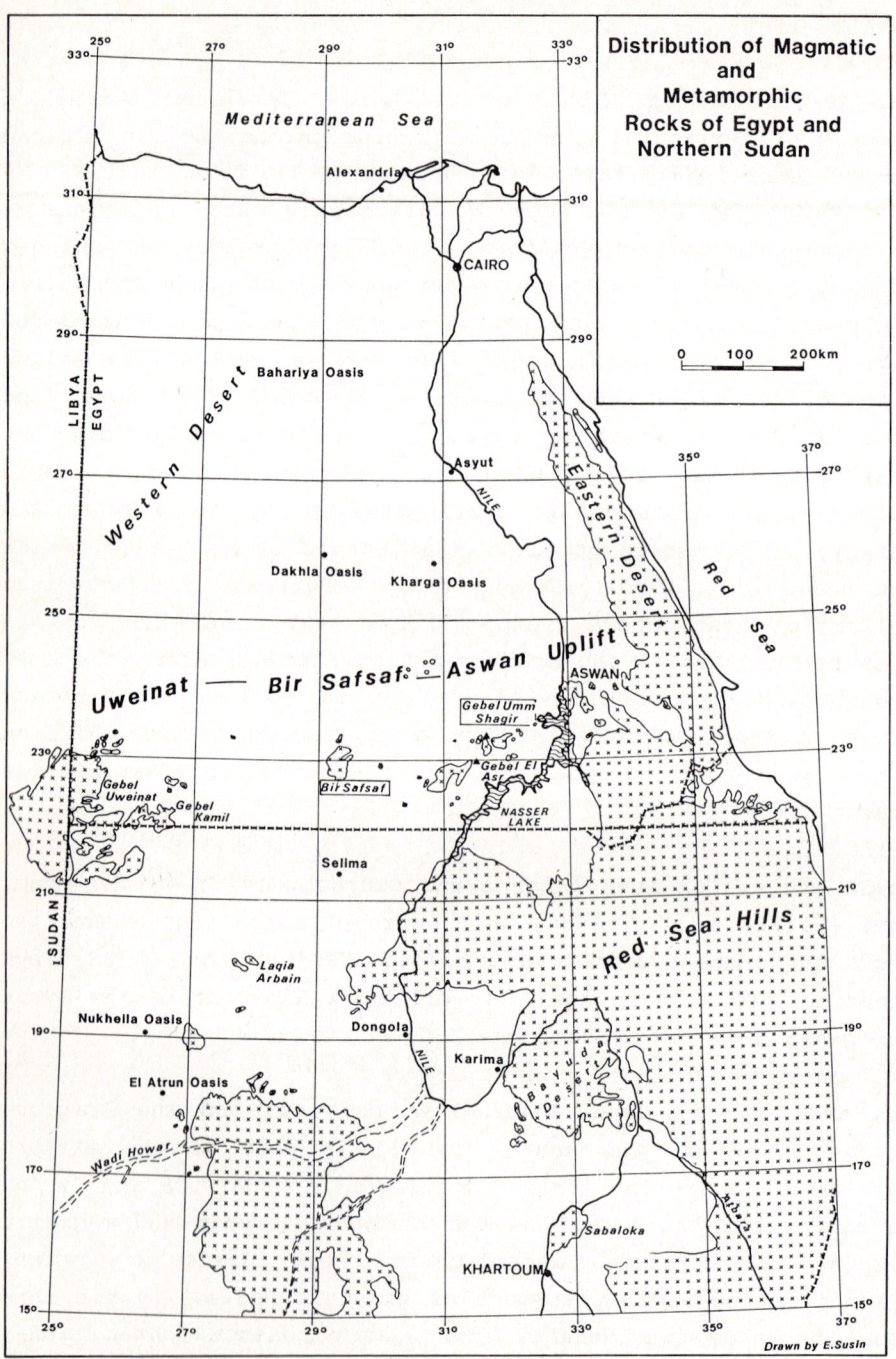

FIGURE 1.
     Distribution of basement rocks (crosses) in Egypt, northern Sudan and SE Libya

(particularly in southern Egypt) is closely related to accretional and collisional events of the Arabian-Nubian Shield with the African plate.

OUTLINE OF GEOLOGY

Unlike the Eastern Desert and the Red Sea Hills in Sudan, the Precambrian of NW Sudan and SW Egypt is not continuously exposed. Extensive areas are covered by generally flat lying, undisturbed sedimentary strata of the Nubian Group (Klitzsch, 1983). Basement rocks occur as structural highs of varying size. The distance between some of them is several hundreds of kilometres and this doubtlessly restricts correlations of rock units.

The areas which were investigated are (see Fig. 2): The Uweinat-Gebel Kamil basement inlier in SW Egypt and NW Sudan, the Bir Safsaf - Aswan uplift in southern Egypt, the Nubian Desert basement area west of the Nile, the Wadi Howar basement area north of Wadi Howar and the minor inliers around Nukheila and Laquiya Arbain, all in northern Sudan. Many smaller outcrops of basement rocks, especially in the Nubian Desert and the Bir Safsaf - Aswan uplift, generally show the same rock associations as in the above mentioned basement inliers.

Up to now these areas were characterized as "undifferentiated basement" (Geological map of the Sudan, 1 : 2 000 000, 1981) or even more confusing, differentiated according to a system which was developed for the Eastern Desert of Egypt (Geologic map of Egypt, 1 : 2 000 000, 1981).

Gebel Uweinat - Gebel Kamil basement inlier

The major rock assemblages of the Gebel Uweinat-Gebel Kamil basement inlier were subdivided by Richter (1986) on a basis of lithology and metamorphic grade into three major units: The

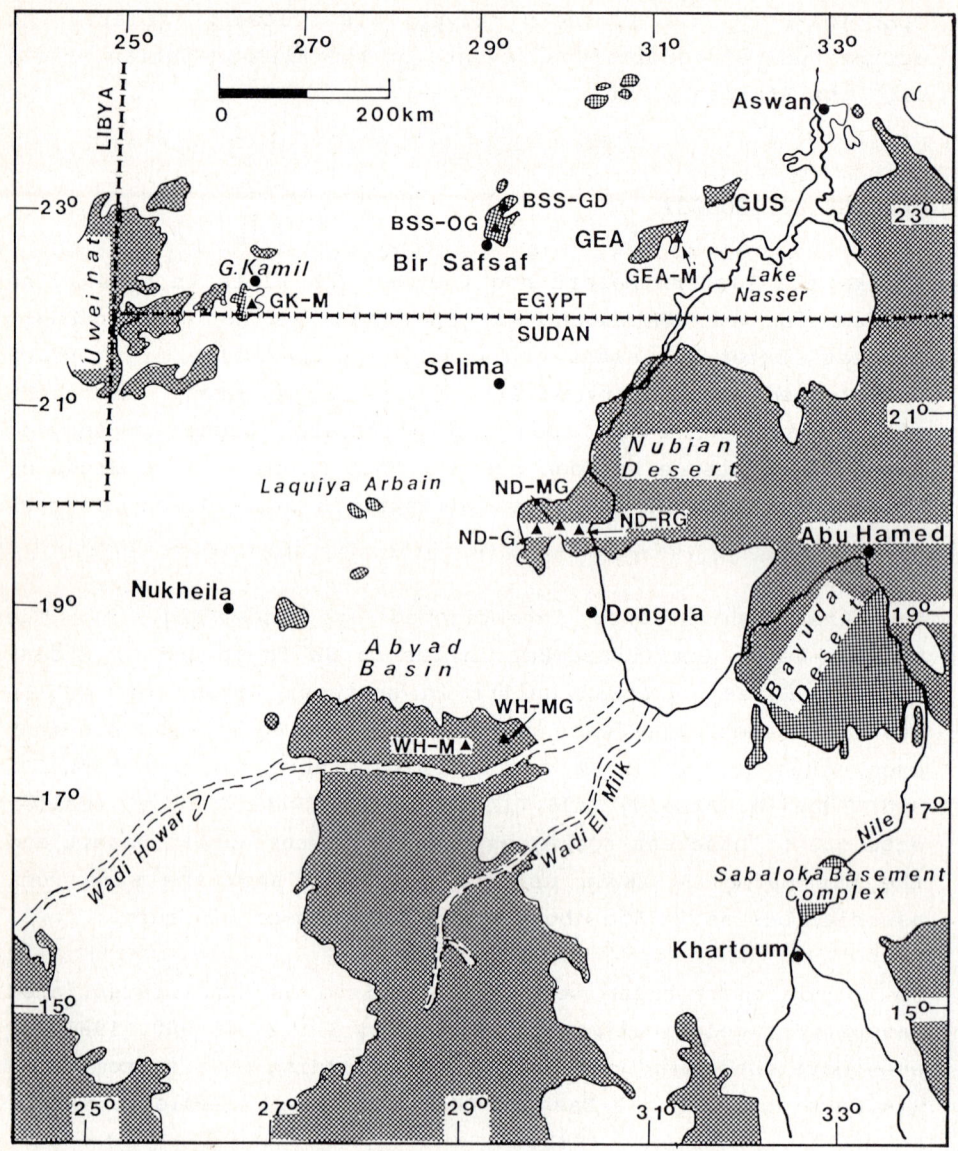

FIGURE 2.
    Basement areas in southern Egypt and northern Sudan
    showing sample locations ( ); for comparison see Table 2.
    GEA = Gebel El Asr area; GUS = Gebel Umm Shagir area

Granoblastite Formation, the Anatexite Formation and the Metasedimentary Formation (exposed in NW Sudan and not shown in Fig. 3). The Granoblastite Formation consists mainly of quartzofeldspathic gneisses, minor metamorphosed basic and ultrabasic rocks and a few intercalated supracrustal rocks. The Anatexite Formation is dominated by migmatites interpreted as anatectic granulites and contains abundant supracrustal intercalations. Both rock series underwent granulite facies metamorphism, HP-HT for the Granoblastite Formation and LP-HT for the Anatexite Formation. The Metasedimentary Formation is a low to medium grade supracrustal sequence of psammitic and semipelitic gneisses and schists; concordant intercalations of bimodal basaltic-rhyolitic rocks are believed to be the source of layered ironquartzites. The tectonic environment of these rocks is interpreted as an intracontinental rift (Richter, 1986).

Intrusive rocks are mainly diorites to granodiorites and rapakivi granites. Foliation indicates for the majority of them at least a Pan-African age, but a number of them might be considerably older. The lack of radiometric age data does not permit a more detailed approach.

The major style of folding of the basement is tight to isoclinal. NNE to NE fold axis trends dominate, but locally they change to E-W. Axial planes are frequently overturned to the east. The deformation of the area occurred in the Mid-Proterozoic, accompanied and succeeded by the regional anatectic event (~ 1800 Ma; Klerks, 1980; Cahen et al., 1984). The characteristic sigmoidal folds, clearly visible on satellite images, may be due to a clockwise rotation of smaller crustal blocks during the culmination of this tectono-thermal event.

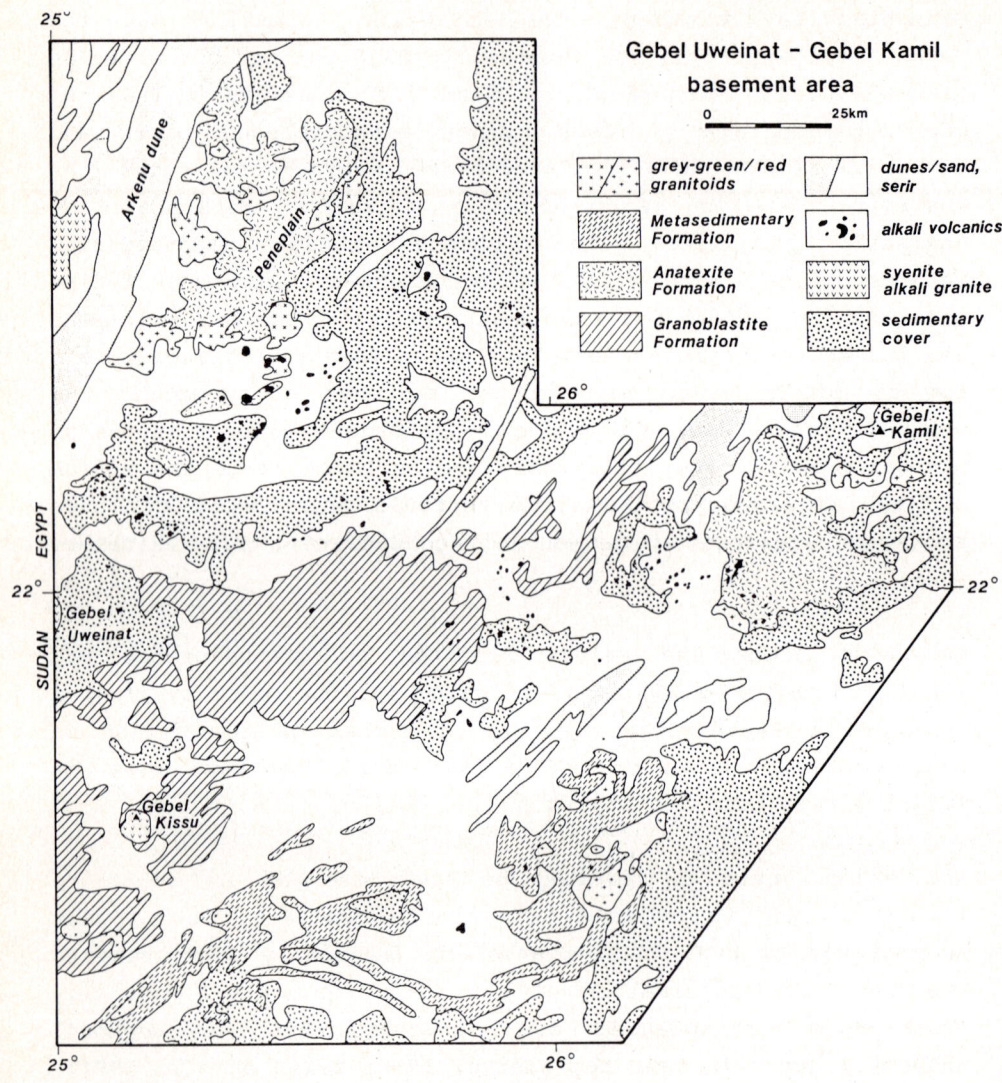

FIGURE 3.
    Geological sketchmap of the Gebel Kamil basement area.

Bir Safsaf - Aswan uplift

The Bir Safsaf - Aswan uplift forms a major basement high in
southern Egypt. Basement is exposed in three areas: Bir Safsaf,
Gebel El Asr and Gebel Umm Shaghir. The Precambrian lithologic
units are similar in all three areas (Bernau et al., 1987).

The oldest unit consists of high grade gneisses of granitic
composition, frequently with migmatitic structures.
Intercalated into this series are amphibolites, marbles and
calcsilicate rocks. Metagranodiorites and metatrondhjemites
also occur within the granitic gneisses. This series could be
the equivalent to Richter's (1986) Anatexite Formation, except
that granulites and granoblastites, like in the western part of
the Gebel Kamil area (Fig. 3) are absent. However, detailed
investigations on the metamorphic development of calcsilicate
rocks from the Bir Safsaf - Aswan uplift revealed P-T
relationships which are very similar to those of the granulites
and granoblastites from the Gebel Uweinat-Gebel Kamil area.

The migmatitic gneisses are intruded by small bodies of
syntectonic S-type granites and disrupted by intrusions of late
tectonic LILE-enriched diorites, granodiorites and biotite-
granites. Dykes and volcanic plugs of various age are common.

The Pre-Pan African deformational history is much more unclear
than in the Gebel Uweinat-Gebel Kamil area. Migmatization and
intensive fracturing and intrusions of granitoids have dis-
located the original gneissic foliation in many places.
Mylonitic shear zones in the Gebel El Asr and Gebel Umm Shaghir
areas, which generally strike E-W, post-date the migmatization.
this can be clearly seen from quartz-c-axis fabrics (Bernau et
al., 1987). The mylonitic shear planes seem to be related to
thrust fault movements in the Late Proterozoic, contemporaneous
with generally SW to W directed thrusting which occured when
the Nubian Shield was accreted onto continental Africa (for
comparison see El-Ramly et al., 1984).

Northern Sudan basement areas:

(Nubian Desert, Wadi Howar, Nukheila, Laquiya Arbain)

All parts of the northern Sudan basement areas are, compared with previously described areas, very poorly known. Knowledge of the petrology and lithostratigraphy of the rock units is still in the reconnaissance stage (Huth et al., 1984). The Nubian Desert area west of the river Nile (Fig. 2) frames the Mesozoic Abyad basin. The westernmost part is dominated by grey granitic gneisses which are intercalated with marbles, calc-silicates, talc schists, chlorite schists, quartzites and quartzgarnet schists. These units strike N-S and dip 60° to 80° to the east. They are isoclinally folded. The central and eastern part of the area is occupied by red granitic gneisses which appear to be folded into a huge N-S trending syncline, at least 30 kilometres wide. The dip of its western limb is steeply (80°) to the east while the eastern limb dips about 40°-50° to the west.

Two major granite bodies intrude the basement gneisses. In the central part is a medium grained biotite granite body of about 20 kilometres in diameter and in the east near the Nile is a more coarse grained, unfoliated, homogeneous biotite granite body of at least 15 kilometres in diameter. The latter is certainly the youngest major intrusive body in the area. Both are chemically very similar LILE-enriched biotite-granites (Bernau et al., 1987).

The eastern and central Wadi Howar area is dominated by exposures of granite. Older units include steeply dipping, isoclinally folded, granitic gneisses with a general NE-SW strike. They are intercalated with metasediments including marbles and calcsilicates. The basement in the western part of the Wadi Howar is very poorly exposed and presently unknown.

The minor basement exposures east of Nukheila (Fig. 2) consist mainly of grey granitic gneisses with intercalations of amphibolites, marbles, calcsilicates, chlorite-schists and quartzites. Minor intrusions of grey and red granites occur in the central and eastern parts, as well as in smaller outcrops near the oasis Laquiya Arbain (Fig. 2).

CHRONOSTRATIGRAPHY

Compared to the vast extension of the area concerned, radiometric age data are extremely scarce. For this reason, correlations of rock units are often exclusively based on geochemical, structural and metamorphic similarities and must be necessarily tentative. Nevertheless, there are sufficient data available now to establish a preliminary chronostratigraphic sequence within the basement of NW Sudan and SW Egypt.

Mid Archean to Early Proterozoic

Although a number of workers (Meinhold, 1979; Almond, 1980; Vail, 1983; El-Ramly et al., 1984; Vail, 1985; Kröner, 1985) agree that the high grade gneiss and migmatite terrane in the basement west of the Nile is part of a Pre-Pan-African continental plate, there is little data to support this interpretation. Apart from relatively unreliable ages which are derived from single sample Rb/Sr and K/Ar analyses (Marholz, 1968; Hunting Geology and Geophysics Ltd., 1974; Schürmann, 1974), the only evidence for Archean crust in NE Africa comes from the Gebel Uweinat area in SE Libya. A seven point Rb/Sr age from pyroxene granulites (Karkur Murr series) yielded 2656 ± 71 Ma (Cahen et al., 1984, recalculated from Klerkx and Deutsch, 1977). The former authors concluded that this was the time of the amphibolite facies metamorphism which affected the granulitic rocks (Klerkx, 1980). Granulite facies metamorphism may have occurred around 2900 Ma as suggested by Rb/Sr model age calculations on one granulite sample. The Archean age is confirmed by Nd model ages of two samples of the Karkur Murr granulites which are 3000 Ma and 3200 Ma (Harris et al., 1984). $_{Nd}$ values at the time of amphibolite facies metamorphism were moderately negative (-5.2 and -6.6) indicating that the 2650 Ma event involved considerable crustal reworking. There are no other rocks from the area which have a uniquely Archean source. Nevertheless, few arguments indicate that crust, at least as old as Mid Proterozoic, was present east of the Gebel Uweinat inlier.

Based on detailed petrological and structural investigations in the Gebel Kamil area (Fig. 3), Richter (1986) concluded, that the Granoblastite Formation in the western part of the Gebel Kamil area might be correlated with the Karkur Murr series of Uweinat and the Anatexite Formation of central and eastern Gebel Kamil area (Fig. 3) with the Ayn Daw series (Klerkx, 1980) of Uweinat. In particular the polymetamorphic history of the Gebel Kamil rocks was similar to that of the Uweinat area (Richter, 1986).

In the Gebel El Asr and Gebel Umm Shaghir areas (Fig. 2), calcsilicate rocks were investigated by Bernau et al. (1987) to determine their metamorphic conditions. Although these rocks are no granulites in the petrological sense, their metamorphic history can be compared with those of the Gebel Uweinat and Gebel Kamil areas. A high T (800° to 1000° C) and high P (P_{tot} = 8 to 9 kbar) granulite facies event was followed by an amphibolite facies event (T = 650° to 700° C; P_{tot} = 6 kbar), responsible for migmatization of the country rocks. A low grade metamorphic overprint was the final event.

Mid Proterozoic

There are no radiometric determinations of Mid Proterozoic age for the crust of this region east of Uweinat. Abdel Monem and Hurley (1977) reported an upper intercept U/Pb zircon age of 1770 ± 40 Ma from arkosic metasediments which were collected from southeast of the Hafafit dome in the Eastern Desert of

Egypt. This age is interpreted by these authors as the probable age of the crustal block that supplied the detritus for the original sediments. Dixon (1981) separated zircons from conglomerate units which occur within sequences of Late Proterozoic metasediments in the Eastern Desert of Egypt. He dertermined ages of 1100 to 2100 Ma and argued that these zircons were derived from adjacent continental areas west of the Nile. There is additional evidence that in Mid Proterozoic times a significant amount of new continental crust formed between the Uweinat-Kamil area and the present Nubian Shield. Harris et al. (1984) reported six Nd model ages between 1600 Ma and 1900 Ma from rocks which are more or less located near the Nile.

The Nubian Desert granitic gneisses are exposed in the western part of the Nubian Desert west of the Nile (Fig. 2). A Rb/Sr whole rock regression line age through four points indicates an Early Pan African resetting of the istope system at 918 ± 40 on these rocks which is supported by the high Sr_i ratio of 0.7161 ± 0.0007 (Schandelmeier et al., in prep.). A Nd model age of 2004 Ma (Table 2, ND G 6) was determined for these gneisses and the $_{Nd}$ value at T = 918 Ma is -15.8, indicating considerable reworking of old continental crust.

A Nd model age from a sample from the Gebel Kamil migmatite is 1913 Ma. This rock which suffered a Late Proterozoic resetting of the Sr isotope system at 673 Ma (Schandelmeier and Darbyshire, 1984) has an $_{Nd}$ value at T = 673 Ma of -18.9 (Table 2, GK M 3), characteristic for reworked older crust. The location of these rocks is more than 600 km west of the oceanic terrane/continental plate boundary. There is no indication of juvenile Pan African material in the Gebel Kamil area which makes a mixed Archean/Pan African source very unlikely. More likely, the Nd model age reflects a source of Mid Proterozoic age.

In the Gebel El Asr and Gebel Umm Shaghir areas (Fig. 2) homogeneous migmatic gneisses are exposed in a few large

	tonalitic gneisses		leucogranites		LILE enriched granites		
	GEA M	a. ton.	GUS G	WH MG	ND RG/MG	BSS RG	BSS GD
SiO_2	68.81 ± 1.16	69.40	70.73 ± 1.23	73.15 ± 4.34	72.14 ± 1.59	74.44 ± 1.20	63.88 ± 2.01
TiO_2	0.28 ± 0.04	0.35	0.35 ± 0.09	0.42 ± 0.45	0.31 ± 0.19	0.26 ± 0.09	0.89 ± 0.10
Al_2O_3	16.11 ± 0,59	15.80	14.86 ± 0.36	13.10 ± 0.75	13.90 ± 0.35	13.26 ± 0.53	15.10 ± 0.74
$Fe_2O_3^*$	2.56 ± 0.22	3.17	2.37 ± 0.39	2.34 ± 1.79	1.91 ± 0.90	1.44 ± 0.50	5.10 ± 0.73
MnO	0.05 ± 0,01	0.04	0.05 ± 0.01	0.05 ± 0.01	0.04 ± 0.01	0.04 ± 0.01	0.07 ± 0.01
MgO	1.14 ± 0.13	1.14	0.72 ± 0.15	0.41 ± 0.34	0.44 ± 0.18	0.45 ± 0.17	2.91 ± 0.53
CaO	4.41 ± 0.34	3.37	2.01 ± 0.36	1.29 ± 0.87	1.38 ± 0.61	1.18 ± 0.21	4.33 ± 0.52
Na_2O	4.41 ± 0.13	4.68	3.34 ± 0.18	3.57 ± 0.24	3.48 ± 0.38	3.74 ± 0.16	4.04 ± 0.21
K_2O	1.34 ± 0.13	1.58	4.52 ± 0.41	4.62 ± 0.70	5.20 ± 0.38	4.53 ± 0.39	2.68 ± 0.54
P_2O_5	0.13 ± 0.02	0.11	0.14 ± 0.02	0.12 ± 0.11	0.11 ± 0.04	0.09 ± 0.04	0.26 ± 0.05
l.o.i.	0.55 ± 0.05	0.54	0.77 ± 0.20	0.69 ± 0.31	0.86 ± 0.12	0.41 ± 0.07	0.70 ± 0.05
Rb	35 ± 9	44	132 ± 21	145 ± 37	212 ± 52	158 ± 22	63 ± 11
Sr	494 ± 34	460	294 ± 17	161 ± 125	206 ± 74	197 ± 55	760 ± 66
Ba	301 ± 60	400	1084 ± 161	942 ± 472	822 ± 274	876 ± 194	1172 ± 132
Y	9 ± 4	-	23 ± 9	36 ± 4	21 ± 12	19 ± 7	18 ± 2
Nb	8 ± 2	-	8 ± 2	29 ± 13	8 ± 2	5 ± 2	14 ± 3
Zr	113 ± 17	175	245 ± 53	230 ± 172	156 ± 119	196 ± 61	243 ± 25
V	21 ± 3	-	28 ± 14	14 ± 12	13 ± 8	15 ± 10	88 ± 12
Cr	5 ± 1	12	10 ± 3	21 ± 14	6 ± 2	9 ± 3	77 ± 30
Ni	3 ± 1	13	4 ± 2	23 ± 11	9 ± 2	6 ± 3	42 ± 13

Major - element oxides in wt. %; trace - element concentrations in ppm; total Fe as Fe_2O_3

TABLE 1

Average chemical composition and standard deviation (1 s) of granitoid rocks from southern Egypt and northern Sudan. For analytical details see Bernau et al. (1987).

GEA M = 11 meta - trondhjemitic to meta - granodioritic rocks from Gebel el Asr, southern Egypt

a ton. = Average of Archean high alumina tonalitic gneisses; data from Condie (1981).

GUS G = 13 leucogranites (S-Types) from Gebel Umm Shagir, southern Egypt

WH MG = 6 leucogranites (S-Types) from Wadi Howar, northern Sudan

ND MG/RG = 8 LILE enriched granites from Nubian Desert, northern Sudan

BSS G = 7 LILE enriched granites from Bir Safsaf, southern Egypt

BSS GD = 14 LILE enriched granodiorites from Bir Safsaf, southern Egypt

TABLE 2
Rb/Sr whole rock ages and Nd model ages.
* Assumed ages (see text)

Lithology Sample No.	Nubian Desert gneiss ND G6	Wadi Howar migmatite WH M3	Gebel Kamil migmatite GK M3	Gebel El Asr migmatite GEA M4	Gebel El Asr migmatite GEA M8
Age (Ma)	918 ± 40	686 ± 26	673 ± 56	670*	670*
MSWD	34.1	8.3	5.3		
$(^{87}Sr/^{86}Sr)_i$	0.7167 ± 7	0.7064 ± 8	0.7049 ± 4		
$^{147}Sm/^{144}Nd$	0.08390	0.10166	0.07826	0.12920	0.09150
$^{143}Nd/^{144}Nd$	0.511152	0.511868	0.511149	0.511585	0.511485
ε_{Nd}	-15.8	-6.7	-18.9	-14.8	-13.5
T_{CHUR} (Ma)	2004	1237	1913	2371	1670

Lithology Sample No.	Nubian Desert micro-granite ND MG3	Bir Safsaf ortho-gneiss BSS OG 6a	Wadi Howar micro-granite WH MG 6	Bir Safsaf grano-diorite BSS GD4	Nubian Desert red granite ND RG4
Age (Ma)	623 ± 37	597*	585 ± 19	564 ± 77	565 ± 8
MSWD	15.5		6.6	0.1	2.0
$(^{87}Sr/^{86}Sr)_i$	0.7053 ± 18		0.7086 ± 8	0.7055 ± 3	0.7065 ± 2
$^{147}Sm/^{144}Nd$	0.09873	0.09961	0.12350	0.10090	0.07223
$^{143}Nd/^{144}Nd$	0.511901	0.511881	0.511952	0.511917	0.511906
ε_{Nd}	-6.6	-7.4	-8.0	-7.2	-5.3
T_{CHUR} (Ma)	1149	1191	1430	1148	899

outcrops. A Rb/Sr whole rock age determination failed since the eleven analysed samples scatter considerably around the regression line. The regression line has a very flat slope which perhaps indicates a young event which disturbed the Sr isotopic system. Field evidence, geochemical and petrographic characteristics suggest that these rocks are clearly older than the homogeneous, late tectonic biotite-granites. Chronostratigraphically they can be probably correlated with the previously mentioned S-type granitoids (Bernau et al., 1987). Assuming an age of 670 Ma for the formation of the rock, one sample (Table 2, GEA MIG 4) yielded a Nd model age of 2371 Ma and an $_{Nd}$ value at T = 670 Ma of -14.8. A second sample (Table 2, GEA MIG 8) yielded a Nd model age of 1670 Ma and an $_{Nd}$ valute at T = 670 Ma of -13.5, both values indicating that Pre-Pan African crustal material contributed to their formation.

Late Proterozoic (Pan African)

The most easily recognizable regional event throughout NW Sudan and SW Egypt is the Pan African episode. The oldest Pan African rocktype in the area is the Nubian Desert gneisses with their Rb/Sr age of 918 ± 40 Ma. Whether this early Pan African event is restricted to the eastern edge of the Proterozoic continent or extends further west, is still uncertain.

A distinct Mid Pan African overprint which affected the whole area, except the Gebel Uweinat inlier, and caused the present-day migmatitic texture of the metamorphic rocks, occurred around 680 Ma (Schandelmeier et al., in prep.; Table 2, Gebel Kamil migmatite, Wadi Howar migmatite, Gebel El Asr migmatite). S-type leucogranites were formed during this migmatization (Bernau et al., in press).

LILE-enriched granitoids (Table 1) were intruded during Late Pan African times. They may be classified as "Caledonian I-types" (Pitcher, 1983) or post-collisional granites (Pearce

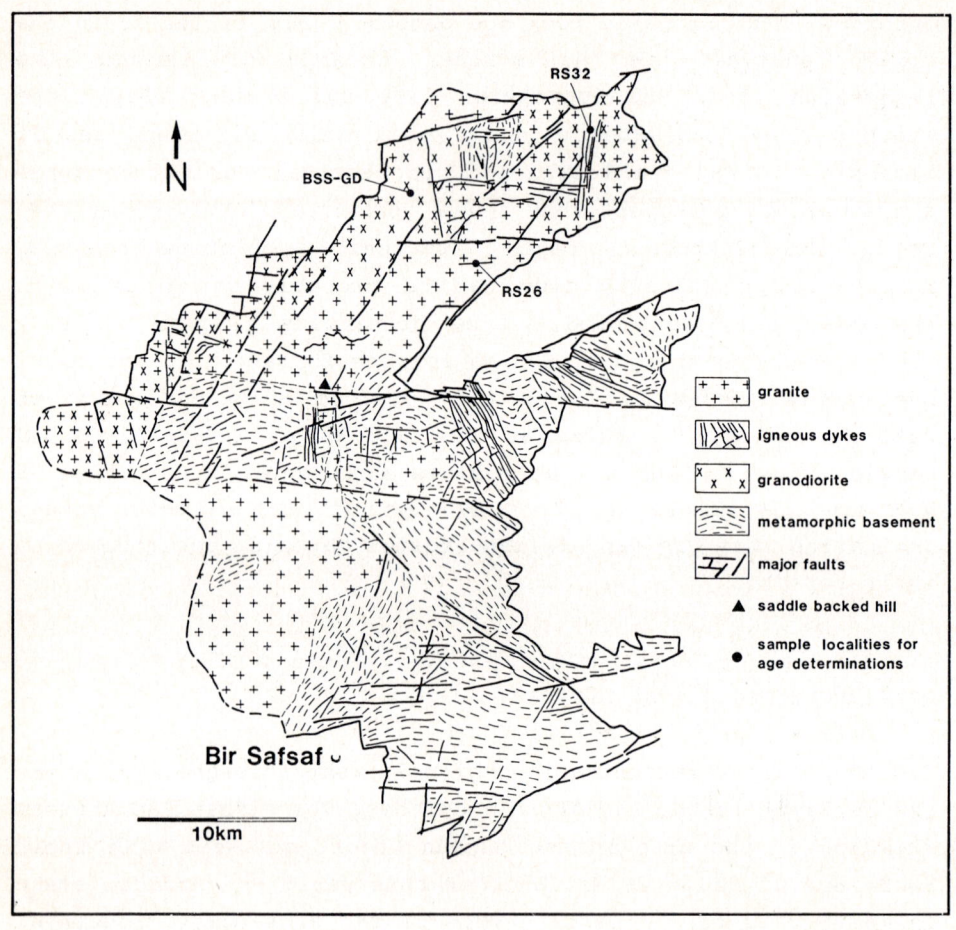

FIGURE 4.
   Geological sketchmap of the Bir Safsaf area (BSS-GD :
   564 ± 77 Ma, Rb/Sr whole rock isochron; RS 26 :
   521 ± 13 Ma; RS 32 : 193 ± 5 Ma, both K/Ar whole rock ages
   from dyke rocks; (data from Bernau et al., 1987)

et al., 1984). Early granodiorites and subordinate diorites were followed by intrusions of huge biotite-granite plutons throughout the whole area (Figs. 2, 3, 4). A granodiorite from the Bir Safsaf area (Fig. 4, BSS GD) yielded a Rb/Sr whole rock age of 564 ± 77 Ma and a biotite-granite from the Nubian Desert area a Rb/Sr whole rock age of 565 ± 8 Ma (Fig. 2, Table 2, ND RG). The same succession of Egyptian granitoids was obtained by Engel et al. (1980) near Aswan with identical ages but slightly lower Sr_i ratios and by Stern and Hedge (1985).

Granites and some gneisses from our working area yielded Rb/Sr whole rock ages which range from 686 Ma to 565 Ma (Schandelmeier et al., in prep.; Table 2). The Nd model ages for these rocks range from 1400 Ma to 900 Ma and there is a close time-equivalence between these rocks and a series of rocks from the Arabian-Nubian Shield (Duyverman et al., 1982; Harris et al., 1984) The rocks reported by Harris et al. (1984) which have Nd model ages between 1900 Ma and 1600 Ma have exclusively Late Proterozoic Rb/Sr ages.

The striking difference, however is the $_{Nd}$ values which are for our samples from the continental plate exclusively negative (Table 2), while the rocks from the Arabian Shield have positive $_{Nd}$ values. This indicates, that within the continental plate of NE Africa, the Pan African episode involved reworking of pre-existing crust, while the process in the Arabian-Nubian Shield was one of rapid crustal growth through extraction of juvenile material directly from the mantle (Duyverman et al., 1982). The location of the samples and their appropriate Nd model ages are shown in Figure 5. Figure 6 shows the initial 143Nd/144Nd ratios relative to the growth line of CHUR.

It will be shown in the following discussion, that magmatic and structural events near the margin of the continental plate are closely related to accretional and collisional tectonic events in the Arabian-Nubian Shield.

DISCUSSION

A series of newly obtained data argue for Kröner's (1979)
approach, that NE Africa west of the Nile was underlain by a
continuous Pre-Pan African crust and was for that reason part
of an older craton. The theory of Rogers et al. (1978) that a
large ocean basin in NE Africa (west of the Nile) was
cratonized in Late Proterozoic times, can be confidently ruled
out.

As discussed earlier, the only radiometric data which supply
evidence for a Mid Archean to Early Proterozoic event were
obtained from granulitic gneisses from the Gebel Uweinat area,
but Richter (1986) has reasonably well documented that the
granulite facies rocks extend as far as into the western Gebel
Kamil area. The Nd Model ages for migmatites from the Gebel
El Asr area and from the Gebel Kamil area and for gneisses from
the Nubian Desert area indicate the formation of new crust in
the Early to Mid Proterozoic. Harris et al.'s (1984) suggestion
that the Archean basement of Gebel Uweinat was rimmed by a Mid
Proterozoic crust of 1900 Ma to 1600 Ma is substantiated by our
investigations (Bernau et al., 1987; Schandelmeier et al., in
prep.). It will be the subject of further geochronological work
to find radiometric ages to document this Mid Proterozoic
event.

The Late Proterozoic tectono-thermal episode obscured these
older structures with various intensity. Only in the Gebel
Uweinat area proper this influence seems to be unimportant. In
the Gebel Kamil, Nubian Desert and Wadi Howar areas it is
demonstrated by migmatization and intrusion of granitic bodies.
In the Bir Safsaf - Aswan uplift the Pan African imprint on the
foreland of the continental plate seems to be related to
accretion events in the Arabian-Nubian Shield.

Nd model ages of the intrusive rocks (Table 2) cluster at
around 1200 Ma, a time when the earliest stage of development

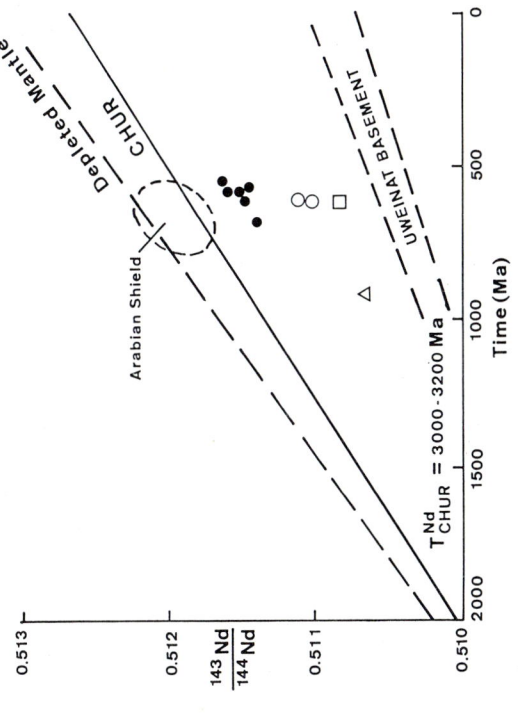

FIGURE 6.
Initial 143Nd/144Nd ratios in samples from the continental plate of NE Africa. Location of the Arabian Shield and the Uweinat basement from Harris et al. (1984)

granitoids and meta-granitoids
● Nubian Desert Gneiss
○ Gebel Kamil migmatite
□ Gebel El Asr migmatite

FIGURE 5.
Sample location map showing the distribution of Nd model ages in NE Africa; Circles are data from HARRIS et al. (1984); squares are data from Schandelmeier et al. (in prep.).

3200 – 3000 Ma   1900 – 1600 Ma   1300 – 800 Ma
>1900 Ma   1900 – 1600 Ma   1400 – 900 Ma

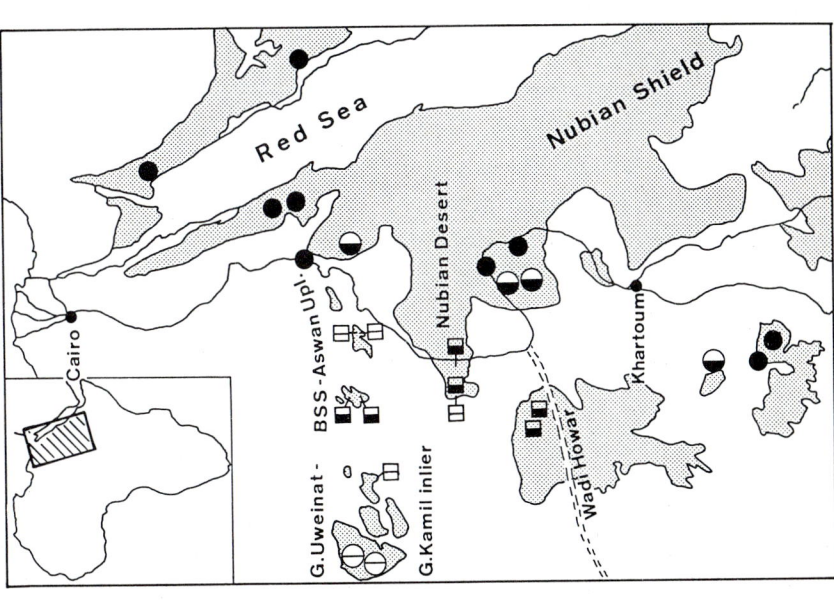

of the Arabian-Nubian Shield was probably initiated through
large-scale rifting of the older NE African plate (Stoeser and
Camp, 1984). At this time, rift related magmas which were
differentiated from the mantle, might have been added to the
continental crust near the rift zone. These LREE-enriched,
highly fractionated rocks might have been part of the source
rocks which contributed to the formation of the Late
Proterozoic granitic intrusions, producing their distinct
negative $_{Nd}$ values.

El-Ramly et al. (1984) have demonstrated in the Hafafit area of
the Eastern Desert of Egypt that Nubian Shield assemblages were
thrust onto the continental plate by collisional processes and
extensive horizontal shortening. Figure 7 shows a schematic
section which extends roughly from the Hafafit area in the
Eastern Desert of Egypt in WSW direction into the Bir Safsaf
uplift.

At the eastern margin of the continental plate, in the Gebel
Umm Shaghir area, S-type granites developed in the lower crust
as a result of crustal thickening during migmatization at
around 680 Ma. This process was approximately time-equivalent
with the closure of the back-arc basin through compression and
early thrusting (El-Ramly et al., 1984; Fig. 7). Bernau et al.
(1987) have shown that the extensive mylonitic shear zones in
the Gebel Umm Shaghir area developed shortly after the
migmatization under decreasing temperatures. The shear planes
are generally flat and are related to thrust movement within
the marginal plate. Isostatic equilibration of the passive
continental margin lead to continuous uplift and LILE-enriched
granitoids ("Caledonian I-type") were generated at 580 Ma.
Reworking of older crust must have been involved into this
process as indicated by the negative $_{Nd}$ values of these rocks
(Table 2).

During the final stage of collision, when Nubian Shield
assemblages were thrust over the marginal plate, an E-W
oriented dextral wrench fault system developed in the

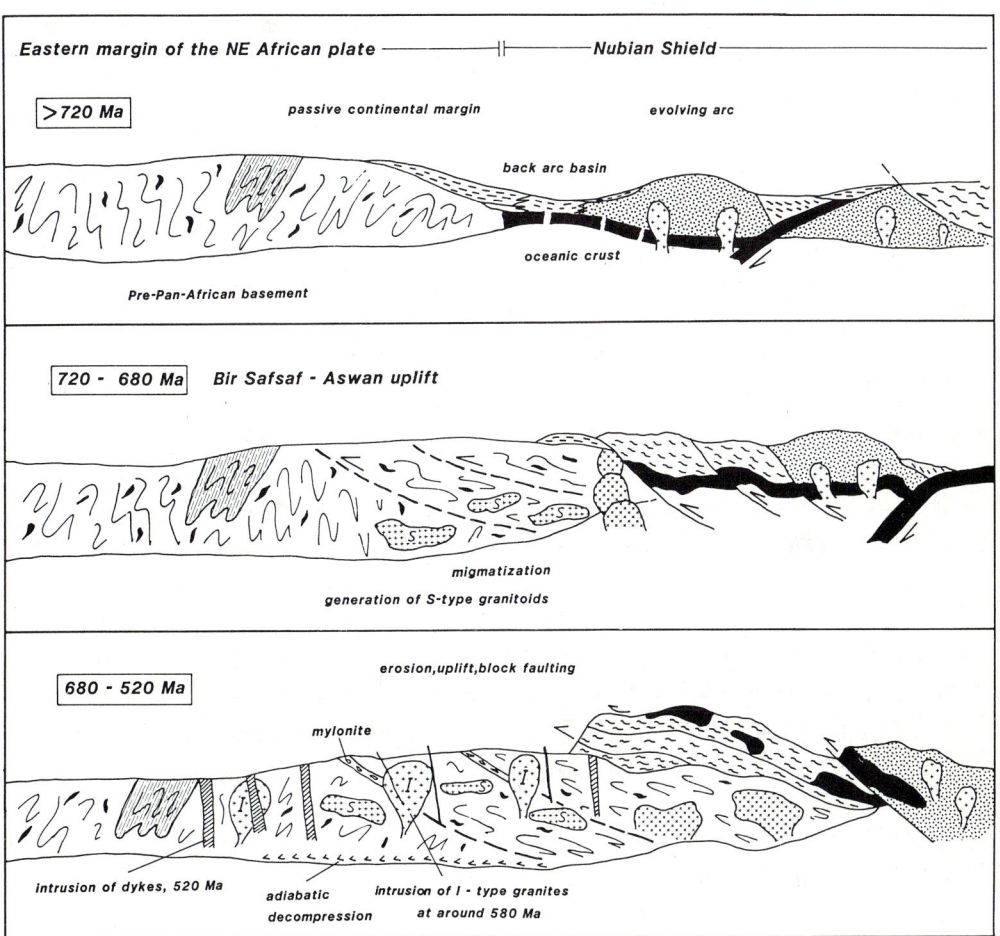

FIGURE 7.
    A proposed model for the tectonic development of the
continental marginal plate of NE Africa in the eastern
Bir Safsaf - Aswan uplift. Eastern part of the section
(Nubian Shield) is from El-Ramly et al. (1984) and refers
to the Hafafit area of the Eastern Desert of Egypt.
Western part of the section refers to the basement area
around Gebel Umm Shaghir (see Fig. 1)

Bir Safsaf - Aswan uplift which might be the conjugate wrench zone to the sinistral Najd fault system of Saudi Arabia (El-Gaby, 1983; Bernau et al., 1987).

After the final consolidation, uplift resulted in extensive vertical block faulting and many of the fractures were invaded by micromonzodioritic dykes. One of these dykes which cuts the LILE-enriched granitoids yielded a K/Ar whole rock, as well as mineral age, of 521 ± 13 Ma (Bernau et al., 1987).

ACKNOWLEDGEMENTS

This study was carried out as part of the "Special Research Project Arid Areas" of the Technical University of Berlin, financed by the German Research Foundation (DFG). The authors would like to express their thanks to a number of colleagues from the Geological Survey of Sudan for their logistic support and their fruitful discussions. R. J. Stern and an anonymous reviewer considerably improved an earlier version of this manuscript. The contribution of D. P. F. Darbyshire is published with the approval of the Director, British Geological Survey (NERC).

REFERENCES

Abdel-Monem, A.A. and P.M. Hurley, 1979. U-Pb dating of zircons from psammitic gneisses, Wadi Abu Rusheid-Wadi Sikait area, Egypt. In: S.A. Tahoun (ed.). Evolution and Mineralization of the Arabian-Nubian Shield. Pergamon Press, Oxford, 2, 165-170.

Almond, D. C., 1980. Precambrian events at Sabaloka, near Khartoum, and their significance in the chronology of the basement complex of North-East Africa. Precambrian Res., 13, 43-62.

Bernau, R., D.P.F. Darbyshire, G. Franz, U. Harms, A. Huth, N. Mansour, P. Pasteels and H. Schandelmeier, 1987. Petrology, Geochemistry and structural development of the Bir Safsaf - Aswan uplift/Southern Egypt. J. Afr. Earth Sci. 6, 79-90.

Cahen, L. and N.J. Snelling, J. Delhal and J.R. Vail, 1984. The geochronology and evolution of Africa. Clarendon Press, Oxford, 512 pp.

Condie, K. C., 1981. Archean greenstone belts. Elsevier, Amsterdam, 574 pp.

Dixon, T. H., 1981. Age and chemical characteristics of some Pre-Pan-African rocks in the Egyptian Shield. Precambrian Res., 14, 119-133.

Duyverman, H.J., N.B.W. Harris and C.J. Hawkesworth, 1982. Crustal accretion in the Pan African: Nd and Sr isotope evidence from the Arabian shield. Earth Planet. Sci. Lett., 59, 315-326.

El-Gaby, S., 1983. Architecture of the Egyptian basement complex (Abstract). 5th Int. Conf. basement tectonics, Cairo.

El-Ramly, M.F., R. Greiling, A. Kröner and A.A.A. Rashwan, 1984. On the tectonic evolution of the Wadi Hafafit area and environs, Eastern Desert of Egypt. Bull. Fac. Earth Sci., King Abdulaziz Univ., 6, 113-126.

Engel, A.E.J., T.H. Dixon and R.J. Stern, 1980. Late Precambrian evolution of Afro-Arabian crust from ocean arc to craton. Geol. Soc. Am. Bull., 91, 699-706.

Gass, I.G., 1981. Pan-African (Upper Proterozoic) plate tectonics of the Arabian-Nubian shield. In: A. Kröner (ed.). Precambrian Plate Tectonics, Elsevier, Amsterdam, 387-405.

Harris, N.B.W., C.J. Hawkesworth and A.C. Ries, 1984. Crustal evolution in northeast and east Africa from model Nd ages. Nature, 309, 773-776.

Hunting Geology and Geophysics Ltd., 1974. Geology of the Jabal Al Uwaynat area, Libyan Arab Republic. Unpublished report, Industrial Research Centre, Tripoli.

Huth, A., G. Franz and H. Schandelmeier, 1984. Magmatic and metamorphic rocks of NW Sudan: A reconnaissance Survey. Berliner geowiss. Abh. (A), 50, 7-21.

Klerkx, J. and S. Deutsch, 1977. Resultats preliminaires obtenus par la méthode Rb/Sr sur làge des formations précambriennes de la région d'Uweinat (Libye). Mus. roy. Afr. centr. Dépt. Géol. et Min. Rapp. Ann. 1976, 83-94.

Klerkx, J., 1980. Age and metamorphic evolution of the basement complex around Jabal al Awaynat. In: M.J. Salem and M.T. Busrewil (eds.). The geology of Libya, 3, 901-906.

Klitzsch, E., 1983. Geological Research in and around Nubia. Episodes, 6, 15-19.

Kröner, A., 1979. Pan African plate tectonics and its repercussions on the crust of northeast Africa. Geol. Rdsch., 68, 565-583.

Kröner, A., 1985. Ophiolites and the evolution and tectonic boundaries in the Late Proterozoic Arabian-Nubian shield of Northeast Africa and Arabia. Precambrian Res., 27, 277-300.

Marholz, W.W., 1968. Geological exploration of the Kufra region, April-May 1965. Bull. Geol. Surv. Libya 8, 1-76.

Meinhold, K.-D., 1979. The Precambrian Basement complex of the Bayuda Desert, Northern Sudan. Rev. Géol. dyn. Géogr. phys., 21, 395-401.

Pearce, J. A., N. B. W. Harris and A. G. Tindle, 1984. Trace element discrimination diagrams for the tectonic interpretation of granitic rocks. J. Petrol. 25, 956-983.

Pitcher, W. S., 1983. Granite type and tectonic environment. In: K. Hsu (ed.). Mountain building processes. Academic Press. London, 19-40.

Richter, A., 1986. Die Geologie der metamorphen und magmatischen Einheiten im Gebiet zwischen Gebel Uweinat und Gebel Kamil - SW Ägypten/NW Sudan. Berliner geowiss. Abh. (A), in press.

Rocci, G., 1965. Essai d'interprétation des mésures geochronologiques. La structure de l'Ouest africain. Sci. Terre, 10, 461-479.

Rogers, J.J.W., M.A. Ghuma, M.R. Nagy, J.K. Greenberg and P.D. Fullagar, 1978. Plutonism in Pan-African belts and the geologic evolution of northeast Africa. Earth Planet. Sci. Lett., 39, 109-177.

Schandelmeier, H. and F. Daryshire, 1984. Metamorphic and magmatic events in the Uweinat-Bir Safsaf uplift (Western Desert/Egypt). Geol. Rdsch. 73, 2, 819-831.

Schürmann, H.M.E., 1974. The Pre-Cambrian in North Africa. E.J. Brill, Leiden.

Stern, R. J. and C. E. Hedge, 1985. Geochronologic constraints on the late Precambrian crustal evolution in the Eastern Desert of Egypt. Am. J. Sci, 285, 97-127.

Stoeser, D. B. and V. E. Camp, 1985. Pan-African microplate accretion of the Arabian Shield. Geol. Soc. Am. Bull. 96, 817-826.

Vail, J.R., 1983. Pan-African crustal accretion in north-east Africa. J. Afr. Earth Sci. 1, 285-294.

Vail, J. R., 1985. Continental and oceanic Terrains in NE Africa and Arabia (Abstract). 13th Coll. Afr. Geol., St. Andrews, Scotland, CIFEG Occ. Publ. 1985/3, 71.

Chapter 4

Is There any Pre-Pan-African (> 950 Ma) Basement in the Eastern Desert of Egypt?

A. Kröner[1] / T. Reischmann[2] / H.-J. Wust[1, 2] / A. A. Rashwan[3]

[1] Institut für Geowissenschaften, Universität Mainz, Postfach 3980, 6500 Mainz, FRG.
[2] Max-Planck-Institut für Chemie, Postfach 3060, 6500 Mainz, FRG.
[3] Egyptian Geological Survey and Mining Authority, 3 Salah Salem Street, Abbasiya, Cairo, Egypt.

Keywords: Proterozoic, Pan-African, Egypt, NE Africa, Arabian-Nubian Shield, Nile Craton, Geochronology, Rb/Sr, U/Pb, Pb/Pb, Sm/Nd

Abstract: The inference of old continental crust underneath the Pan-African island-arc assemblage in the Eastern Desert (ED) of Egypt is largely based on a few geochronological dates that indicate ages in excess of ~1000 Ma. We assess these ages and the analytical data and conclude that there is as yet no conclusive evidence for the existence of early Proterozoic or even Archaean crust east of the River Nile.

U-Pb ages for zircons in granitoid clasts as well as detrital zircons from early Pan-African clastic metasediments in the central and southern ED reflect an early to middle Proterozoic detrital component that we relate to the Nile craton of which remnants are now exposed at Ouweinat, in western Egypt and in NW Sudan. However, these sediments also reveal a significant input of early Pan-African crust as shown by zircons in the age

range ~770 - 830 Ma; they are therefore derived from at least
two distinct source regions.

We interpret the published data arrays derived from whole-rock
Rb-Sr dating of early Pan-African clastic metasediments at
Abu Swayel, El Shalul and El Bakriya as reflecting mixing of
old continental crust with newly-generated, juvenile crust of
~ 800 Ma age, and this is supported by U-Pb zircon and whole-
rock Pb-Pb and Nd-data. "Ages" calculated from such mixing
lines in the range 1130-1300 Ma are geologically meaningless
and neither date the time of deposition nor metamorphic
equilibration in these rocks.

Although we must consider an evolutionary scenario for the ED
that involves incorporation of exotic continental fragments
within a collage of juvenile terranes, on present evidence we
favour the hypothesis of accretion of new crust along a highly
irregular active plate margin. The available isotopic data
provide no evidence for the suggestion that the entire ED and
Sinai are floored by pre-Pan-African crust.

INTRODUCTION

The recognition of island-arc assemblages and ophiolitic
remnants in the Pan-African (~950-450 Ma) complex of the
Arabian shield (Greenwood et al., 1976; Al-Shanti and Mitchell,
1976; Roobol et al., 1983) and in the Red Sea Hills of the
Sudan (Neary et al., 1976; Vail, 1983; Reischmann and Kröner,
1984) has led to a re-assessment of the basement "stratigraphy"
in the Eastern Desert of Egypt that was previously based on the
classical geosynclinal concept of pre-plate tectonic days
(Hume, 1934; El Ramly, 1972; Akaad and Noweir, 1980; Geologic
Map of Egypt, 1981). Shackleton et al. (1980) and Ries et al.
(1983) provided structural evidence to show that the scattered
mafic-ultramafic complexes are remnants of giant nappes with
fragments of ophiolites and that thrusting and shearing has
largely obliterated any original stratigraphy. Geochemical and
isotopic data have also shown that the Eastern Desert and Sinai
assemblages belong to the same general arc-ophiolite
association as now recognized in Arabia and in the Sudan
(Bentor, 1985; Beyth et al., 1978; Engel et al., 1980; Furnes
et al., 1985; Stern, 1981; Shimron , 1980) and that most of
these rocks must have evolved in an oceanic environment prior
to their accretion onto the African craton (Duyverman et al.,
1982; Stern and Hedge, 1985; Harris et al., 1984).

However, the discovery of zircons older than ~1700 Ma in rocks
originally interpreted as clastic metasediments (Abdel-Monem
and Hurley, 1979) and in granite clasts contained in sediments
(Dixon, 1981) as well as the recognition of old crustal
components in the Pb-isotopic systematics of the Aswan granite
(Gillespie and Dixon, 1983; Stacey and Stoeser, 1984) has led
to the assumption that the "Nile Craton" (Rocci, 1965) with
early Proterozoic and Archaean rocks extends at least as far
east as the River Nile. Several authors have therefore
interpreted the clastic sediments found structurally below the
arc-ophiolite assemblage, such as at Hafafit and Meatiq, as a
passive continental margin sequence that fringed the ancient
craton in early Pan-African times (~800-950 Ma) and onto which

the arc-ophiolite nappes were thrust, from E to W, during the
subsequent accretion event (El-Ramly et al., 1984; Ries et al.,
1983; El-Gaby et al., 1984; Elbayoumi and Greiling, 1984;
Kröner, 1985; Kröner et al., 1987a). El-Gaby et al. (1987) have
taken the extreme view that the early Proterozoic-Archaean
crust extends underneath the Pan-African cover almost to the
present Red Sea coast and that is is presently exposed in the
cores of large, medium- to high-grade gneiss domes in the
Eastern Desert and Sinai. They regard these migmatites and
granite-gneisses   as   a   "remobilized   infrastructure"   but
recognize the presently observed structures and fabrics in
these rocks as having resulted from Pan-African overprinting.

The common view that high-grade and complexly deformed rocks
must be very old has also prevailed in the interpretation of
gneisses in Arabia (e.g. Delfour, 1981) and in the Sudan (e.g.
Sabaloka granulites, Almond, 1980), but was shown to be
erroneous after reliable isotopic ages became available (Kröner
et al., 1987b). We therefore critically review the available
isotopic data for rocks of the Eastern Desert that indicate
ages in excess of 1000 Ma in order to constrain present models
for the evolution of this part of the Arabian-Nubian Shield. We
recalculated some of the published data using the decay
constants of Steiger and Jäger (1977) and York (1969)
regression (if not otherwise stated in the text). The errors in
age and Sr_i (87Sr/86Sr initial ratio) are 2 sigma. We discuss
the analytical quality of the data and their chronological and
geological significance.

PRE-PAN-AFRICAN AGES IN THE EASTERN DESERT AND ARABIA

The oldest reliably dated rocks of the Pan-african assemblage
in the Arabian-Nubian Shield occur in SW Arabia and yield ages
in the range ~900-950 Ma. They belong to mafic to intermediate
metavolcanics of the Jiddah Group as well as to tonalitic,
dioritic and granodioritic plutons that intrude the volcano-
sedimentary Baish-Baha sequences (Fleck et al., 1980; Marzouki

et al., 1982) and that are interpreted as a "primitive arc assemblage" of an intra-oceanic environment (Roobol et al., 1983). An earlier Rb-Sr age of 1165 ± 110 Ma on three samples of metabasalt from the SW Arabian Shield (Fleck et al., 1980) could not be substantiated (Fleck, Reischmann and Kröner, unpubl. data), and it is therefore likely that the Pan-African magmatic evolution in Arabia began less than 1000 Ma ago (Calvez et al., 1984; Kröner, 1985). Much older ages recently reported from the Shield (e.g. Calvez et al., 1984; Stacey and Hedge, 1984; Stacey et al., 1984; Stacey and Stoeser, 1984; Pallister et al., 1987) are now considered to define small crustal blocks that may represent exotic or displaced terranes (microcontinents), probably accreted to the African continent together with the arcs and similar to the present situation in the SW Pacific (Kröner, 1985; Stoeser and Camp, 1985; Kröner et al., 1987a).

U-Pb zircon ages

In the eastern Desert of Egypt pre-950 Ma ages were reported from several localities. Dixon (1981) dated two granite cobbles from a conglomerate lens within a metagreywacke unit associated with arc volcanics in the Wadi Mobarak region NNW of Marsa Alam (locality 1 in Fig. 1) and obtained highly discordant U-Pb zircon ages. Four size fractions of one cobble (78-D) can be fitted to a regression line with concordia intercepts at 1120+230/-90 Ma and 580+65/-90 Ma respectively. Dixon (1981) inferred from this a maximum age of 1120 Ma for the enclosing metasediment.

This interpretation is questionable, however, since the above data array, if indeed geologically meaningful, requires severe Pb-loss of the dated zircons at about 580 Ma ago. The cobbles do not show significant metamorphic recrystallization, and the enclosing metagreywacke is of lower greenschist grade. Furthermore, there are no igneous rocks in this region that reflect the age of 580 Ma. To the contrary, the Wadi El Mia

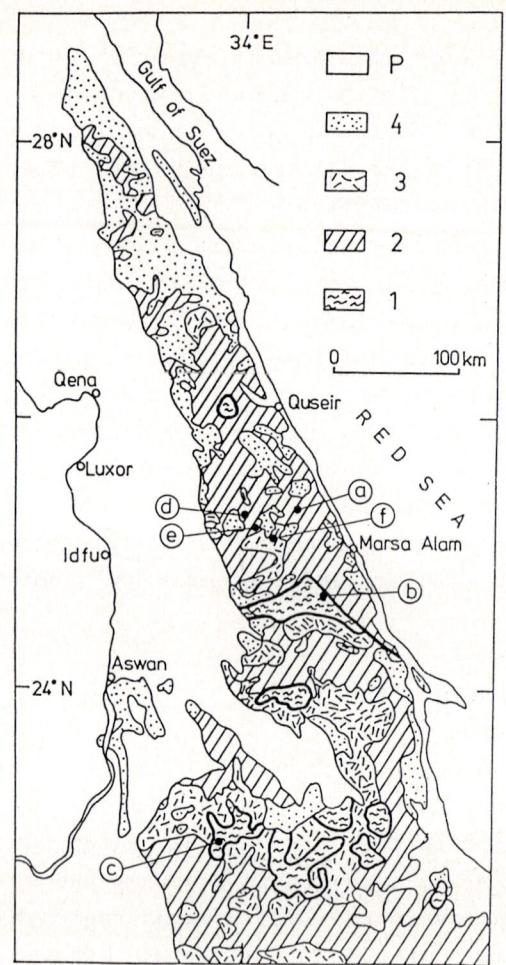

FIGURE 1.

Simplified geological map of the Eastern Desert of Egypt showing location of samples discussed in this paper. Modified from Stern and Hedge (1985).

1 = Gneisses and migmatites, 2 = Mafic to intermediate metavolcanics and related immature metasediments, gabbro, serpentinite, 3 = Quartz diorite and granodiorite, 4 = Posttectonic granite and granodiorite, P = Phanerozoic sedimentary cover. (a) Wadi Mobarak for granite cobbles 78-D and 78-L of Dixon (1981), (b) Wadi Abu Rosheid-Wadi Sikait (Abdel Monem and Hurley, 1979), (c) Abu Swayel Mine (El Shazly et al., 1973; Wust et al., 1987b), (d) El Shalul (El Manharawy, 1977), (e) El Bakriya (El Manharawy, 1977), (f) Wadi El Miya (El Manharawy, 1977; Stern and Hedge, 1985).

granodiorite, exposed W of the above cobble locality, has a 4-point Rb-Sr isochron age of 674±13 Ma (Stern and Hedge, 1985) and shows no effect of disturbance at about 580 Ma ago.

Zircons are not normally affected by low-grade metamorphism as also demonstrated by Dixon (1981) for a ~710 Ma old quartz diorite in the SE Desert, and we suggest that the data array for cobble 78-D is also compatible with Pb-loss of ~885 Ma old zircons at unspecified later time (Fig. 2). If the 207Pb/206Pb age of ~885 Ma for the most concordant size fraction of hand-picked, clear, euhedral grains is taken as the crystallization age of the parent rock, the more discordant data points are displaced to the left of the discordia line through the origin (Fig. 2) and would thus reflect Pb-loss after ~885 Ma ago. Kinny (1986) and Kröner and Compston (1987) have shown from comparative single grain ion microprobe and bulk zircon dating that such Pb-loss patterns are common and that it is hazardous to rely on data arrays where a minority of U-rich grains, which are prone to radiogenic Pb-loss, can strongly influence the mean U-Pb isotopic composition of the composite sample. It is thus equally possible to assign an age of ~885 Ma to the granite cobble. This is compatible with an age of ~830 Ma for a euhedral, detrital zircon in a meta-greywacke at El Bakrya and a granite pebble age of 775±28 Ma from the Atud Conglomerate in the Southern ED (Wust et al., 1987a) as well as early Pan-African granite ages reported from Jordan (Jarrar et al., 1983) and the Red Sea Hills farther S (Manton, Stern and Kröner, unpubl. data; Kröner et al., 1987a). It documents that at least some igneous activity in the Nubian segment is as old as in Arabia.

The second granite cobble, 78-L. dated by Dixon (1981) has even more discordant zircons, and the 4 measured fractions do not define a good linear array (MSWD=85, upper intercept at 1870±440, lower intercept at 490±180 Ma, see Fig. 3). Dixon (1981) considered only the most discordant fraction to have been affected by recent Pb-loss and calculated a discordia line with intercepts at 2060+100/-40 Ma and 575+36/-18 Ma

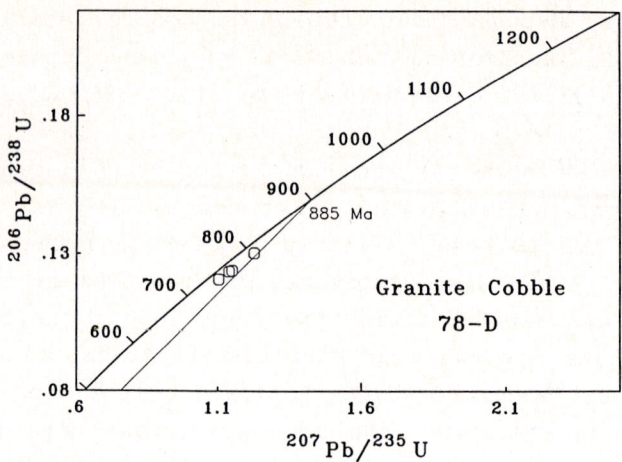

FIGURE 2.

Concordia diagram showing analytical data of granite cobble 78-D dated by Dixon (1981) and our interpretation, suggesting a granite source age of ~ 885 Ma.

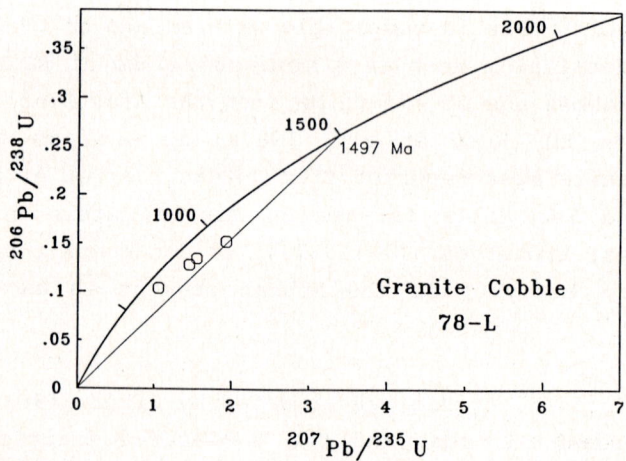

FIGURE 3.

Concordia diagram showing analytical data of granite cobble 78-L dated by Dixon (1981) and our interpretation, suggesting a detrital component at ~ 1497 Ma.

respectively from the remaining 3 fractions. As in the previous case, an equally valid interpretation for the 78-L zircon data is that the 207Pb/206Pb age of 1497 Ma for the least discordant, handpicked fraction of clear, euhedral grains reflects the time of crystallization of the parent rock and that the remaining fractions display the sums of variable Pb-losses of U-rich and U-poor grains at unspecified younger times. It is particularly noteworthy that the clear fraction has the lowest U-content while the fraction with highest U is the one with lowest apparent 207Pb/206Pb age, in line with the Pb-loss pattern observed by Kinny (1986) and Kröner and Compston (1987).

Although an age of ~2060 Ma is possible for cobble 78-L, we prefer the safer interpretation that the minimum cobble age is ~1500 Ma. This is still pre-Pan-African and does not invalidate the conclusion of Dixon (1981) that this clast is derived from an old continental source area to the W of the Eastern Desert. Schandelmeier et al. (1987) report Nd model ages from the Western Desert of Egypt that are also compatible with this interpretation. The second region from which old U-Pb zircon ages were reported is situated adjacent to the Hafafit gneissic terrain in Wadis Abu Rosheid and Sikait (locality 2 in Fig. 1). Abdel-Monem and Hurley (1979) dated two samples of what they described as "psammitic gneiss" from localities that were about 3 km apart. they considered the gneiss to be a metamorphosed arkosic sediment and reported an unusual abundance of accessory minerals with zircon reaching up to 0.2 % of the total rock volume. The zircon morphology is described as euhedral to subhedral, but subrounded to rounded types are also common. These variable shapes are suggestive of a mixture of magmatic zircons and detrital and/or metamorphic grains as frequently found in clastic Pan-African sediments of the Eastern Desert (Wust et al., 1987) and seems to support a sedimentary origin for the host rock. Hassan et al. (1985) provided a detailed petrographic and geochemical study of the psammitic gneiss from the area where the above zircon samples were collected, and their data also favour a sedimentary origin. However, Elbayoumi

and Greiling (1984), El-Gaby (in press) and El-Gaby et al.
(1987) interpret the rocks at the sample localities as
gneissose granite of clear igneous derivation that, if correct,
must have an unusual trace element and accessory mineral
composition for an igneous rock. The isotopic data reported by
Abdel-Monem and Hurley (1979) are ambiguous and difficult to
interpret. The 206Pb/204Pb ratios for all zircon fractions are
so low (between 44 and 226) that common Pb correction has a
significant effect on the resulting age. The common Pb
correction applied by the authors (laboratory lead) is
implausible since it assumes the zircons to be entirely free of
Pb. This is certainly not the case as shown by the frequent
feldspar inclusions. The result of the (incorrect) Pb
correction may therefore be responsible for some of the
spurious and anomalously low ages the more so since some sample
points show reverse discordancy. Considering the data as
published, several zircon fractions from the Wadi Abu Rosheid
sample form a tight cluster (Fig. 4). The 207Pb/206Pb ages of
these fractions scatter between 232 Ma and 540 Ma while
2 fractions plot away from this cluster and have 207Pb/206Pb
ages of 729 Ma and 781 Ma respectively. It is difficult to
attach any significance to these ages in view of the poor
analytical data. All that can be said is that the Wadi Abu
Rosheid zircons may be Pan-African in age and that, if the
parent rock is indeed a granite, this pluton could have a
crystallization age of ~750 Ma. Abdel Monem and Hurley (1979)
attribute the scatter "around 400 Ma" to a hydrothermal event
but any reliable interpretation of these data is meaningless in
view of the data quality and uncertain common Pb correction.

Only one composite, non-magnetic zircon fraction was analyzed
from the Wadi Sikait sample, and this yielded a ~50 %
discordant data point with a 207Pb/206Pb age of 1582 ± 91 Ma
(Fig. 4). The U-content of this sample is almost 10 times lower
than that of the Wadi Abu Rosheid fractions, and there is no
a-priori reason why this fraction should be genetically related
to the others. Nevertheless, Abdel-Monem and Hurley (1979)
constructed a best-fit line through all points that yields an

(El-Ramly et al., 1984), one of which yielded a Nd(DM) model age of ~800 Ma (Harris et al., 1984). If a (realistic) emplacement age of ~750 Ma is assigned to this volcanic clast, a Nd(T) of +7.1 can be calculated (Harris et al., 1984). Clearly, this rock cannot be intruded by a 1580 Ma old granite.

Alternatively the 1580 Ma old zircon fraction may be interpreted as a xenocrystic component. There is increasing evidence for the presence of inherited, xenocrystic zircons in crustally-derived granitoids as demonstrated by conventional and single-grain dating (e.g. Williams et al., 1983; Kinny et al., 1987). Such grains are probably derived from the lower crust and are typically low in U. If they are mixed with other, high-U zircons that crystallized during emplacement of the granite magma, the isotopic composition of this mixture would represent a point on a mixing line in the concordia diagram, defined by the source age of the xenocrysts and the newly-formed magmatic zircons. The best-fit line constructed by Abdel-Monem and Hurley (1979) could be interpreted in this way. Thus, if the psammitic gneiss was a granite, the true emplacement age may be between 730 and 780 Ma as argued above. Such a date is compatible with the regional geological evidence and the granite pebble age of 775 ± 28 of Wust et al. (1987a) from the Atud Conglomerate and would place the Hafafit gneiss assemblage into the age bracket ~730-800 Ma. Nevertheless, the remarkable variations in major and trace element contents in the psammitic gneiss as reported by Hassan et al. (1985) and the rounded shape of some zircons in the samples of Abdel-Monem and Hurley (1979) convince us that this rock was a metasediment, and we therefore favour a detrital origin for the 1580 Ma zircon fraction.

Rb-Sr whole-rock ages

The remaining pre-1000 Ma ages reported from the Eastern Desert are Rb-Sr whole-rock data of uncertain quality and significance. We recalculated these ages on the basis of the

upper intercept age with concordia at 1770 ± 40 Ma. Our
Yorkfit-recalculation of the same data (MSWD = 37) gives
1790 ± 140 Ma (2-sigma error).

The practice of regressing data points together where a genetic
relationship is not established must be regarded as
questionable, and the upper intercept "age" thus obtained for
the above mixed zircon population is probably geologically
meaningless. If the source rock of the samples is a granitized
metasediment, as argued by Abdel-Monem and Hurley (1979), the
source region may be as old as 1582 ± 91 Ma, rather similar,
incidentally, to the 207Pb/206Pb age of Dixon`s (1981) cobble
78-L. In this case the ~730-780 Ma zircons of the Abu Rosheid
sample may reflect input from the Pan-African arc assemblage
while the 232-540 Ma grains were disturbed during post-
crystallizational events.

If, however, the zircon host rock was a granite, the crystallization age of this rock could be around 1580 Ma and would thus
be clearly pre-Pan-African. Ries et al. (1983), Elbyoumi and
Greiling (1984) and Greiling et al. (1987) assert that all
contacts of the psammitic gneiss in the Wadi Abu Rosheid area
with the other rock types are tectonic and that the gneisses
are part of a stacked nappe complex in which they occur as
tectonic slices. If this is correct, the age of the psammitic
gneiss cannot be used to constrain the ages of the neighbouring
rock types, and it may well be possible that the gneiss is an
exotic fragment, perhaps detached from its original base, that
signifies the presence of microcontinental crustal blocks in
the Eastern Desert as has now also been proposed for Arabia
(Stoeser and Camp, 1985) and the Sudan (Kröner et al., 1987a).

The psammitic gneiss, if a granite and ~1580 Ma old, is
unlikely to have intruded the layered early Pan-African
supracrustal sequences of the Hafafit-Wadi Nugrus area prior to
nappe formation for the following reason. The psammitic gneiss
at Hafafit contains irregular and flattened clasts of dacitic
composition that are interpreted as volcanic fragments

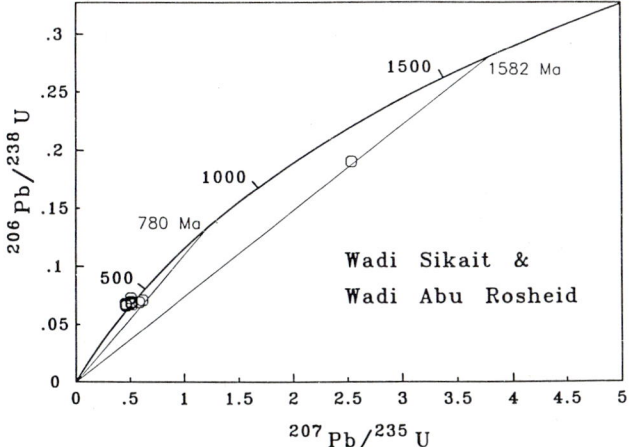

FIGURE 4.

Concordia diagram showing analytical data of Abdel-Monem and Hurley (1979) for psammitic gneiss from Wadi Abu Rosheid-Wadi Sikait and our interpretation, suggesting a detrital or xenocrystic component at ~ 1582 ± 91 Ma and a possible granite emplacement age of ~ 780 Ma.

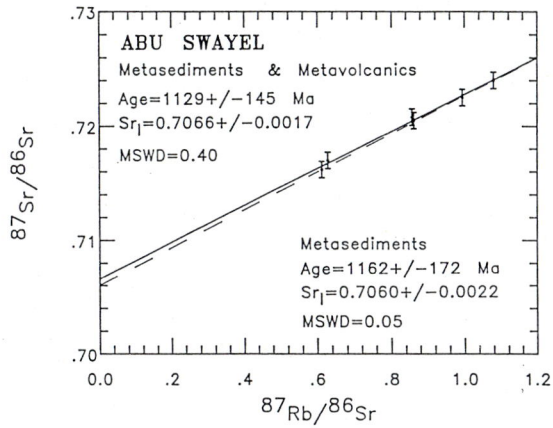

FIGURE 5.

Plot showing distribution of Rb-Sr data for interlayered metasediments and metavolcanics from the Abu Swayel Mine, SE Desert, as determined by El Shazly et al. (1973). The two regression lines are for metasediments only and for the combined data respectively. Note that age and error for the two regressions overlap, suggesting apparent isotopic homogenization at the sampling scale.

published data, the new decay constant and at a 95 % confidence
level. El Shazly et al. (1973) dated 4 whole-rock samples of
amphibolite-grade metasediments from the Abu Swayel Mine SSE of
Aswan (Locality 3 in Fig. 1). We have regressed their data
using the published errors (4 % for 87Rb/86Sr, 0.1 % for
87Sr/86Sr) and obtained a good linear fit (Fig. 5) suggesting
an age of 1162 ± 172 Ma (MSWD = 0.40, Sr = 0.7060 ± 22).
El Shazly et al. (1973) interpret this age as reflecting the
time of regional metamorphism and also suggest that the
original sediments had a virtually uniform 87Sr/86Sr ratio at
the time of deposition and were thus derived from a very
homogeneous source area.

It is important to note that the Abu Swayel metasediments are
interlayered with metavolcanics, as also oberseved by El Shazly
et al. (1973), and two amphibolites of this suite measured by
these authors can be fitted to the same regression line as the
associated metasediments (Fig. 5). The combined age is
1129 ± 145 Ma (MSWD = 0.40, Sr = 0.7066 ± 17) and would suggest
that Sr-isotopic homogenization within the error level of the
analyses has apparently taken place between sediments and
volcanics over distances of several metres. The above age may
either date the time of volcanic activity and Sr-isotopic
homogenization during sediment deposition or diagenesis, or
homogenization took place during a regional metamorphic event
as suggested by El Shazly et al. (1973).

Stern and Hedge (1985) have dated acid metavolcanics from a
locality some 20 km S of the Abu Swayel Mine and obtained a
Rb-Sr whole-rock isochron age of 768 ± 31 Ma with remarkably
low Sr_i of 0.7019 ± 3. If this metavolcanic sequence is
equivalent to the succession at the old mine, as seems apparent
from the published geological maps, the age of El Shazly et al.
(1973) cannot reflect the time of sediment deposition.

Wust et al. (1987b) have also analyzed metasediments and
metavolcanics from the Abu Swayel Mine that contain garnet- and
staurolite-bearing metamorphic assemblages and obtained a good

linear array by the Pb-Pb whole-rock method, suggesting an apparent age of ~ 2 Ga. These authors demonstrate, however, that this alignment is a mixing line between an old crustal component of Archaean to early Proterozoic age and new, juvenile crust formed in early Pan-African times and that the apparent "age" has no geochronological significance.

Furthermore, Wust et al. (1987b) have analyzed Nd-isotopes in both metasediments and metavolcanics from Abu Swayel Mine. The metasediments have strongly negative Nd(T) values, in line with their presumed derivation from an old continental source, while the metavolcanics cannot be older than ~ 800 Ma as shown by Nd(DM) model age. The entire Abu Swayel sequence is therefore clearly of early Pan-African age. Sr-isotopes analyzed by Wust et al. (1987b) display considerable scatter and are not compatible with the data array of El Shazly et al. (1973). These data document open system behaviour and significantly different 87Sr/86Sr initial ratios in the metasediments and metavolcanics. These samples have not been in isotopic equilibrium at about 800 Ma ago. This example is a good demonstration that Rb-Sr whole-rock data for clastic sediments are difficult to interpret and often reflect the input of the source terrain. Ages derived from such assemblages may be geologically meaningless if the data arrays represent mixing lines.

El Manharawy (1977) reported Rb-Sr whole-rock "isochron" ages for two metasedimentary suites at El Shalul and El Bakriya in the central Eastern Desert N of Barramiya (locality 4 in Fig. 1). Our regression of the analytical data (3 % error in 87Rb/86Sr, 0.01 % error in 87Sr/86Sr) shows considerable scatter outside the large analytical error (Fig. 6) that indicates significant isotopic heterogeneity, and the "ages" obtained do not meet isochron criteria. The 5 data points for the El Shalul suite scatter about a regression line (MSWD=8.4), and a Model 3 age (McIntyre et al., 1973) of 1188 ± 360 Ma can be calculated. The Sr$_i$ of 0.7015 ± 77 is implausibly low for a metasediment. Wust et al. (1987a) have dated single zircons from sediments of the same localities and found well rounded grains with ages up to 2.65 Ga as well as a clear, euhedral population with an age of ~ 830 Ma. Clearly, this indicates mixing of old continental material, presumably from the Nile craton, with early Pan-African juvenile magmatic crust, and the sediment composition now found is a mixture of these two components. The Rb-Sr data for the metasediments from the El Bakriya locality show even greater scatter (MSWD=79, Fig. 7), and a Model 3 "age" of 1302 ± 424 Ma can be calculated (Sr_i = 0.7031 ± 59). El Manharawy (1977) excluded sample MS-37 of this suite from his regression in view of visible alteration, but the scatter of the remaining 6 samples is still large (MSWD=43), and a Model 3 "age" (McIntyre et al., 1969) is 1173 ± 325 Ma (Sr_i = 0.7040 ± 42). The above Sr-isotopic arrays provide no meaningful geological age information, and the age and metamorphism of these two sedimentary suites thus remains unconstrained.

We are surprised that detrital zircons of the 1000 - 1200 Ma age range have so far not been detected in the Eastern Desert, and crust of this age is thus unlikely to be present. The occurrence of clear, euhedral, little-transported, zircons of magmatic origin and ~ 830 Ma age in the El Bakriya metasediments (Wust et al., 1987a) clearly shows that these strata are not as old as their Rb-Sr "ages" indicate.

El Manharawy (1977) has also analyzed 6 samples of a "Grey Granodiorite" from Wadi El Miyah in the central Eastern Desert (locality 5 in Fig. 1). This rock belongs to the "Older Granites" of El-Ramly (1972) and intrudes metasediments and metavolcanics. The data points correlate extremely well (MSWD=0.12), and our regression yields on isochron age of 1015 ± 59 Ma with Sr_i = 0.7058 ± 4. Stern and Hedge (1985) have dated a granodiorite from the same pluton (in their publication referred to as located in Wadi Mia), also using the Rb-Sr method and obtained an isochron age of 674 ± 13 Ma (Sr_i = 0.7027 ± 2, MSWD=0.8). This latter age is compatible with ages for other granitoids of the region and appears to

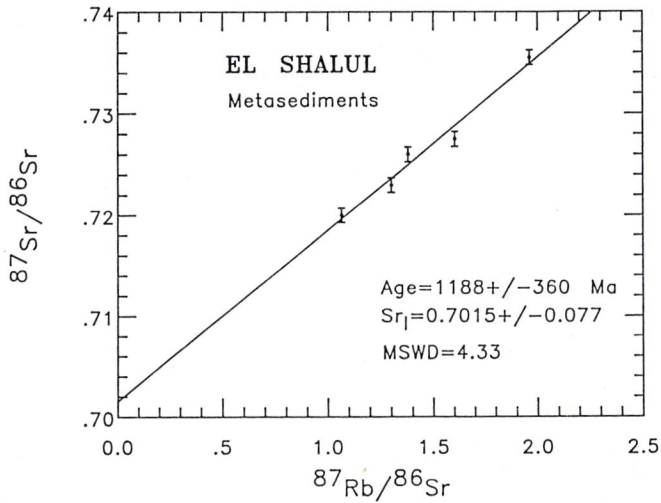

FIGURE 6.

Plot showing poorly correlated distribution of Rb-Sr data for metasediments from El Shalul, central Eastern Desert, as determined by El Manharawy (1977).

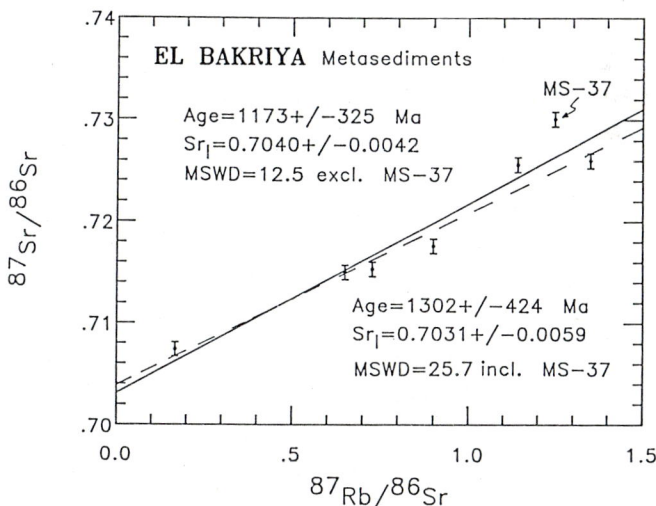

FIGURE 7.

Plot showing poorly correlated distribution of Rb-Sr data for metasediments from El Bakriya, central Eastern Desert, as determined by El Manharawy (1977). The two regression lines are for all data points and for 6 samples excluding MS-37 respectively.

reflect the time of intrusion (Stern and Hedge, 1985) rather than metamorphic overprinting since the rock is only slightly foliated (own field observations), and the Sr_i is too low for the granodiorite to be drived from a ~ 1000 Ma old crustal precursor. We are unable to explain the age discrepancy and the good linear fit of the El Manharawy (1977) data. One possibility would be that the rocks analyzed by El Manharawy (1977) have isotopic systematics inherited from their source. Roddick and Compston (1977) and Allsopp et al. (1980) have described such cases from Australia and Namibia and could show that whole-rock data from granites may define "isochrons" that reflect the mean age of the crustal reservoir from which they were derived. However, we do not favour this interpretation in view of the high Sr_i of 0.7058 ± 8 for the ~ 1000 Ma data array and suggest to discard the ~ 1000 Ma age until precise zircon data are available.

Lastly, we discuss the gneisses and granites in southern Sinai that El-Gaby et al. (1987) consider as pre-Pan-African "infrastructure". These rocks were discussed by Bentor (1985) who correlated them with the early Pan-African (~ 900 - 650 Ma) arc assemblages in Egypt and Arabia. All these gneisses and associated metasediments and metavolcanics have low Sr_i-values around 0.703 (see Table VII in Bentor, 1985), and some of the migmatitic rocks of the ~ 650 Ma age generation can be interpreted as anatectic products of remelted early Pan-African crust (Bielski, 1982). There is no indication of older continental crust in this region. Rb-Sr data for plutonic rocks E of Wadi Araba in southern Jordan, that constitute the extension of the Sinai suite, also have low Sr_i between 0.7032 and 0.7046, and the oldest granite in this area yielded a U-Pb zircon concordia intercept age of 820 ± 54/-25 Ma that is seen as dating a synmetamorphic granitic event in the NW Arabian Shield (Jarrar et al., 1983).

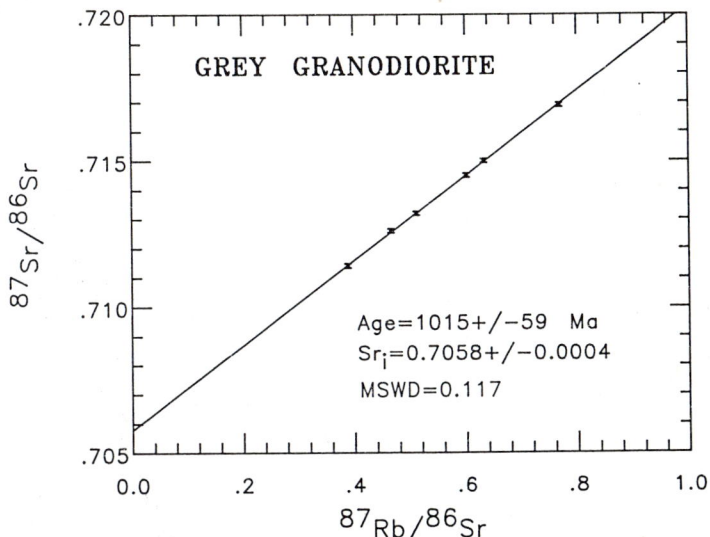

FIGURE 8.

Isochron diagram showing excellent correlation of Rb-Sr data for grey granodiorite from Wadi El Miya, central Eastern Desert, as determined by El Manharawy (1977). Note slightly elevated Sr_i in comparison to plutonic rocks from Saudi Arabia.

CONCLUSIONS

There is as yet no conclusive evidence for the presence of pre-Pan-African crust in the Eastern Desert of Egypt. The granite cobbles of Dixon (1981) were derived partly from an old continental source to the W, S or SW and partly from early Pan-African granitoids and thus reflect a similar heterogeneity of clastic components in the metasediments of the Eastern Desert as documented by detrital zircons (Wust et al., 1987a).

The zircon ages of Abdel-Monem and Hurley (1979) for gneisses SW of Hafafit are inconclusive because of poor analytical data, uncertain common Pb correction and uncertain derivation of these rocks. One single zircon fraction suggests a mid Proterozoic age, and this may either be interpreted as reflecting the detrital source as we favour it (if the gneiss originated from a sediment) or the zircons are xenocrysts derived from older but allochthonous crust (if the gneiss originated from a granite).

The Rb-Sr data of El Shazly et al. (1973) and El Manharawy (1977) for metasediments at Abu Swayel, El Shalul and El Bakriya are equally inconclusive, and the data arrays may represent mixing lines between old continental crust and juvenile additions of early Pan-African age. At least for the Abu Swayel rocks an upper age limit of ~ 800 Ma is given by the Nd isotopic systematics (Wust et al., 1987). These sediments are thus comparable to the early Pan-African Baish-Baha sequences in Arabia.

Evidence for early Pan-African ca. 800 - 880 Ma old granitoid plutonism and regional metamorphism as known from Saudi Arabia (e.g. Bentor, 1985) is also gradually emerging from the Red Sea Hills, the southern Eastern Desert and SW Jordan and suggests that arc formation and accretion occurred virtually synchronously over almost the entire area now occupied by the Arabian-Nubian Shield.

Although the available isotopic data do not support the contention of El-Gaby et al. (1987) that large regions of the Eastern Desert consist of pre-Pan-African gneisses, they nevertheless substantiate the incorporation of some old continental crust at least in parts of the Eastern Desert. If the 1582 ± 91 Ma zircons at Abu Rosheid are xenocrysts they must be derived from mid-Proterozoic crust underneath the Hafafit sedimentary assemblage that El Ramly et al. (1984) and Kröner et al. (1987a) associated with a passive continental margin. However, it is difficult to demonstrate continuity of this old crust underneath the Pan-African assemblage of the Eastern Desert from the River Nile towards the present Red Sea the more so since many of the volcano-plutonic suites in the ~ 700 - 570 Ma age range have low Sr_i (Stern and Hedge, 1985) and oceanic-type Pb-isotopic ratios (Gillespie and Dixon, 1983). It is therefore possible that the ancient African continental margin was either highly irregular, with promontories and embayments, or that the old crustal components suspected below the Eastern Desert represent small crustal blocks or microcontinents that were accreted onto the African margin as has been postulated for Arabia and the Sudan (Kröner et al., 1987a).

The inferred incorporation of old crust into the Pan-African assemblage in the Eastern Desert may have a significant effect on crust-production rates during the ~ 950 - 450 Ma Pan-African episode, since the crust-formation rates estimated by Reymer and Schubert (1986) may be much too high as already suspected by Reischmann (1986).

ACKNOWLEDGEMENTS

We thank the German Federal Ministry of Research and Technology (BMFT), the German Science Foundation (DFG) and the Egyptian Geological Survey and Mining Authority (EGSMA) for support during fieldwork, and we acknowledge the excellent cooperation with our Egyptian colleagues A. Dardir and M.F. El-Ramly during this collaborative project.

REFERENCES

Abdel-Monem, A.A. and P.M. Hurley, 1979. U-Pb dating of zircons from psammitic gneisses, Wadi Abu Rusheid-Wadi Sikait area, Egypt. Inst. Applied Geol., Univ. Jeddah, Bull., 3, 2, 45-55.

Akaad, M.K. and A.M. Noweir, 1980. Geology and lithostratigraphy of the Arabian Desert orogenic belt of Egypt between Lat. 25 35'and 26 30'. Inst. Applied Geol., Univ. Jeddah, Bull., 3, 3, 127-134.

Allsopp, H.L., E.S. Barton, A. Kröner, H.J. Welke and A.J. Burger, 1983. Emplacement versus inherited isotopic age patterns: A Rb-Sr and U-Pb study of Salem-type granites in the central Damara belt. Trans. geol. Soc. S. Afr., 11, 281-287.

Almond, D.D., 1980. Precambrian events at Sabaloka, near Khartoum, and their significance in the chronology of the basement complex of NE Africa. Precambrian Res., 13, 43-62.

Al-Shanti, A.M.S. and A.H.G. Mitchell, 1976. Late Precambrian subduction and collision in the Al Amar-Idsas region, Arabian shield, Kingdom of Saudi Arabia. Tectonophysics, 30, 41-47.

Bentor, Y.K., 1985. The crustal evolution of the Arabo-Nubian massif with special reference to the Sinai Peninsula. Precambrian Res., 28, 1-74.

Bielski, M., 1982. Stages in the evolution of the Arabian-Nubian massif in Sinai. Unpubl. Ph.D. thesis, Hebrew Univ., Jerusalem, 155 p.

Calvez, J.Y., C. Alsac, J. Delfour, J. Kemp and C. Pellaton, 1984. Geological evolution of the western, central and eastern parts of the northern Precambrian shield, Kingdom of Saudi Arabia. Fac. Earth Sci., Univ. Jeddah, Bull., 6, 24-48.

Delfour, J., 1981. Geologic, tectonic and metallogenic evolution of the northern part of the Precambrian Arabian shield (Kingdom of Saudi Arabia). Bull. BRGM (2), II, 1-2, 1-19.

Depaolo, D.J., 1981. Neodymium isotopes in the Colorado Front Range and crust-mantle evolution in the Proterozoic. Nature, 291, 193-196.

Dixon, T.H., 1981. Age and chemical characteristics of some pre-Pan-African rocks in the Egyptian shield. Precambrian Res., 14, 119-133.

Duyverman, H.J., N.B.W. Harris and C.J. Hawkesworth, 1982. Crustal accretion in the Pan African: Nd and Sr isotope evidence from the Arabian shield. Earth Planet. Sci. Lett., 59, 315-326.

Elbayoumi, R.M.A. and R. Greiling, 1984. Tectonic evolution of a Pan-African plate margin in southeastern Egypt - a suture zone overprinted by low angle thrusting? In: Klerkx, J. and Michot, J. (eds.) African Geology-Géologie africaine. Mus. roy. Afr. centr., Tervuren, Belgium, 47-56.

El-Gaby, S., in press. Architecture of the Egyptian basement complex. Proc. 5th Int. Conf. Basement Tectonics, Cairo.

El-Gaby, S., F.K. List and R. Tehrani, 1987. Geology, evolution and metallogenesis of the Pan-African belt in Egypt. In: El-Gaby, S. and Greiling, R. O. (eds.). The Pan-African belt of NE Africa and adjacent areas. Earth evol. sci. Vieweg, Wiesbaden.

El-Manharawy, M.A., 1977. Geochronological investigations of some basement rocks in the central Eastern Desert, Egypt, between lat. 25 and 26 North. Unpubl. Ph.D. thesis, Cairo University, Egypt, 211 p.

El-Ramly, M.F., 1972. A new geological map for the basement rocks in the Eastern and Southwestern Deserts of Egypt. Ann. Geol. Surv. Egypt, 2, 1-18.

El-Ramly, M.F., R. Greiling, A. Kröner and A.A.A. Rashwan, 1984. On the tectonic evolution of the Wadi Hafafit area and environs, Eastern Desert of Egypt. Fac. Earth Sci., Univ. Jeddah, Bull., 6, 113-126.

El-Shazly, E.M., A.H. Hashad, T.A. Sayyah and F.A. Bassyuni, 1973. Geochronology of Abu Swayel area, South Eastern Desert. Egypt. J. Geol., 17, 1-18.

Engel, A.E.J., T.H. Dixon and R.J. Stern, 1980. Late Precambrian evolution of Afro-Arabian crust from ocean arc to craton. Geol. Soc. America Bull., 91, 699-706.

Fleck, R.J., W.R. Greenwood, D.G. Hadley, R.E. Anderson and D.L. Schmidt, 1980. Rubidium-strontium geochronology and plate-tectonic evolution of the southern part of the Arabian shield. U.S. Geol. Surv., Prof. Paper 1131, 38p.

Furnes, H. A.E. Shimron and D. Roberts, 1985. Geochemistry of Pan-African volcanic arc sequences in sotheastern Sinai Peninsula and plate tectonic implications. Precambrian Res., 29, 359-382.

Geological Map of Egypt Scale 1 : 2 000 000, 1981. Geol. Surv. Egypt, Cairo.

Gillespie, J.G. and T.H. Dixon, 1983. Lead isotope systematics of some igneous rocks from the Egyptian shield. Precambrian Res., 20, 63-77.

Greenwood, W.R., D.G. Hadley, R.E. Anderson, R.J. Fleck and D.L. Schmidt, 1976. Late Proterozoic cratonization in southwestern Saudi Arabia. Phil. Trans. R. Soc. Lond., A280, 517-527.

Harris, N.B.W., C.J. Hawkesworth and A.C. Ries, 1984. Crustal evolution in northeast and east Africa from model Nd ages. Nature, 309, 773-776.

Hassan, M.A., M.M. Ali and A.S. Eid, 1985. Petrographic and geochemical studies on the radioactive psammitic gneiss of Wadi Abu Rusheid, south Eastern Desert, Egypt. Annals Geol. Surv. Egypt, 13 (for 1983), 143-155.

Jarrar, G., A. Baumann and H. Wachendorf, 1983. Age determinations in the Precambrian basement of the Wadi Araba area, southwest Jordan. Earth Planet. Sci. Lett D., 1986.

Kinny, P.D., 1986. 3820 Ma zircons from a tonalitic Amitsoq gneiss in the Godthab district of southern West Greenland. Earth Planet. Sci. Lett., 79, 337-347.

Kinny, P.D., I.S. Williams, D.O. Froude, T.R. Ireland and W. Compston, 1987. Early Archaean zircon ages from orthogneiss and anorthosites at Mount Narryer, Western Australia. Precambrian Res., in press.

Kröner, A., 1985. Ophiolites and the evolution of tectonic boundaries in the late Proterozoic Arabian-Nubian shield of northeast Africa and Arabia. Precambrian Res., 27, 277-300.

Kröner, A., R. Greiling, T. Reischmann, I.M Hussein, R.J. Stern, S. Dürr, J. Krüger and M. Zimmer, 1987a. Pan-African crustal evolution in the Nubian segment of northeast Africa. Am. Geophys. Union, Geodynamics Series, 17, in press.

Kröner, A., R.J. Stern, A.A. Dawoud, W. Compston and T. Reischmann (1987). 1987b. The Pan-African continental margin in NE Africa: Evidence from a geochronological study of granulites at Sabaloka, Sudan. Earth Planet. Sci. Lett., in press.

Kröner, A. and W. Compston, 1987. Ion microprobe zircon ages for early Archaean granite pebbles and greywacke, Barberton greenstone belt, southern Africa. Precambrian Res., in press.

Marzouki, F.M.H., N.J. Jackson, C.R. Ramsay and D.P. Darbyshire, 1982. Composition, age and origin of two Proterozoic diorite-tonalite complexes in Arabian shield. Precambrian Res., 19, 31-50.

McIntyre, G.A., C. Brooks, W. Compston and W. Turek, 1966. The statistical assessment of Rb-Sr isochrons. J. geophys. Res., 71, 5459-5468.

Neary, C.R., I.G. Fass, and B.J. Cavanagh, 1976. Granitic association of northeastern Sudan. Geol. Soc. America Bull., 87, 1501-1512.

Pallister, J.S., J.S. Stacey, L.B. Fischer and W.R. Premo, 1987. Precambrian ophiolites of Arabia: U-Pb geochronology, Pb isotopic characteristics, and implications for continental accretion. Precambrian Res., in press.

Reischmann, T., 1986. Geologie und Genese spätproterozoischer Vulkanite der Red Sea Hills, Sudan. Unpubl. Ph.D. thesis, Univ. Mainz, Germany, 202p.

Reischmann, T. and A. Kröner, 1984. Geochemistry of late Proterozoic metavolcanic rocks in the Red Sea Hills, NE Sudan. Terra Cognita. 4, 93.

Reymer, A. and G. Schubert, 1986. Phanerozoic and Precambrian crustal growth. Am. Geophys. Union, Geodynamics Series, 15, in press.

Ries, A.C., R.M. Shackleton, R.H. Graham and W.R. Fitches, 1983. Pan-African structures, ophiolites and mélange in the Eastern Desert of Egypt: a traverse at 26 N. J. geol. Soc. London, 140, 75-95.

Rocci, G., 1965. Essai d'interprétation des mésures géochronologiques. La structure de l'Ouest africain. Sci. Terre, 10, 461-479.

Roddick, J.C. and W. Compston, 1977. Stontium isotopic equilibration: a solution to a paradox. Earth Planet. Sci. Lett., 34, 238-246.

Roobol, M.J., C.R. Ramsay, N.J. Jackson and D.P.F. Darbyshire, 1983. Late Proterozoic lavas of the central Arabian shield-evolution of an ancient volcanic arc system. J. geol. Soc. London, 140, 185-202.

Schandelmeier, H., D.P.F. Darbyshire, U. Harms and A. Richter, 1987. The East Sahara craton: evidence for pre-Pan African crust in NE Africa west of the Nile. In El-Gaby, E. and Greiling, R. O. (eds.) The Pan-African belt of NE Africa and adjacent areas. Earth evol. sci., Vieweg, Wiesbaden.

Shackleton, R.M., A.C. Ries, R.H. Graham and W.R. Fitches, 1980. Late Precambrian ophiolite melange in the Eastern Desert of Egypt. Nature, 285, 472-474.

Shimron, A.E., 1980. Proterozoic island arc volcanism and sedimentation in Sinai. Precambrian Res., 12, 437-458.

Stacey, J.S. and C.R. Hedge, 1984. Geochronologic and isotopic evidence for early Proterozoic crust in the eastern Arabian shield. Geology, 12, 310-313.

Stacey, J.S. and D.B. Stoeser, 1984. Distribution of oceanic and continental leads in the Arabian-Nubian shield. Contrib. Mineral. Petrol., 84, 91-105.

Stacey, J.S., D.B. Stoeser, W.R. Greenwood and L.B. Fischer, 1984. U-Pb zircon geochronology and geologic evolution of the Halaban-Al Amar region of the eastern Arabian shield, Kingdom of Saudi Arabia. J. geol. Soc. London, 141, 1043-1055.

Stern, R.J., 1981. Petrogenesis and tectonic setting of late Precambrian ensimatic volcanic rocks, central Eastern Desert of Egypt. Precambrian Res., 16, 195-230.

Stern, R.J. and C.E. Hedge, 1985. Geochronologic and isotopic constraints on late Precambrian crustal evolution in the eastern Desert of Egypt. Am. J. Sci., 285, 97-127.

Stoeser, D.B. and V.E. Camp., 1985. Pan-African microplate accretion of the Arabian shield. Geol. Soc. America Bull., 96, 817-826.

Vail, J.R., 1983. Pan-African crustal accretion in northeast Africa. J. afr. Earth Sci., 1, 285-294.

Williams, I.S., W. Compston and B.W. Chappell, 1983. Zircon and monazite U-Pb systems and the histories of I-type magmas, Berridale Batholith, Australia. J. Petrol, 24, 76-97.

Wust, H.J., W. Todt, and A. Kröner, 1987a. Conventional and single grain zircon ages for metasediments and granite clasts from the Eastern Desert of Egypt: Evidence for active continental margin evolution in Pan-African times. Terra Cognita, 7, 333-334.

Wust, H.J., T. Reischmann, A. Kröner and W. Todt, 1987b. Conflicting Pb-Pb, Sm-Nd and Rb-Sr systematics in late Precambrian metasediments and metavolcanics from the Eastern Desert of Egypt. Terra Cognita, 7, 333.

Chapter 5

Structural Relationships between the Southern and Central Parts of the Eastern Desert of Egypt: Details of a Fold and Thrust Belt

R. O. Greiling[1, 4] / A. Kröner[1] / M. F. El-Ramly[2] / A. A. Rashwan[3]

[1] Institut für Geowissenschaften, Johannes Gutenberg-Universität, Postfach 3980, 6500 Mainz, FRG.
[2] Geology Department, Faculty of Sciences, Ain Shams University, Abbasiya, Cairo, Egypt.
[3] Egyptian Geological Survey and Mining Authority, 3 Salah Salem Street, Abbasiya, Cairo, Egypt.
[4] present address: Department of Geology, University College, P. O. Box 78, Cardiff CF1 1XL, U.K.

Keywords: Proterozoic, Pan-African, Egypt, Arabian-Nubian Shield, fold and thrust belt, duplex, stretching lineation

Abstract: The central part of the Pan-African basement complex of the Eastern Desert of Egypt is dominated by low grade ophiolitic mélanges and associated calc-alkaline rocks. Structurally, these rocks are part of a fold and thrust belt, which formed during broadly E-W directed crustal shortening at the margin of a pre-Pan-African craton in the west. Towards the south gneissic successions are exposed in tectonic windows beneath the low grade rocks. In contrast to earlier views these gneisses do not represent an autochthonous basement but form thrust units, which are an integral part of the fold and thrust belt. Such a situation is particularly well documented for the Migif-Hafafit gneisses exposed in a window at the margin between the central and southern parts of the Eastern Desert of Egypt. Therefore, geometry, evolution and transport direction of thrust units in the Migif-Hafafit area are shown here as examples for the structure of the Eastern Desert of Egypt and fold and thrust belts in general.

The Migif-Hafafit gneisses are dissected into three major horses, which form the 'Migif-Hafafit antiformal stack' above the 'Migif-Hafafit (basal) thrust'. In the footwall of the Migif-Hafafit thrust, the Shaitian granite represents the lowermost structural unit exposed. The 'Nugrus thrust', which separates the Migif-Hafafit gneisses from overlying low grade successions forms the roof of the Migif-Hafafit antiformal stack. The uppermost (major, 'first order') horse of the Migif-Hafafit antiformal stack is composed of several minor 'second order' horses, which form a WSW-, foreland dipping duplex, the 'Wadi Sikait duplex'. At least one of the 'second order' horses of the Wadi Sikait duplex is again composed of minor ('third order') horses. An example of overlying thrust units in the low grade successions (in the hangingwall of the Nugrus thrust) is the 'Gabal Lewiwi duplex', which is WSW-, foreland dipping.

Whereas the allochthoneity of both the low grade sequences and the Migif-Hafafit gneisses is well documented, information on the direction of tectonic transport is still equivocal (NE-SW or SE-NW). Relevant information comprises geometry of thrust units, orientation of ramps in thrust surfaces, orientation of fault-bend folds and trend of stretching lineations. Cross-cutting relationships of stretching lineations imply NE-SW stretching, restricted to the Migif-Hafafit gneisses, to predate SE-NW stretching related to regional nappe transport. Stretching lineation in SE-NW orientation is the regionally most persistent feature and probably provides the strongest argument for a tectonic transport direction in a regional scale.

INTRODUCTION

In the basement complex of the Eastern Desert of Egypt (EDE), ophiolitic and calcalkaline igneous rocks together with their isotopic characteristics (e.g. Harris et al. 1984) point to an ensimatic tectonic evolution during the early stages of the Pan-African event (about 800-600 Ma), leading to ophiolite subduction and development of magmatic arcs (e.g. Ries et al. 1983, Kröner 1985). Consumption of ocean basins terminated with the collision of magmatic arcs with each other and with their accretion onto the margin of a pre-Pan-African craton in the west, beneath the present Phanerozoic cover (Fig. 1). The final tectonic activity, subsequent to molasse-type sedimentation, comprises low angle thrusting, strike-slip faulting and the intrusion of late-tectonic granite plutons (600-580 Ma; Ries et al. 1983, Stern 1985). Thus, in the central and southern parts of the EDE extensive low angle thrusting, relatively weak folding and a generally low grade of metamorphism (only locally high T/low-medium P, spatially related to tectonic windows; Ries et al. 1983, El-Ramly et al. 1984) led to the view of a foreland thrust and fold belt at the margin of the pre-Pan-African craton in the west, which developed during and after the accretion of the magmatic arcs onto this ancient continental margin (Ries et al. 1983, Shackleton 1986, Kröner et al. 1987a). Whereas E-W convergence during early accretion is generally accepted, the direction of transport on low angle thrusts is still controversial. Both northwest-directed (Ries et al. 1983, Sturchio et al. 1983) and southwest-directed thrusting (El-Gaby et al. 1984, El-Ramly et al. 1984) is assumed in the central parts of the EDE.

At the boundary between the southern part of the EDE (SED) and its central part (CED, Fig. 1; for definition see below) the Migif-Hafafit gneisses (El-Ramly et al. 1984) of Wadi Hafafit area (Fig. 2), previously regarded as the structurally lowest, autochthonous unit of the SED have been thrusted over the Shaitian gneissic granites at Wadi Shait (Greiling 1985, Kröner et al. 1987a; Fig. 7) and are therefore allochthonous.

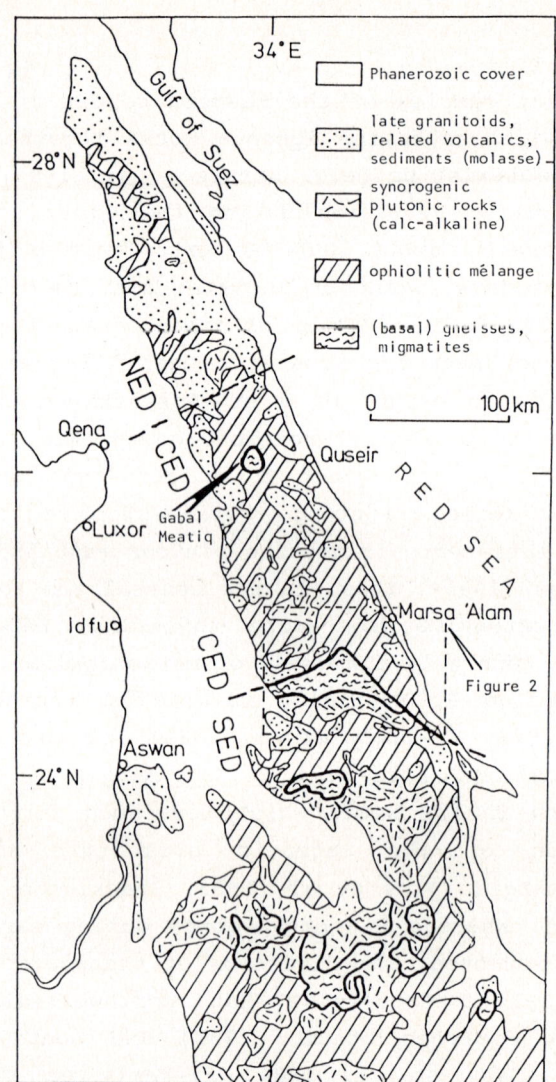

FIGURE 1.

Precambrian rocks of the Eastern Desert of Egypt and their
subdivision into South Eastern Desert (SED), Central
Eastern Desert (CED) and North Eastern Desert (NED),
modified from Stern and Hedge (1985). Gneisses and
migmatites generally occupy the lowermost structural
levels exposed and show thrust contacts with the overlying
units, shown by thick lines. One of these thrusts is taken
here as the boundary between SED and CED. It is shown in
more detail in Figure 2 (position of Fig. 2 indicated by
dashed lines).

Consequently, the shear zones at the northern margin of the Wadi Hafafit area which form the boundary between the SED and the CED can no longer be regarded as delimiting allochthonous, low grade units in the north from a basement culmination of autochthonous gneisses in the south.

These new observations imply a reinterpretation of the structural evolution along the SED/CED margin and we therefore present here further evidence on the tectonic position of the Migif-Hafafit gneisses and discuss the geometry, evolution and transport direction of associated thrust units.

REGIONAL CONTEXT AND LITHOLOGY

Based on lithology and age, the basement rocks of the Eastern Desert of Egypt (EDE) have been divided by Stern and Hedge (1985) into three domains, namely the South Eastern Desert (SED), the Central Eastern Desert (CED) and the North Eastern Desert (NED; Fig. 1).

The SED is characterized by basal units of medium-grade gneisses including continental/shelf facies metasediments, exposed in some tectonic windows beneath low grade successions of ophiolitic mélange, suites of calc-alkaline igneous rocks and related metasediments (e.g. Shackleton et al. 1980, El-Ramly et al. 1984, El-Gaby et al. 1987, Church 1987). The most intensely studied and probably largest of these windows occurs around Wadi Hafafit (about 24° 30-50′ N, Fig. 1) and is given the name Wadi Hafafit culmination (WHC; El-Ramly et al. 1984). Its northern margin has been taken as boundary between SED and CED (Stern and Hedge 1985). Towards the north, the CED is almost exclusively built up of ophiolitic mélange and associated calc-alkaline (island arc) rocks, together with subordinate clastic, molasse-type sediments and late-tectonic bimodal volcanics and granitoid intrusives. There only one small window at Gabal Meatiq (about 26° N, Fig. 1), where gneisses are exposed beneath the low grade thrust units (Ries

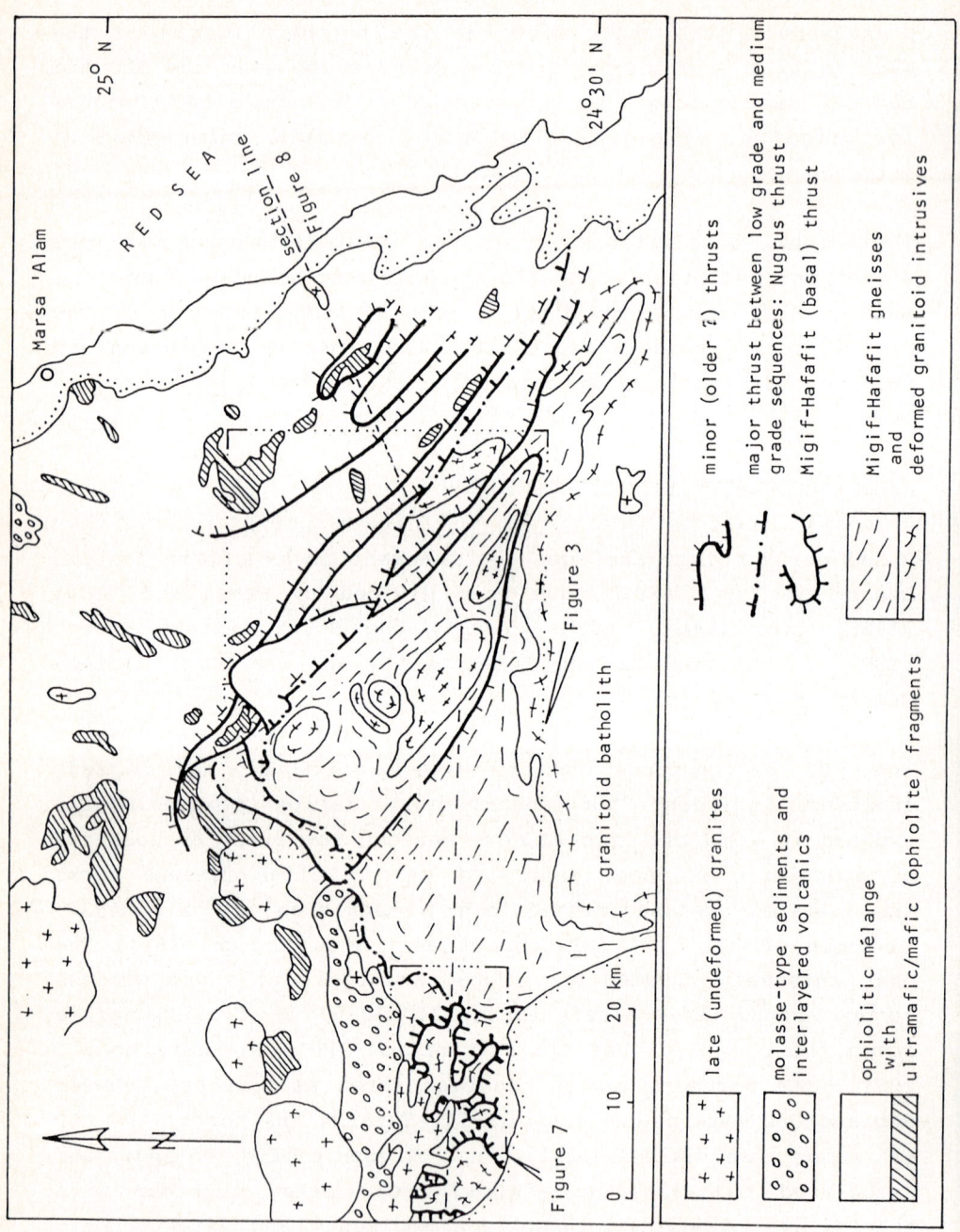

FIGURE 2.

Geological sketch map of the marginal area between SED and CED showing Migif-Hafafit gneisses in the south that dip beneath low grade rocks towards the northwest and the northeast (see Fig. 1 for location). Compiled and revised from El-Ramly and Hermina (1978a), El-Ramly et al. (1984), El-Bayoumi and Greiling (1984). The roof thrust of the Migif-Hafafit gneisses is named Nugrus thrust and represents the margin between SED and CED. The floor thrust of the Migif-Hafafit gneisses is called Migif-Hafafit thrust. Granitoids beneath the Migif-Hafafit thrust represent the 'Shaitian granites' (e.g. Hussein et al. 1982) at their type locality.

The lowermost unit of the Migif-Hafafit gneisses enclosing five granitoid-cored domal structures in the centre of the map area was termed Wadi Hafafit culmination (WHC) by El-Ramly et al. (1984). Minor thrusts in the ophiolitic mélange are shown only where studied in detail. Such thrusts also cut molasse-type rocks (Greiling and El-Ramly 1985). Dip of thrusts is shown schematically. Note general NW – SE structural trend at the WHC and towards E, which terminates abruptly to the NW of the Migif-Hafafit gneissic domain.

Location of detailed maps (Figs. 3, 7) and section (Fig.8) is indicated. Phanerozoic cover is shown with stippled margin.

et al. 1983, Sturchio et al. 1983, El-Gaby et al. 1984). Local,
post-tectonic granite intrusions, same as in the SED, define
the end of the Pan-African event (Ries et al. 1983, El-Ramly
and Hermina 1978a, b).

In the NED and Sinai basement, late orogenic acidic plutons are
widespread (e.g. Stern 1985, Stern and Hedge 1985), while
ophiolitic mélanges and associated rocks form only minor
remnants (e.g. Garson and Shalaby 1976, Shimron 1984, Clark
1985).

MAJOR THRUST UNITS

The major thrust separating hangingwall low grade sequences
towards north from the Migif-Hafafit gneisses in its footwall
towards south can be followed through the Pan-African basement
complex from the Red Sea in the east to the Phanerozoic cover
in the west (Fig. 2). It represents the SED/CED margin (Stern
and Hedge 1985) and is called here 'Nugrus thrust' since it
forms the basal thrust of the unit containing the Nugrus
granite, which is building up Gabal Nugrus (Fig. 3). The basal
thrust of the Migif-Hafafit gneisses is named here 'Migif-
Hafafit thrust'. It is blind at Wadi Hafafit but is exposed
farther west (Figs. 2, 7). Several minor thrusts dissect the
Migif-Hafafit gneisses into horses, which compose a duplex with
the Migif-Hafafit thrust as its floor and the Nugrus thrust as
its roof respectively (Fig. 8). According to the geometry and
relative position of horses (Figs. 6, 8), this major, 'first
order' duplex may be characterized as an antiformal stack
(Boyer and Elliot 1982) and it is named here 'Migif-Hafafit
antiformal stack'. The Migif-Hafafit antiformal stack gave rise
to the regional antiformal structure, which controls the
present map pattern.

The lowermost horse exposed in the core of the regional
antiform is also dominated by antiformal structures (Fig. 3)
and was therefore called Wadi Hafafit culmination (WHC,

El-Ramly et al. 1984). The major folds in the WHC are two antiforms trending NE-SW and NW-SE respectively and terminating at their junction with each other (Fig. 3). These geometric relationships of the two folds preclude an origin by simple lateral compression and the antiforms are therefore interpreted as fault-bend folds (Suppe 1983), which originated above ramps developed in the underlying thrust (Greiling and El-Ramly 1985), named here as Migif-Hafafit thrust.

The NW-SE trending of these antiforms shown in Figure 3 is associated with a steeply inclined northeastern limb, which contains the NW-SE oriented part of the Nugrus thrust together with over- and underlying thrust surfaces. As a consequence of the steep dip of the thrust surfaces, the map pattern reveals a section across different successive thrust units. Details of this pattern are shown in maps (Figs. 4, 5) and a section normal to strike (Fig. 6).

EXAMPLES OF MINOR THRUST UNITS

In lower Wadi Nugrus, both a simple thrust sheet and an overlying duplex structure composed of Migif-Hafafit gneisses are exposed above the WHC (Figs. 3, 4). The duplex is named here Wadi Sikait duplex since it is dissected by Wadi Sikait, where duplex structures have first been observed in the Eastern Desert (Ries et al. 1983). The Wadi Sikait duplex is of 'second order', since it forms a horse of the major ('first order') Migif-Hafafit duplex (Fig. 8).

The lowermost horse exposed of the Wadi Sikait duplex is built up exclusively of granite and gneissic granite. The next overlying horse is dominated by metasediments, mostly quartzites, quartz-biotite and biotite schists. The latter sometimes contain emeralds, which were mined both in Wadi Sikait and in Wadi Nugrus. Subordinate slices of ultramafic and mafic rocks are imbricated within the metasediments, making the

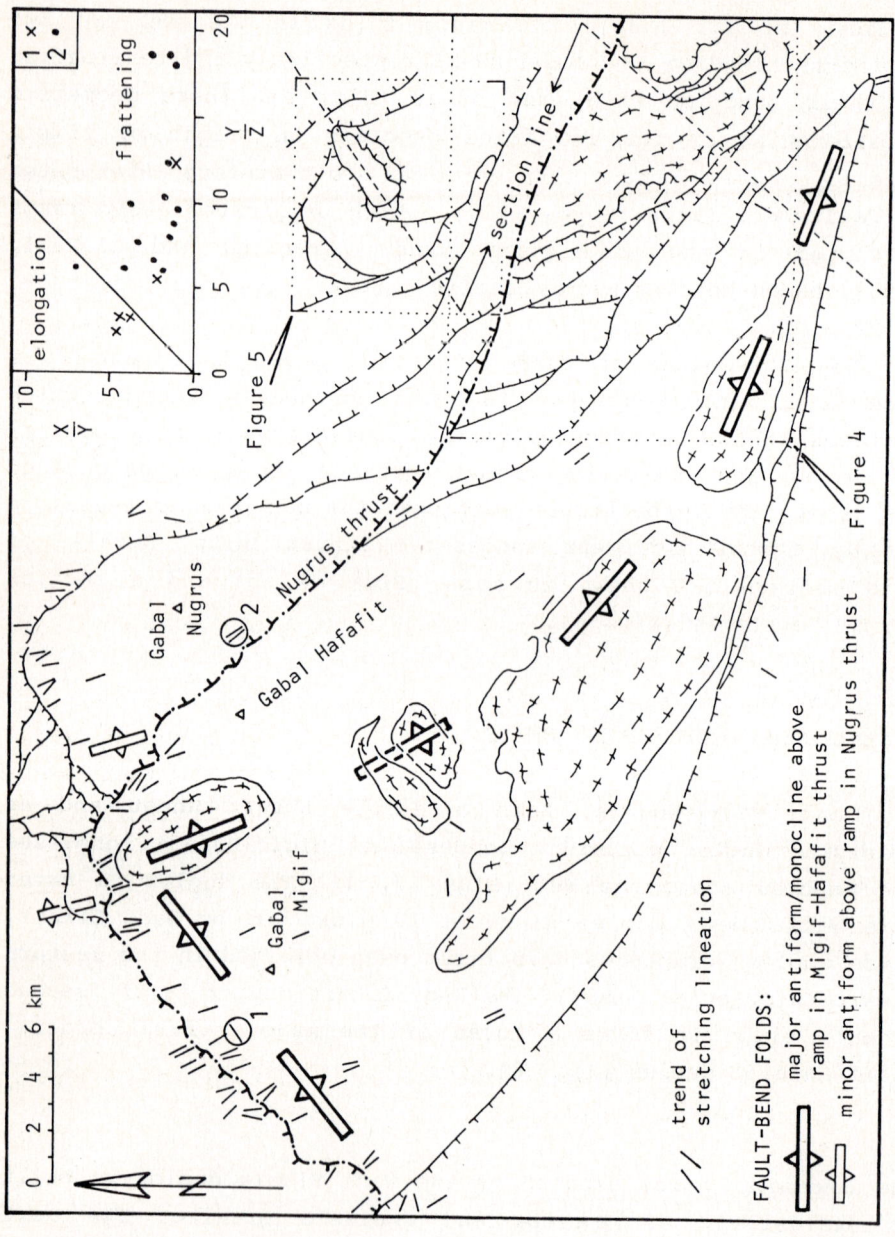

FIGURE 3.

Structural sketch map of the Wadi Hafafit culmination (WHC) with overlying units of Migif-Hafafit gneisses and low grade successions. Granitoid gneisses are shown with 'x' pattern, unornamented areas represent calc-alkaline igneous rocks, mainly of mafic to intermediate composition and rarely acidic, minor ultramafic bodies and subordinate metasediments. Thrusts as in Figure 2.

The WHC, characterized by major antiformal folds, is the lowermost unit built up of Migif-Hafafit gneisses and is containing five granitoid-cored domal structures (El-Ramly et al. 1984). The WHC is delimited by the Nugrus thrust in the northwest and the northeast and by a minor thrust in the south and southeast. The Nugrus thrust separates the Migif-Hafafit gneisses in its footwall from low grade successions in its hangingwall. Details of thrust units are shown in Figures 4 and 5. 'Section line' refers to the cross section in Figure 6.

Local ramps and associated minor antiforms are exposed in the Nugrus thrust to the west of Gabal Nugrus. Major anti-forms are related to ramps in the footwall of the Migif-Hafafit thrust. The Migif-Hafafit thrust is blind in the map area but is exposed farther west (see Figs. 2, 7, 8).

The inset Flinn plot shows shapes of deformed grains in the footwall (locality 1, NW of Gabal Migif) and in the hangingwall of the Nugrus thrust (locality 2, SW of Gabal Nugrus) respectively. Particles shown from locality 1 are elongated in NE-SW direction. Stretching lineations and X axes at locality 2 are oriented SE-NW but particles show flattening-type deformation.

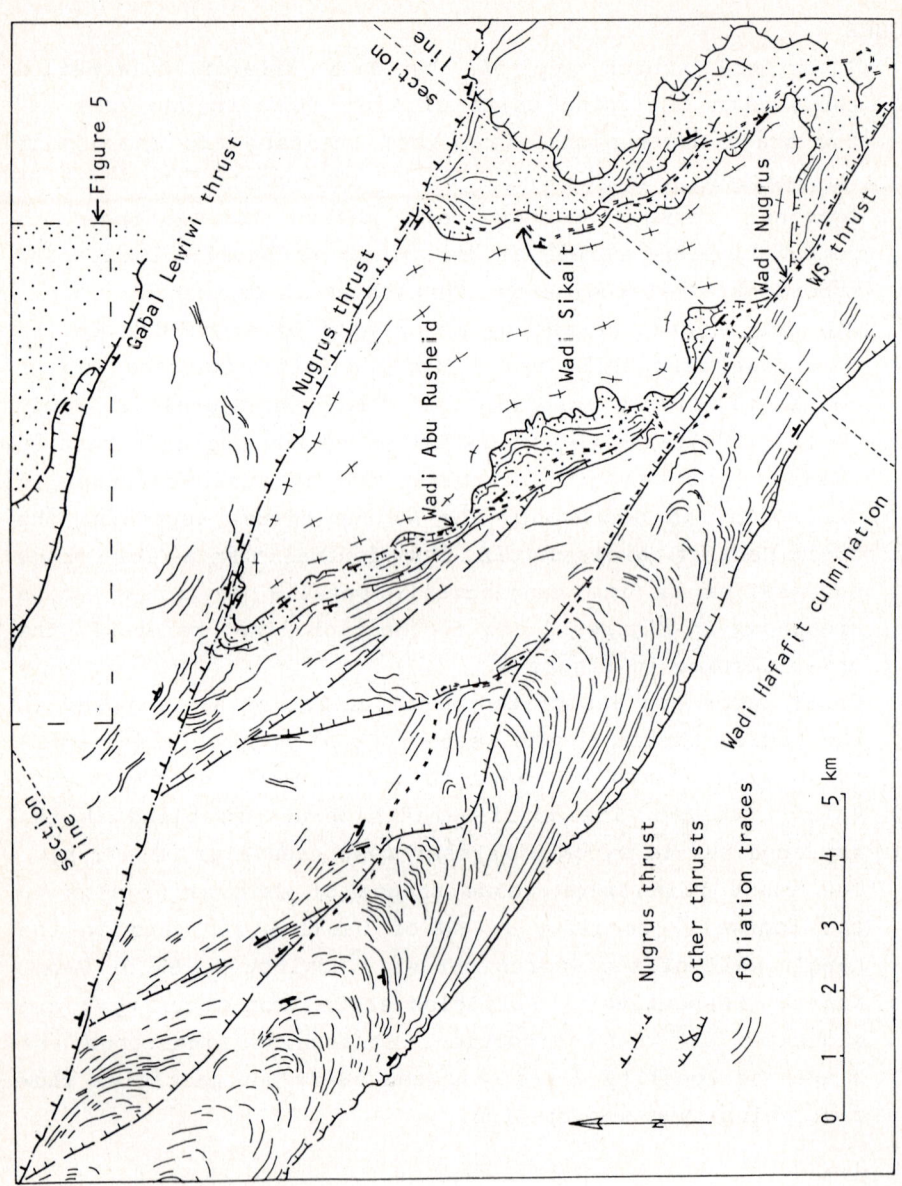

FIGURE 4.

Structural sketch map showing thrust units in the hanging-wall and northeast of the Wadi Hafafit culmination (WHC) in the Wadi Nugrus area (see Fig. 3 for location). Units containing abundant metasediments are stippled, other symbols as in Figure 3.

The thrust unit between the Wadi Sikait and Nugrus thrusts is the ('second order') Wadi Sikait duplex, containing, i.a., horses of metasedimentary and igneous rocks. In lower Wadi Sikait Ries et al. (1983) have shown the meta-sedimentary unit itself to form a ('third order') duplex composed of quartzites, pelites and ultramafic igneous rocks (their Fig. 8 covers the SE corner of the present Fig.).

The granitic unit between Wadis Sikait and Abu Rosheid has structural (thrust) contacts with the overlying rocks. It cores an irregular antiform, apparently unrelated to the regional structure, which is interpreted as due to gravitative doming. Above the Wadi Sikait duplex and the Nugrus thrust an apparently simple thrust sheet is exposed, which is overlain by the Gabal Lewiwi duplex in the hangingwall of the Gabal Lewiwi thrust. This duplex is shown in more detail in Figure 5. Section lines indicate the position of the section in Figure 6.

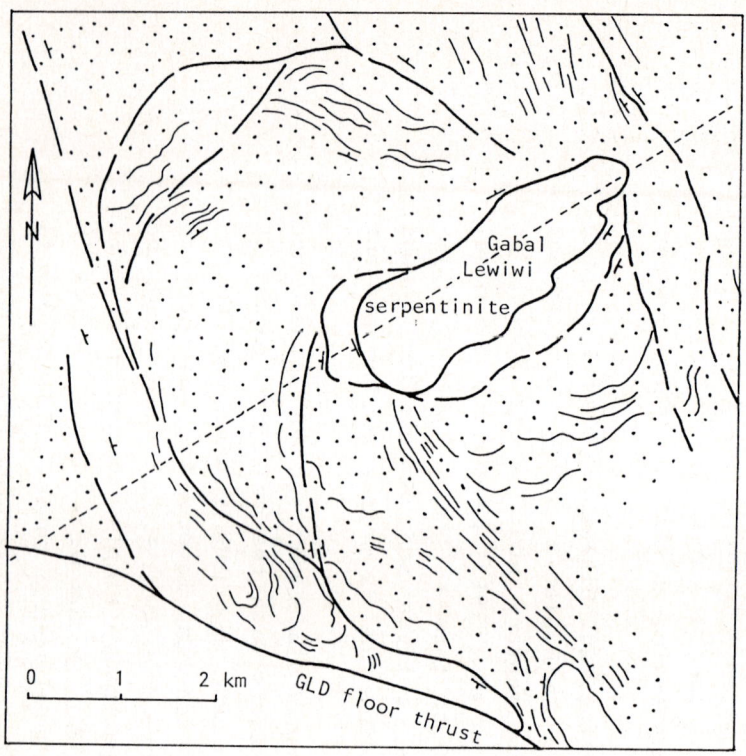

FIGURE 5.

Structural sketch map showing part of the Gabal Lewiwi
duplex (GLD), its floor and roof thrusts (northeasternmost
thrust shown). For location see Figure 3, legend as in
Figure 4.

The irregular domal structure of the Gabal Lewiwi
serpentinite is not related to regional folds and is
therefore interpreted as a gravitative dome (comp.
Shackleton 1987). The NE - SW trending dashed line refers
to the section in Figure 6.

metasediment-dominated horse itself a 'third order' duplex, which is composed of minor horses, as has been shown by Ries et al. (1983, their Fig. 8). Both the horses composed of metasediments and the underlying horse of granite extend from Wadi Sikait to Wadi Abu Rusheid (Fig. 4), where radiometric dating on zircons revealed a pre-Pan-African age (Abdel Monem and Hurley 1979). Since the structural context implies an allochthonous position of the horses constituting the Wadi Sikait duplex, this age does not prove the existence of an (autochthonous) pre-Pan-African basement in the area (in contrast to El-Gaby et al. 1987). Kröner et al. (1987b) further discuss the significance of this age. Structurally overlying horses of the Wadi Sikait duplex are composed of mafic to intermediate intrusive and extrusive igneous members of the Migif-Hafafit gneisses. The duplex roof thrust, the Nugrus thrust, separates the Migif-Hafafit gneisses from overlying low grade sequences.

The conditions in the hangingwall of the Nugrus thrust are shown in Figures 5 and 6, where a simple thrust sheet composed of low grade mafic and intermediate igneous rocks is overlain by a duplex. The most prominent feature of the duplex is the ultamafic complex of Gabal Lewiwi and it is therefore named here 'Gabal Lewiwi duplex'. The Gabal Lewiwi serpentinite complex forms the lowermost horse of the duplex exposed and is overlain by small horses of gabbroic composition, which are, in turn, overlain by relatively larger horses composed of turbiditic metasediments.

GEOMETRY OF THRUST UNITS AND CRITERIA FOR NAPPE TRANSPORT DIRECTION

Possible gravitative overprint

An overprint of regional structures by gravitative doming of granitoid gneisses has been documented for the five granitoid domes in the WHC (Fig. 3; Greiling et al. 1984). In the Wadi

Sikait duplex, the map (Fig. 4) and the section (Fig. 6) show
that the antiformal structure of the granite and gneissic
granite is not compatible with a generally sigmoidal shape of
horses. Therefore, gravitative doming is assumed to have overprinted the horse shape and formed the structural culmination
in the Wadi Sikait duplex between Wadis Sikait and Abu Rosheid
(Fig. 4). Similarly, the Gabal Lewiwi ultramafic complex is
piercing the overlying horses of the Gabal Lewiwi duplex in a
culmination, which is elongated in NE-SW direction and
geometrically unrelated to the NW-SE trending regional strike
of structural elements (Fig. 5). The origin of this culmination
may therefore be related to gravitative doming of the
ultramafic rocks due to a decrease in specific gravity during
serpentinization as envisaged by Shackleton (1987).

Shape of horses

The section (Fig. 6) shows the geometry of the thrust units
normal to the regional strike direction. The shape of horses in
the Gabal Lewiwi duplex together with the generally westward
directed transport known from the regional context implies a
transport from ENE to WSW towards the cratonic foreland. The
dip of the horses towards the cratonic foreland suggests a
foreland dipping duplex (Boyer and Elliot 1982). Apart from the
granitic horse in the Wadi Sikait duplex, also the Wadi Sikait
duplex is a foreland dipping duplex. A similar geometry has
also been observed in the minor horses of Wadi Sikait and taken
as indicating a transport towards WSW (Ries et al. 1983).

Stretching lineations

A WSW or SW directed tectonic transport may be supported by
stretching lineations observed within the Migif-Hafafit
antiformal stack, which trend NE-SW (Fig. 3). However,
stretching lineations associated with the Nugrus thrust
consistently trend SE-NW (Fig. 3). Since stretching is

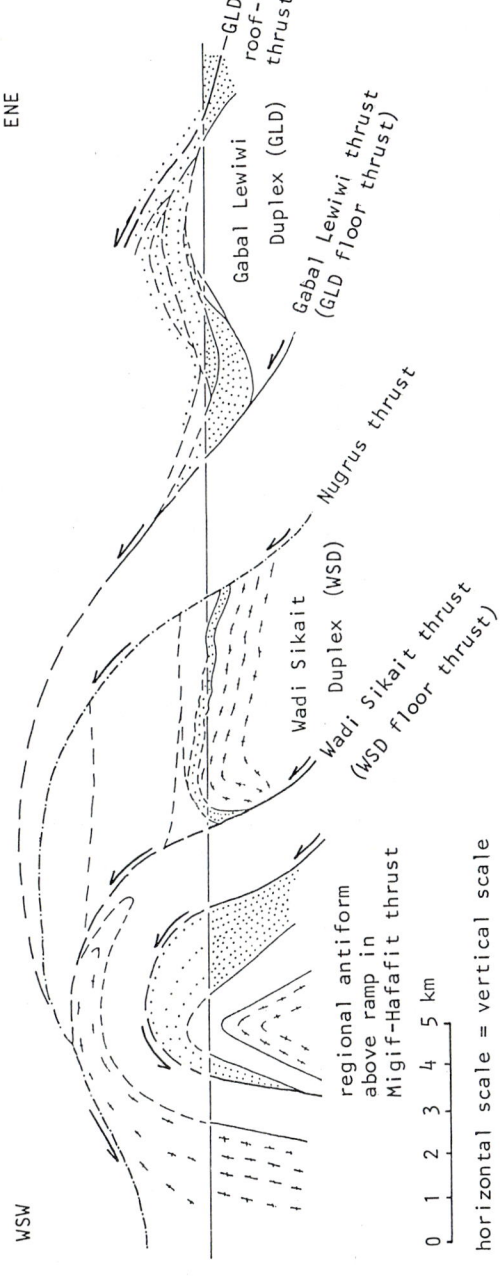

horizontal scale = vertical scale

FIGURE 6.

Structural section from Wadi Hafafit culmination (WHC) towards northeast, about normal to regional strike and approximately parallel to possible ENE-WSW nappe transport direction (Greiling and El-Ramly 1985). Section is compiled from Figures 4 and 5 and extended towards southwest to include the WHC (based on Fig. 3 and Greiling et al. 1984). Legend as in Figure 4. Note over-printing of duplex geometry by gravitative doming in the Wadi Sikait duplex (granitoid rocks) and in the Gabal Lewiwi duplex (serpentinites).

generally assumed to parallel tectonic transport (see discussion by, e.g., Coward 1984, Shackleton and Ries 1984) this orientation implies NW directed movement.

Crosscutting relationships of stretching lineations north of Gabal Migif (locality 1 in Fig. 3) show SE-NW stretching related to thrusting in the footwall of the Nugrus thrust to overprint older, NE-SW directed stretching documented by elongated breccia particles (inset in Fig. 3). Such cross-cutting relationships are also observed elsewhere in the Migif-Hafafit antiformal stack beneath the Nugrus thrust, where NE-SW stretching lineations are relatively frequent besides SE-NW trending lineations.

In contrast, stretching lineations in the low grade successions, in the hangingwall of the Nugrus thrust are oriented exclusively SE-NW.

Ramps and fault-bend folds

The Nugrus thrust displays local ramp-flat geometries north of Gabal Migif and north of Gabal Hafafit. The ramps give rise to local antiformal folds in the hangingwall (low grade) thrust units (Fig. 3; Greiling and El-Ramly 1985). In a larger scale the Nugrus thrust itself is folded by fault-bend folds (Suppe 1983; Fig. 3). The NE-SW trending major fault-bend fold in the WHC (Fig. 3) implies a ramp with NW dip in the footwall of the Migif-Hafafit thrust. However, the present dip angle is distinctly higher than that observed at ramps in general (e.g. Boyer and Elliot 1982, Butler 1982, Suppe 1983). It is therefore assumed that the steep inclination is due not only to rotation above a simple ramp (even if some gravitative overprint is taken into account) but above an antiformal stack beneath the NW-SE trending antiform. This situation implies a passive roof behaviour (Banks and Warburton 1986) of the Migif-Hafafit gneisses in the hangingwall of the Migif-Hafafit thrust. Thus, the major fault-bend folds shown in Figure 3 may also be interpreted as originated in a passive roof of a buried duplex in the footwall of the Migif-Hafafit thrust.

In a smaller scale, the smooth trajectory of the Nugrus thrust relative to the Wadi Sikait duplex does not reflect any 'passive roof' deformation but implies that the Nugrus thrust was active during (Wadi Sikait) duplex stacking (McClay and Insley 1986) or perhaps even later.

DISCUSSION

The Migif-Hafafit gneisses are dissected into three major, allochthonous horses, composing the Migif-Hafafit antiformal stack. This structural situation documented above (e.g. Figs. 6, 8) precludes that the Migif-Hafafit gneisses represent an autochthonous basement. Whereas the allochthoneity of the Migif-Hafafit gneisses and the overlying low grade sequences is hardly questionable, information on the direction of nappe transport is still equivocal.

Apparently contrasting pieces of evidence on tectonic transport directions at the SED/CED margin are derived from the geometry of thrust units, the orientation of ramps in thrust surfaces, the orientation of fault-bend folds and the trend of stretching lineations.

The following models may be relevant to explain the (apparent ?) variations in the directions of tectonic transport towards NW and SW respectively:

1. Primary complexities during duplex formation (e.g. out-of-section thrusts) as observed by McClay and Insley (1986) in the Lewis thrust sheet.
   Such features have not been observed in the actual area.

2. Simple fault-bend folds above a ramp (Suppe 1983) as contrasted to complex fault-bend folds in a passive roof of

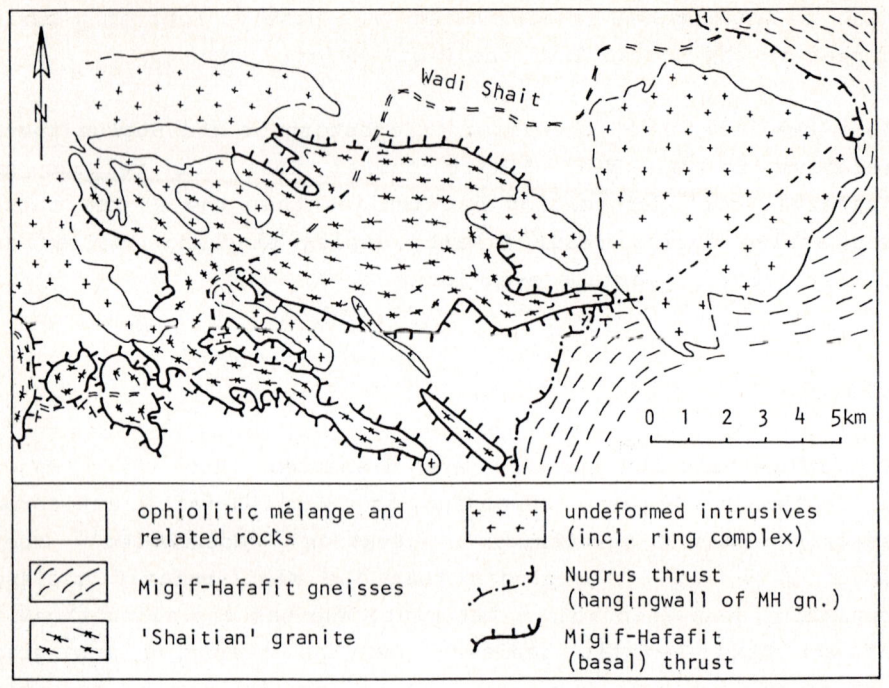

FIGURE 7.

    Structural sketch map showing the Shaitian granite at its type locality in structural windows beneath the Migif-Hafafit gneisses and the ophiolitic mélange. The Migif-Hafafit thrust forms the roof of the Shaitian granite and the floor of the Migif-Hafafit gneisses, as can be seen at the easternmost tip of the Shaitian granite (compare section, Fig. 8).

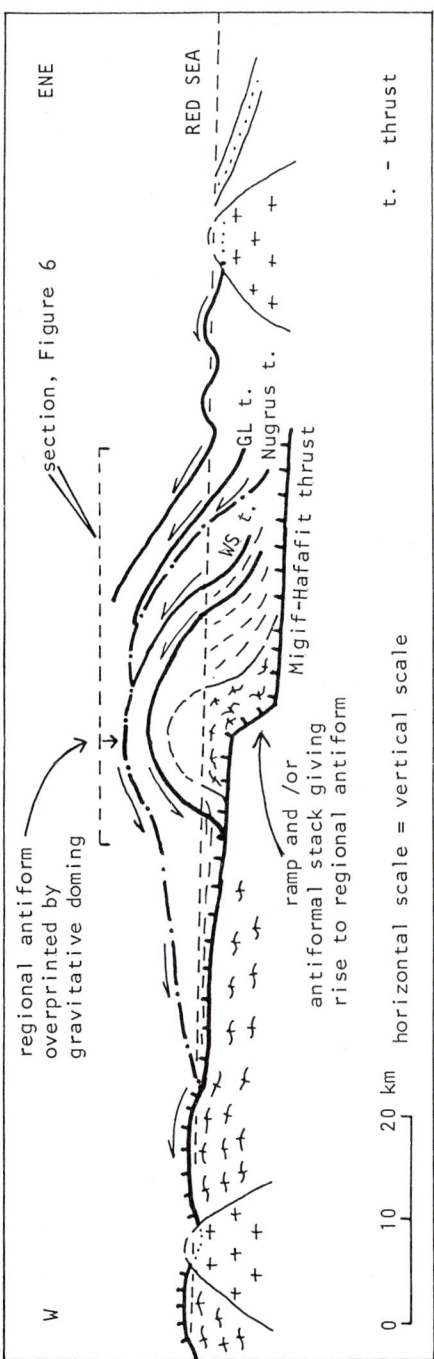

FIGURE 8.

Structural section through marginal area between SED and CED, compiled from Figure 6 and Figure 2, partly based on El-Bayoumi and Greiling (1984) and Kröner et al. (1987a). For location and legend see Figure 2. The section is normal to regional strike but not necessarily parallel to a tectonic transport direction. The horses between Migif-Hafafit and Nugrus thrusts form an antiformal stack, the Migif-Hafafit antiformal stack, situated above a ramp and/or an antiformal stack in the footwall of the Migif-Hafafit thrust. The Migif-Hafafit antiformal stack is overprinted by gravitative doming of granitoid gneisses (Greiling et al. 1984).

a duplex (Banks and Warburton 1986). For example, the orientation of a NW-dipping ramp at the northern margin of the WHC (Fig. 3) apparently precludes thrusting towards NW, if it is accepted that thrusts do not cut steeply down-section in their transport direction (Elliot and Johnson 1980, Butler 1985).

As argued above, the major fault-bend folds (Fig. 3) are more likely related to a passive roof than a ramp. Therefore, a complex relationship between the orientation of these folds and the direction of nappe transport has to be considered. Consequently, the orientation of fault-bend folds may be inconclusive in the actual example.

3. A sequence of separate thrusting events with different transport directions. Crosscutting relationships of stretching lineations imply NE-SW stretching in the Migif-Hafafit gneisses to predate SE-NW stretching related to the Nugrus thrust. Thus, NE-SW stretching obviously is an early feature, restricted to the Migif-Hafafit gneisses (Fig. 3), whereas SE-NW stretching is related to the (late) thrusting event in a regional scale.

Considering the available information, the stretching lineation in SE-NW orientation is the most persistent feature and probably provides the strongest argument for the tectonic transport direction in a regional scale.

ACKNOWLEDGEMENTS

This paper is a contribution to IGCP project 215 'Proterozoic fold belts' and a cooperative project between the Egyptian Geological Survey and Mining Authority (EGSMA), Cairo, Egypt and the Institute of Geosciences, Johannes Gutenberg-Universität, Mainz, W-Germany supported by the Ministry of Research and Technology (BMFT) through KFA Jülich, Internationales Büro, Jülich, W-Germany. This project is also part of the research effort 'Accretion and Differentiation of the Earth' funded by the German Research Council (DFG), Bonn, W.Germany.

We thank Chairman and staff of EGSMA for their continuous interest and their support of our field work, in particular Dr. A.A. Hassan and Dr. A.A. Dardir. Dr. R.W.H. Butler (Durham, England) and Dr. P.N. Mosley (Nottingham, England) reviewed an early draft of the manuscript and suggested substantial improvements.

REFERENCES

Abdel-Monem, A.A. and P.M. Hurley, 1979. U-Pb dating of zircons from psammitc gneisses, Wadi Abu Rosheid-Wadi Sikait area, Egypt. - Inst. Applied Geol., King Abdulaziz Univ., Bull. 3(2), 165-170.

Banks, C.J. and J. Warburton, 1986. 'Passive roof' duplex geometry in the frontal structures of the Kirthar and Sulaiman mountain belts, Pakistan. - J. struct. Geol. 8, 229-237.

Boyer, S.E. and D. Elliot, 1982. Thrust systems. - Amer. Assoc. Petrol. Geol. Bull. 66, 1196-1230.

Butler, R.W.H., 1982. The terminology of structures in thrust belts. - J. struct. Geol. 4, 239-245.

Butler, R.W.H., 1985. Thrust tectonics: a personal view. - Geol. Mag. 122, 223-232.

Church, W.R., 1987. Ophiolites, sutures, and micro-plates of the Arabian-Nubian Shield: a critical comment. - In: El-Gaby, S. and Greiling, R.O. (eds.) The Pan-African Belt of NE Africa and adjacent areas. Earth evol. sci., Vieweg.

Clark, M.D., 1985. Late Proterozoic crustal evolution of the Midyan region, northwestern Saudi Arabia. Geology 13, 611-615.

Coward, M.P., 1984. Major shear zones in the Precambrian crust; examples from NW Scotland and southern Africa and their significance. - In: Kröner, A. and Greiling, R. (eds.) Precambrian tectonics illustrated. Schweizerbart'sche Verlagsbuchhandlung, 205-233.

El-Bayoumi, R.M.A. and R. Greiling, 1984. Tectonic evolution of a Pan-African plate margin in southeastern Egypt - a suture zone overprinted by low angle thrusting? In: J. Klerkx and J. Michot (eds.) African geology, Tervuren, 47-56.

El-Gaby, S., O. El-Nady and A. Khudeir, 1984. Tectonic evolution on the basement complex in the central Eastern Desert of Egypt. Geol. Rdschau. 73, 1019-1036.

El-Gaby, S., F.K. List and R. Tehrani, 1987. Geology, evolution and metallogenesis of the Pan-African Belt in Egypt. - In: El-Gaby, S. and Greiling, R.O. (eds.) The Pan-African Belt of NE Africa and adjacent areas. Earth evol. sci., Vieweg.

Elliot, D. and M.R.W. Johnson, 1980. Structural evolution in the northern part of the Moine thrust belt, NW Scotland. - Transact. R. Soc. Edinburgh, Earth Sci. 71, 69-96.

El-Ramly, M.F. and M.H. Hermina (eds.), 1978a. Geologic map of the Aswan quadrangle, Egypt. Egyptian Geol. Surv. Mining Author.

El-Ramly, M.F. and M.H. Hermina (eds.), 1978b. Geologic map of the Qena quadrangle, Egypt. Egyptian Geol. Surv. Mining Author.

El-Ramly, M.F., R. Greiling, A. Kröner and A.A. Rashwan, 1984. On the tectonic evolution of the Wadi Hafafit area and environs, Eastern Desert of Egypt. Bull. Fac. Earth Sci., King Abdulaziz Univ. 6, 113-126.

Garson, M.S. and I.M. Shalaby, 1976. Precambrian - Lower Palaeozoic plate tectonics and metallogenesis in the Red Sea region. Geol. Assoc. Can., Spec. Pap. 14, 573-596.

Greiling, R. 1985. Thrust tectonics in Pan-African rocks of SE Egypt. Terra cognita 5.

Greiling, R.O. and M.F. El-Ramly, 1985. Thrust tectonics in the Pan-African basement of SE Egypt. - CIFEG, Publ. Occ. 1985/3, 73-74.

Greiling, R., A. Kröner and M.F. El-Ramly, 1984. Structural interference patterns and their origin in the Pan-African basement of the southeastern Desert of Egypt. - In: Kröner, A. and Greiling, R. (eds.) Precambrian tectonics illustrated. Schweizerbart'sche Verlagsbuchhandlung, 401-412.

Harris, N.B.W., C.J. Hawkesworth and A.C. Ries, 1984. Crustal evolution in north-east and east Africa from model Nd ages. - Nature 309, 773-776.

Hussein, A.A., M.M. Aly and M.F. El-Ramly, 1982. A proposed new classification of the granites of Egypt. - J. Volcanolog. and Geothermal Res. 14, 187-198.

Kröner, A., 1985. Ophiolites and the evolution of tectonic boundaries in the Late Proterozoic Arabian-Nubian Shield of northeast Africa and Arabia. - Precambrian Res. 27, 277-300.

Kröner, A., R. Greiling, T. Reischmann, I.M. Hussein, R.J. Stern, J. Krüger, S. Dürr and M. Zimmer, 1987a. Pan-African crustal evolution in the Nubian segment of north-east Africa. - Am. Geophys. Union, Spec. Publ. 17, 235-257.

Kröner, A., T. Reischmann, H.J. Wust and A.A. Rashwan, 1987b. Is there any pre-Pan-African (>950 Ma) basement in the Eastern Desert of Egypt? In: El-Gaby, S. and Greiling, R.O. (eds.) The Pan-African Belt of NE Africa and adjacent areas. Earth evol. sci., Vieweg.

McClay, K.R. and M.W. Insley, 1986. Duplex structures in the Lewis thrust sheet, Crowsnest Pass, Rocky Mountains, Alberta, Canada. - J. struct. Geol. 8, 911-922.

Ries, A.C., R.M. Shackleton, R.H. Graham and W.R. Fitches, 1983. Pan-African structures, ophiolites and mélange in the Eastern Desert of Egypt: a traverse at 26° N.-J. Geol. Soc. London 140, 75-95.

Shackleton, R.M., 1986. Precambrian collision tectonics in Africa. - In: Collision tectonics, edited by Coward, M.P. and Ries, A.C., Spec. Publ. Geol. Soc. London 19, 329-349.

Shackleton, R.M., 1987. Contrasting structural relationships of Proterozoic ophiolite in Northeast and Eastern Africa. - In: Pan-African Belt of NE Africa and adjacent areas, El-Gaby, S. and Greiling, R.O. (eds.), Earth evol. sci., Vieweg.

Shackleton, R.M. and A.C. Ries, 1984. The relation between regionally consistent stretching lineations and plate motions. - J. struct. Geol. 6, 111-117.

Shackleton, R.M., A.C. Ries, R.H. Graham and W.R. Fitches, 1980. Late Precambrian ophiolitic mélange in the eastern desert of Egypt. - Nature 285, 472-474.

Shimron, A.E., 1984. Evolution of the Kid Group, southeast Sinai peninsula: thrusts, mélanges and implications for accretionary tectonics during the Late Proterozoic of the Arabian-Nubian Shield. - Geology 12, 242-247.

Stern, R.J., 1985. The Najd fault system, Saudi Arabia and Egypt: a Late Precambrian rift system? Tectonics 4, 497-511.

Stern, R.J. and C.E. Hedge, 1985. Geochronologic and isotopic constraints on Late Precambrian crustal evolution in the Eastern Desert of Egypt. Am. J. Sci. 285, 97-127.

Sturchio, N.C., M. Sultan and R. Batiza, 1983. Geology and origin of Meatiq dome, Egypt: a Precambrian metamorphic core complex? Geology 11, 72-76.

Suppe, J., 1983. Geometry and kinematics of fault-bend folding. - Amer. J. Sci. 283, 684-721.

Chapter 6

Bimodal Dike Swarms in the North Eastern Desert of Egypt:
Significance for the Origin of Late Precambrian "A-Type" Granites in Northern Afro-Arabia

Robert J. Stern[1] / George Sellers[2] / David Gottfried[2]

[1] Programs in Geosciences, The University of Texas at Dallas, Box 830688, Richardson, Texas 75083-0688, U.S.A.
[2] United States Geological Survey, Reston, VA, 22092, U.S.A.

Keywords: Proterozoic, Pan-African, Egypt, NE Africa, bimodal dike swarm, geochemistry, A-type granite

Abstract: Compositionally bimodal dike swarms are an important but poorly known aspect of the 575-600 Ma extensional event in northern Afro-Arabia. The "mafic" suite consists of basalts and andesites with 49 - 66 % SiO_2 and a mean of ~ 58 % silica. Porphyritic members contain phenocrysts of plagioclase and amphibole or clinopyroxene. Approximately 25 % of the mafic dikes are mantle-derived alkali basalts (up to 9 % MgO, 560 ppm Cr, 180 ppm Ni) that are enriched in incompatible elements (1.5 - 2.4 % TiO_2, .8 - 1.8 % K_2O, 290 - 650 ppm Ba, 210 - 270 ppm Zr, $(La/Yb)_n$ = 5.7 - 7.2). The rest of the mafic suite is composed of andesites enriched in Large Incompatible Lithophile (LIL) and Light Rare Earth Elements (LREE). The andesites formed by hydrous melting of garnet periodite or eclogite. The felsic suite consists of rhyolites; porphyritic variants contain phenocrysts of K-feldspar and quartz, biotite, and/or amphibole. The felsic dikes contain 70 - 78 % SiO_2 with a mean of 75 % SiO_2, and are largely metaluminous. The full range of compositions observed in the Eastern Desert "Pink

Granites" is recorded in the dikes. Compositional differences between low silica rhyolites (70 - 73 % SiO_2) and high silica rhyolites (77 - 78 % SiO_2) cannot be explained by fractional crystallization of the former to produce the latter. Instead it seems likely that liquid-liquid or vapor-liquid fractionation processes such as thermogravitational diffusion or halide-/carbonate-complexing was responsible.

The field and chemical data indicate that the mafic dikes were hypabyssal feeders for the Dokhan Volcanics while the felsic dikes fed epizonal sills and magma cushions emplaced at the base of the Dokhan. The large variations in felsic compositions indicate that the processes responsible for much of the chemical variability of Pink Granite was established at depth and did not occur, for the most part, in the presently exposed epizonal plutons. Compositional variability may either reflect differences in the crustal source or in deeper-seated magmatic fractionation processes.

Finally, the composition of the mafic dikes provides a <u>caveat</u> for interpreting tectonic environments from the chemical composition of igneous rocks. These rift-related magmas show some chemical fingerprints of Andean-type calc-alkaline suites. this supports the suggestion of other scientists that igneous rocks of rapidly extending, highly volcanic continental rifts show a large degree of compositional overlap with Andean margin magmatic suites.

INTRODUCTION

The latest stage in the formation of the continental crust of
Afro-Arabia was dominated by the widespread invasion of
abundant K-rich granites. These are the "Alkali Granites" of
Rogers and Greenberg (1981) and are indistinguishable from the
"A-type" granites intruded elsewhere in the closing stages of
cratonization (Collins et al., 1982). These granites were
emplaced during the interval 680 - 520 Ma in the Arabian Shield
(Fleck et al., 1980; Stoeser, 1986) and 600 - 540 Ma in Egypt
(Bielski et al., 1979; Fullagar, 1980; Halpern and Tristan,
1981; Stern and Hedge, 1985). These and similar suites
elsewhere in Afro-Arabia comprise one of the most extensive
concentrations of alkali granites on earth (Bentor, 1985) and
clearly manifest critical aspects of crustal evolution in Afro-
Arabia.

The origin of these melts is still unresolved. A major obstacle
to successful reconstruction of granite petrogenesis results
from the fact that much of the chemical evolution of these
bodies may result from processes that do not reflect crystal-
melt equilibrium. Especially the latest stages in the evolution
of the plutons may involve liquid-liquid and vapor-liquid
fractionation processes such as thermogravitational diffusion
(Hildreth, 1981) or halide- or carbonate-complexing and
transport (Harris and Marriner, 1980; Harris et al., 1986).
These processes can be especially important in redistributing
otherwise diagnostic trace elements such as the Rare Earth
Elements (REE) and so can strongly overprint the chemical
compositions that resulted from earlier, crystal-liquid
fractionation. Where such processes have occurred it becomes
very difficult to decipher the melt's petrogenetic history.

One objective of this paper is to attempt to circumvent this
obstacle and further examine the petrogenesis of the alkali
granites of Afro-Arabia. We approach this by focussing on the
late Precambrian dike swarms in the North Eastern Desert of
Egypt, one of the densest areas of continental dike swarms on

earth (Vail, 1970). They are closely related in space and time
with the other members of the late Precambrian bimodal suite,
the alkali Pink Granites and the Dokhan Volcanics. This
approach has the further advantage that the felsic dikes
represent liquid compositions, not cumulus "mushes" as are many
of the Pink Granites (Sultan et al., 1986). A second objective
is to document the composition and tectonic affinity of one
part of the late Precambrian dike swarms of Afro-Arabia.

GEOLOGICAL SETTING

Late Precambrian crustal evolution in the North Eastern Desert
of Egypt occurred in a tectonic environment of strong crustal
extension. Five principal lithologies dominate the crust of
this region (Stern et al., 1984). The oldest rocks exposed are
Granodiorites, 610 - 680 Ma in age (Stern and Hedge, 1985),
although 780 Ma granodiorites are found in Sinai (Stern and
Manton, in press). The tectonic setting of these older units is
poorly understood, although terranes immediately to the south
were experiencing accretion of juvenile arc terranes during
this interval (Ries et al., 1983).

Most of the crust exposed in the North Eastern Desert formed in
the interval 575 - 600 Ma. Units that formed during this
episode include the Dokhan Volcanics, a bimodal suite of alkali
basalts to medium-K andesites and rhyolites (Basta et al.,
1980; Ressetar and Monrad, 1983; Stern and Gottfried, 1986).
The Dokhan has been dated at 583 - 594 Ma (Stern and Hedge,
1985). These volcanics are intimately related with first cycle
clastic sediments of the Hammamat Group which were deposited in
intermontane basins and grabens (Grothaus et al., 1979). The
Dokhan and the Hammamt are preserved in E-W or NE-SW-trending
basins (Fig. 1) and are the principal supracrustal expressions
of the 575 - 600 Ma rifting episode.

The Pink Granites are the most important intrusive bodies of
this episode. They are largely discordant epizonal plutons,

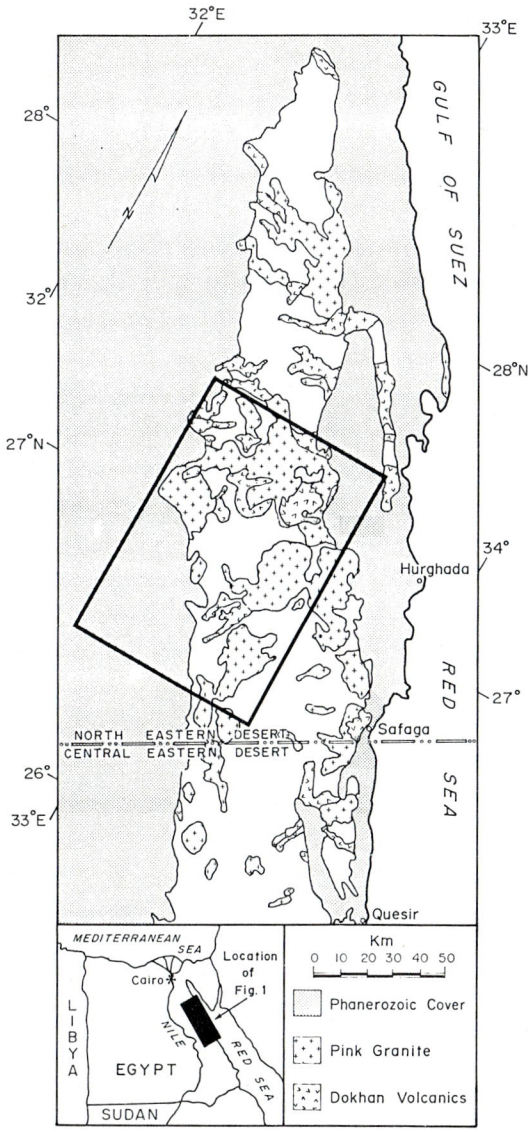

FIGURE 1.

Precambrian basement in the North Eastern Desert of Egypt showing the distribution of the Dokhan Volcanics and Pink Granites as well as the location of Figure 2 (rectangle). Figure modified after the map of Ghanem et al., 1973.

peraluminous to peralkaline in composition, typically with
70 - 78 % SiO_2 and 4 - 5 % K_2O (e.g. El-Gaby, 1975). The Pink
Granites are mostly quartz monzonites and granites according to
the modal classification scheme of Streckeisen (1976). Pink
Granites throughout Sinai and the Eastern Desert range in age
from 568 to 597 Ma (Fullagar, 1980; Bielski et al., 1979;
Halpern and Tristan, 1981; Stern and Hedge, 1985). They are
found wherever basement is exposed but become increasingly
common to the north (El Ramly, 1972). Granites of similar age
and composition are also common in the basement of Jordan and
northern and central Arabia (Jarrar et al., 1983; Stuckless et
al., 1985; Stoeser, 1986). Granites of this age are not common
in the Sudan (Cavanagh, 1979; Stern and Kröner, unpublished
data).

Abundant subparallel <u>dike swarms</u>, which are the main focus of
this study, are found throughout the North Eastern Desert
(Fig. 2). These range in age from 543 - 589 Ma (Stern and
Hedge, 1985), although most predate the cooling and
consolidation of the granites. The dikes predominantly strike
NE-SW to E-W; N-S trending sets are much less common. The dikes
are compositionally bimodal, with subequal volumes and numbers
of melanocratic, aphyric to porphyritic mafic dikes and
leucocratic, quartz- and feldspar-rich felsic dikes. The dikes
are important for two principal reasons. First, they
unequivocally demonstrate that extensional tectonics dominated
North Eastern Desert crustal evolution when the Pink Granites
were emplaced. Second, the dikes were hypabyssal feeders to the
surficial and near-surface expressions of late Precambrian
bimodal igneous activity. Geochemical studies of felsic dikes
may allow us to determine the compositonal heterogeneity that
existed in the felsic melts prior to the late-stage magmatic
modifications associated with the development of large magma
bodies. Intrusive relationships between the mafic and felsic
dikes may allow us to determine the relative timing of the
different components of the bimodal suite, and compositional
comparisons may permit us to evaluate possible petrogenetic
relationships and tectonic settings.

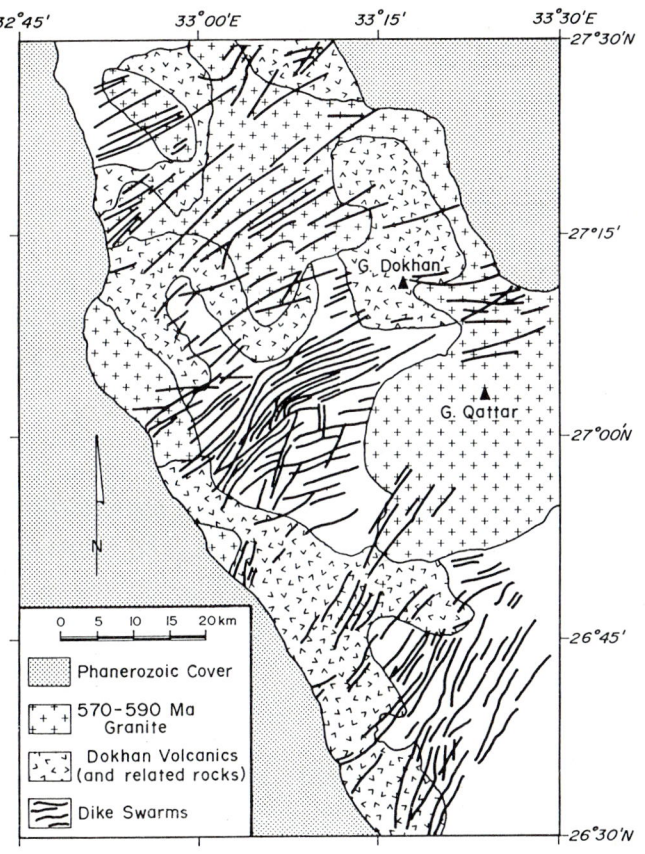

FIGURE 2.

Distribution of dike swarm in a portion of the N.E.
Desert. This figure was generated by visual interpretation
of LANDSAT MSS bands 4, 5, 7 color composite image in
conjunction with field studies. Resolution of this image
is ~ 80 m; dikes are typically 3 - 10 m wide. What appears
as a single dike on the image may actually consist of
several closely-spaced subparallel dikes. For example, in
the area between G. Dokhan and G. Qattar, about 20 "dikes"
are identified on the LANDSAT image. Field and aerial
photographic studies of this area permit the
identification of several hundred dikes. Note also that
discrimination of dikes intruded into units of similar
lithology such as mafic dikes in the Dokhan Volcanics, is
extremely difficult. The samples studied here were
collected from the area within 15 km of G. Dokhan.

The region around Gebels Qattar and Dokhan was chosen for this study (Fig. 2). This area was selected because the relationships between the 575 - 600 Ma crustal units are spectacularly exposed, because this area is very accessible, and because field relations, geochronology, and geochemistry have already been studied (Ghobrial and Lotfi, 1967,; Stern and Hedge, 1985; Massey, 1984; Stern and Gottfried, 1986). Samples from 35 different dikes were collected from exposures east, west, and south of G. Dokhan. Some of these samples were analyzed in the geochronological study of Stern and Hedge (1985), where they gave a Rb-Sr whole rock age of 589 ± 8 Ma and initial $^{87}Sr/^{86}Sr$ of .7030 ± 2 (2 sigma). Many of the dikes are composite, with mafic to intermediate margins and felsic cores. One of these, clearly indicating that mafic and felsic components were simultaneously liquid, has been studied in detail (Voegeli, 1985).

PETROGRAPHY

The dikes are readily separable in the field into mafic and felsic members. The mafic dikes are dark grey or green to black and can be aphyric or porphyritic. Prophyritic varieties always contain phenocrysts of plagioclase and sometimes clinopyroxene or amphibole. The mafic dikes are variably altered to a lower greenschist mineral assemblage (chlorite-calcite-quartz-epidote-sericitized feldspar). Nevertheless, relict whole-grains and cores of plagioclase, clinopyroxene, and amphibole are common.

The felsic dikes are black or pink to orange and are aphyric or porphyritic. Potassium feldspar (orthoclase) is always a phenocryst in porhyritic varieties; it is also common as mantles on plagioclase. Quartz is the second most common phenocryst. Plagioclase and mafic minerals, biotite or amphibole, are less common. The felsic rocks are generally very fresh with a minimum of low-grade metamorphic minerals.

How well the original chemical compositions of the igneous
rocks has been preserved in spite of the low-grade metamorphism
cannot be quantitatively assessed. We note, however, that
whole-rock Rb-Sr isochrons on these samples have not been
obviously disturbed (Stern and Hedge, 1985; Voegeli, 1985),
suggesting that mobile elements have not migrated on a scale
larger than a hand specimen. There are very few veins or
segregations of secondary minerals, indicating that relatively
little fluid was added to or subtracted from the dikes during
metamorphism. For this reason we believe that the present
chemical compositions of the dike rocks closely approximate the
original igneous compositions.

GEOCHEMISTRY AND PETROGENESIS

Specimens of 16 mafic and 19 felsic dikes were analyzed for
major and selected trace elements at USGS facilities in Reston.
Major elements were determined by x-ray fluorescence with the
exceptions of MgO and Na_2O which were determined by rapid-rock
wet chemical techniques. Major element analyses listed in
Tables 1 and 2 have been recalculated to a volatile-free basis
with losses upon heating to $925^{O}C$ (L.O.I.) listed separately.
Trace elements Cr, Ni, Cu, Zn, Rb, Sr, Y, Zr, Nb, and Ba were
also determined by XRF, with typical precisions of ± 5 %. Some
samples were previously analyzed for Rb and Sr by isotope
dilution. Nine samples, chosen to encompass the compositional
range, were analyzed for REE by isotope dilution at UTD, using
techniques outlined by Stern and Gottfried (1986). To ensure
dissolution of minor phases, felsic samples were dissolved in
sealed pressure vessels. These results are listed in Table 3.

The chemical analyses confirm the field and petrographic
observation that the dikes represent a bimodal suite (Fig. 3).
The mafic dikes range from 49 - 66 % SiO_2 with a mean of
56.6 % SiO_2. The felsic dikes range from 70 to over 78 % SiO_2
with a mean of 75.1 % SiO_2. The "Daly Gap" at 66 % to 70 % SiO_2

Table 1

Major and Trace Element Data, Mafic Dikes

	195A	195B	195D	195E	195H	196A	196B	198C	199	201A	204B	204G	204H	204J	204M	205D
SiO_2	63.67	54.31	61.12	52.79	58.20	62.92	59.46	51.92	50.53	61.95	55.74	58.56	59.37	49.05	65.90	56.23
TiO_2	1.14	1.54	1.26	1.54	1.50	1.51	1.11	2.03	2.01	1.40	1.71	1.09	1.31	2.40	1.05	1.28
Al_2O_3	15.52	14.80	15.12	13.84	15.99	15.21	16.32	15.86	15.12	15.03	16.06	16.21	16.51	15.90	15.63	17.90
$Fe_2O_3^*$	6.54	10.92	7.22	9.83	8.61	7.24	8.01	11.92	12.44	8.21	9.33	7.85	7.70	13.13	5.17	8.53
MgO	2.15	6.94	3.65	8.90	3.57	1.99	3.81	5.47	6.81	1.64	4.31	3.89	2.96	5.36	2.67	3.28
MnO	0.14	0.16	0.10	0.15	0.16	0.11	0.11	0.16	0.22	0.12	0.10	0.12	0.10	0.21	0.07	0.10
CaO	3.42	7.32	5.28	7.50	5.21	3.57	5.04	8.10	8.75	4.01	5.93	6.38	5.00	9.75	3.98	6.53
Na_2O	3.96	2.69	3.29	3.16	3.86	4.49	3.64	3.34	2.65	4.06	4.30	3.68	4.55	2.89	3.20	4.04
K_2O	3.11	0.97	2.59	1.77	2.47	2.54	2.24	0.76	1.15	3.14	2.06	1.95	2.16	0.87	2.05	1.79
P_2O_5	0.35	0.35	0.37	0.52	0.43	0.42	0.26	0.44	0.32	0.44	0.46	0.27	0.34	0.44	0.28	0.32
Total	100.00	100.00	100.00	100.00	100.00	100.00	100.00	100.00	100.00	100.00	100.00	100.00	100.00	100.00	100.00	100.00
L.O.I.	1.30	2.45	6.02	4.31	1.72	1.99	2.39	3.02	1.25	2.03	3.75	2.89	2.46	5.02	1.60	1.21
Trace Elements (ppm)																
Cr	<20	286	131	556	48	<20	115	98	157	<20	123	109	<20	23	93	<20
Ni	<5	143	46	176	33	6	43	33	82	<5	55	41	11	67	37	13
Cu	9	31	20	46	9	18	26	20	99	8	30	23	22	31	14	4
Zn	103	112	106	94	105	106	103	118	133	111	117	106	105	112	110	71
Rb	65+	23+	61+	27+	83+	44	49	14+	30	89	65	62	59	18+	59	67
Sr	489+	552+	632+	626+	573+	473	629	717+	539	488	953	569	705	614+	1164	1036
Y	34	26	22	26	28	35	27	30	27	38	18	23	20	28	12	21
Zr	300	251	231	265	279	351	236	273	205	413	235	208	255	257	250	235
Nb	10	11	<10	11	11	13	11	11	12	21	<10	<10	13	13	<10	<10
Ba	687	759	1130	649	755	742	505	369	294	697	1048	435	626	288	690	474
Element Ratios																
$Fe_2O_3^*/MgO$	3.0	1.6	2.0	1.1	2.4	3.6	2.1	2.2	1.8	5.0	2.2	2.0	2.6	2.5	1.9	2.6
K_2O/Na_2O	0.79	0.36	0.79	0.56	0.64	0.57	0.62	0.23	0.43	0.77	0.48	0.53	0.47	0.30	0.64	0.44
K/Rb	397	350	352	544	246	479	379	441	318	293	263	261	304	401	288	222
K/Ba	38	11	19	23	27	28	37	17	32	37	16	37	29	25	25	31
Rb/Sr	0.13	0.04	0.10	0.04	0.14	0.09	0.08	0.02	0.06	0.18	0.07	0.11	0.08	0.03	0.05	0.06

$Fe_2O_3^*$ = total iron as Fe_2O_3

L.O.I. = % wt. loss on ignition

+ = analysis by isotope dilution

Table 2

Major and Trace Element Data, Felsic Dikes

	195F	195I	196C	196E	198B	200C	200F	201C	201D	202G	203A	203B	203E	203F	204D	204E	204F	204I	204L
SiO_2	71.00	74.89	77.51	77.30	77.41	77.42	77.80	75.81	72.45	69.94	72.63	76.91	70.34	78.35	73.96	77.80	77.33	76.58	71.98
TiO_2	0.53	0.17	0.12	0.14	0.13	0.18	0.08	0.15	0.41	0.64	0.55	0.13	0.49	0.08	0.31	0.11	0.12	0.23	0.51
Al_2O_3	14.61	14.14	12.26	12.84	12.82	12.45	12.55	12.47	13.75	14.13	14.03	11.99	14.19	11.68	13.85	11.99	12.6	12.48	14.71
$Fe_2O_3^*$	2.08	0.95	1.02	0.46	0.40	0.73	0.86	1.75	2.48	3.26	2.21	1.40	2.85	0.93	1.83	0.91	0.88	1.01	1.91
MgO	0.45	0.14	0.10	0.08	0.06	0.08	0.08	0.03	0.04	0.06	0.38	0.08	0.36	0.01	0.56	0.09	0.06	0.08	0.38
MnO	0.09	0.04	0.02	0.02	0.02	<0.01	0.01	0.03	0.04	0.06	0.09	0.03	0.08	0.03	0.03	0.02	0.03	0.01	0.11
CaO	1.20	0.67	0.38	0.72	0.55	0.46	0.63	0.79	2.03	2.22	1.30	0.40	1.27	0.48	1.15	0.49	0.50	1.27	0.94
Na_2O	5.85	4.90	3.90	3.87	3.98	3.96	4.83	4.30	4.10	4.63	4.97	4.08	4.40	3.89	3.63	3.83	3.88	3.13	5.59
K_2O	4.12	4.01	4.60	4.49	4.60	4.64	3.16	4.53	3.94	3.92	3.70	4.88	5.88	4.46	4.57	4.71	4.54	5.12	3.77
P_2O_5	0.07	0.09	0.09	0.08	0.03	0.08	0.07	0.10	0.14	0.18	0.14	0.10	0.14	0.09	0.11	0.05	0.04	0.09	0.10
Total	100.00	100.00	100.00	100.00	100.00	100.00	100.00	100.00	100.00	100.00	100.00	100.00	100.00	100.00	100.00	100.00	100.00	100.00	100.00
L.O.I.	0.59	0.48	0.61	0.46	0.45	0.77	0.59	0.73	0.79	1.17	0.74	0.69	0.72	0.57	1.81	0.90	0.72	1.82	0.51
Type**	M	M	M	M	M	M	M	M	M	M	M	A	M	M	P	M	M	M	M
Trace Elements (ppm)																			
Cr	<20	<20	<20	<20	<20	<20	<20	<20	<20	<20	<20	<20	<20	<20	<20	<20	<20	<20	<20
Ni	<5	<5	5	8	<5	12	<5	<5	7	13	9	7	<5	7	8	<5	<5	<5	6
Cu	7	<2	9	14	3	15	26	6	4	15	14	10	<2	10	10	<2	17	2	9
Zn	56	48	65	60	43	60	83	95	73	85	118	84	88	84	64	44	77	42	99
Rb	55+	98+	223+	99	82.8+	211	125+	194	150	110	80	167	124	380	206	256	212+	219	228
Sr	183+	70.3+	37.6+	120	92.5+	43	122+	53	296	394	458	20	89	5	118	55	22+	70	44
Y	39	25	68	11	13	26	118	58	28	25	38	45	48	103	34	74	76	46	44
Zr	413	221	241	104	109	191	252	310	277	389	428	379	783	237	238	248	242	202	459
Nb	16	15	57	11	12	39	98	33	15	10	16	23	24	87	21	60	62	36	17
Ba	1076	849	63	743	851	50	43	443	710	809	1268	66	481	18	572	56	45	116	973
Element Ratios																			
$Fe_2O_3^*/MgO$	4.6	6.8	10.2	5.8	6.7	9.1	10.8	58.3	62.0	54.3	5.8	17.5	7.9	93.0	3.3	10.1	14.7	12.6	5.0
K_2O/Na_2O	0.70	0.82	1.18	1.16	1.16	1.17	0.65	1.05	0.96	0.85	0.74	1.20	1.34	1.15	1.26	1.23	1.17	1.64	0.67
K/Rb	622	340	171	376	461	183	210	194	218	296	384	243	394	97	184	153	177	194	406
K/Ba	32	39	606	50	45	770	610	85	46	40	24	614	101	2060	66	698	837	366	32
Rb/Sr	0.30	1.39	5.9	0.83	0.90	4.9	1.0	3.7	0.51	0.28	0.17	8.4	1.4	76	1.7	4.7	9.6	3.1	0.34

$Fe_2O_3^*$ = total iron as Fe_2O_3

L.O.I. = % wt. loss on ignition

Type**: A = Peralkaline, M = Metaluminous, P = Peraluminous

(Note: Peraluminous-Metaluminous cutoff at molar $Al_2O_3/Na_2O+CaO+K_2O$ = 1.05 (Stuckless et al., 1986))

+ = Analysis by isotope dilution

Table 3

Rare Earth Element Data (ppm)

	------------Mafic Dikes---------					-------Felsic Dikes------			
	195B	199	204J	204M	205D	195F	195I	203E	204E
La	21.9	16.3	21.5	20.6	21.4	41.3	35.1	23.5	24.6
Ce	51.4	41.0	54.3	47.3	49.3	95.5	71.3	66.5	105.8
Nd	29.3	25.4	32.3	24.6	26.3	48.2	26.8	34.0	35.5
Sm	6.30	5.86	7.23	4.76	5.22	9.55	4.48	10.6	9.50
Eu	1.81	1.90	2.15	1.40	1.64	2.35	0.74	0.04	0.11
Gd	6.35	5.76	6.72	3.96	4.52	9.00	3.73	12.1	8.78
Dy	4.54	4.75	5.31	2.59	3.52	6.50	3.37	13.6	9.04
Er	2.43	2.29	2.51	1.12	1.79	3.56	1.50	8.12	5.24
Yb	2.06	1.91	1.97	0.79	1.51	2.81	1.40	7.70	4.73
$(La/Yb)_n$	7.1	5.7	7.2	17.5	9.4	9.7	16.7	1.9	3.5
Ce/Ce^*	0.99	1.01	1.04	1.00	0.99	1.01	0.96	1.18	1.78
Eu/Eu^*	0.85	0.99	0.93	0.97	0.99	0.77	0.51	0.01	0.04
$(Yb/Er)_n$	0.85	0.85	0.81	0.72	0.86	0.81	0.97	1.02	0.92

Ratios with subscript "n" refer to chondrite-normalized abundances.

Ratios with asterisk refer to ratios of observed to expected concentrations.

corresponds closely to the break at 69 % SiO_2 observed between the Dokhan Volcanics and Pink Granites of Qattar-Dokhan area (Stern and Gottfried, 1986). We will hereafter discuss the geochemistry and petrogenesis of the mafic and felsic suites separately.

a) The Mafic Dikes

The mafic dikes are similar to the lavas of the Qattar-Dokhan Area in being dominated by andesites with medium to high contents of K_2O (Fig. 4). The andesitic dike samples are virtually indistinguishable from the andesitic lavas of this region (Stern and Gottfried, 1986). This oberservation coupled with the field and geochronologic data further compels the conclusion that these dikes were feeders for the Dokhan Volcanics of the region. The dike suite differs from the Dokhan Volcanics sampled to date in containing more basalts (¼ of the mafic suite contain less than 53 % SiO_2). These basalts are especially significant because one contains up to 8.9 % MgO, 556 ppm Cr, and 176 ppm Ni, indicating that this sample must represent a primary mantle melt. The basalts are enriched in Fe and Ti, with $Fe_2O_3^*$ = 9.8 to 13.1 %, $Fe_2O_3^*/MgO$ = 1.1 to 2.5, and TiO_2 = 1.5 to 2.4 %. They are also enriched in K_2O (.8 - 1.8 %), Ba (290 - 650 ppm), Zr (210 - 270 ppm) and LREE $(La/Yb)_n$ = 5.7 - 7.2 (Fig. 5b). These are alkalic basalts that were derived from partial melting of enriched upper mantle.

The relationships between the basalts and the andesites is problematic. A fractionation relationship is not indicated because incompatible elements, which would be expected to be enriched in the residual liquid, are commonly as low or lower in the andesites as compared to the basalts. This is especially true for the high field strength cations (HFSC) such as Ti, P, Y, Nb, Zr, and the REE. There is an overall decrease in HSFC such as Y with increasing SiO_2 (Fig. 6) that is largely inconsistent with a low-P fractionation relationship existing between the basalts and andesites. These lithologies nevertheless are petrogenetically related as shown by their

FIGURE 4.

K₂O vs. SiO₂ classification scheme for andesitic rocks (Gill, 1981), with analyses of mafic suite dikes plotted. Also shown are contours enclosing 100, 90, 75, and 50 % of the analyses of Dokhan Volcanics (Basta et al., 1980; Monrad and Ressetar, 1983; Stern and Gottfried, 1986). Note that one dike sample with less than 50 % SiO₂ is not plotted (E-204J).

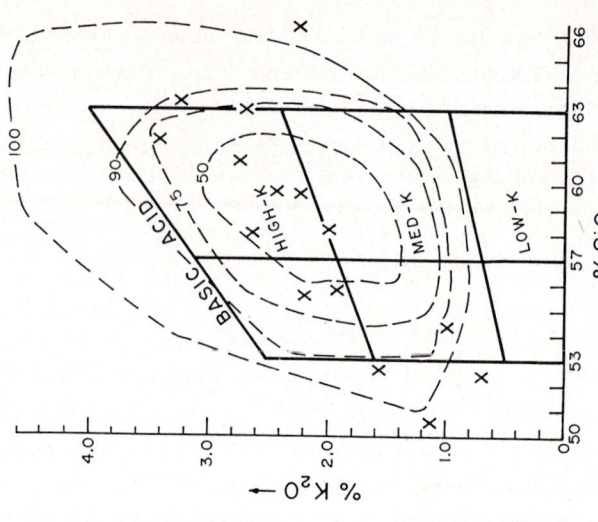

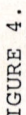
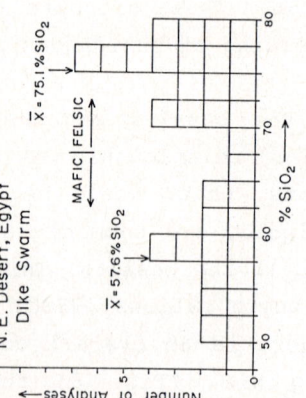

FIGURE 3.

Silica histogram for the North Eastern Desert dikes. The upper part of the Figure shows a generalized frequency diagram for similar rift-related hypabyssal intrusions from the Oslo Rift (Watt, 1966, reported in Windley, 1977). Note the strongly bimodal composition of both rift-related igneous suites.

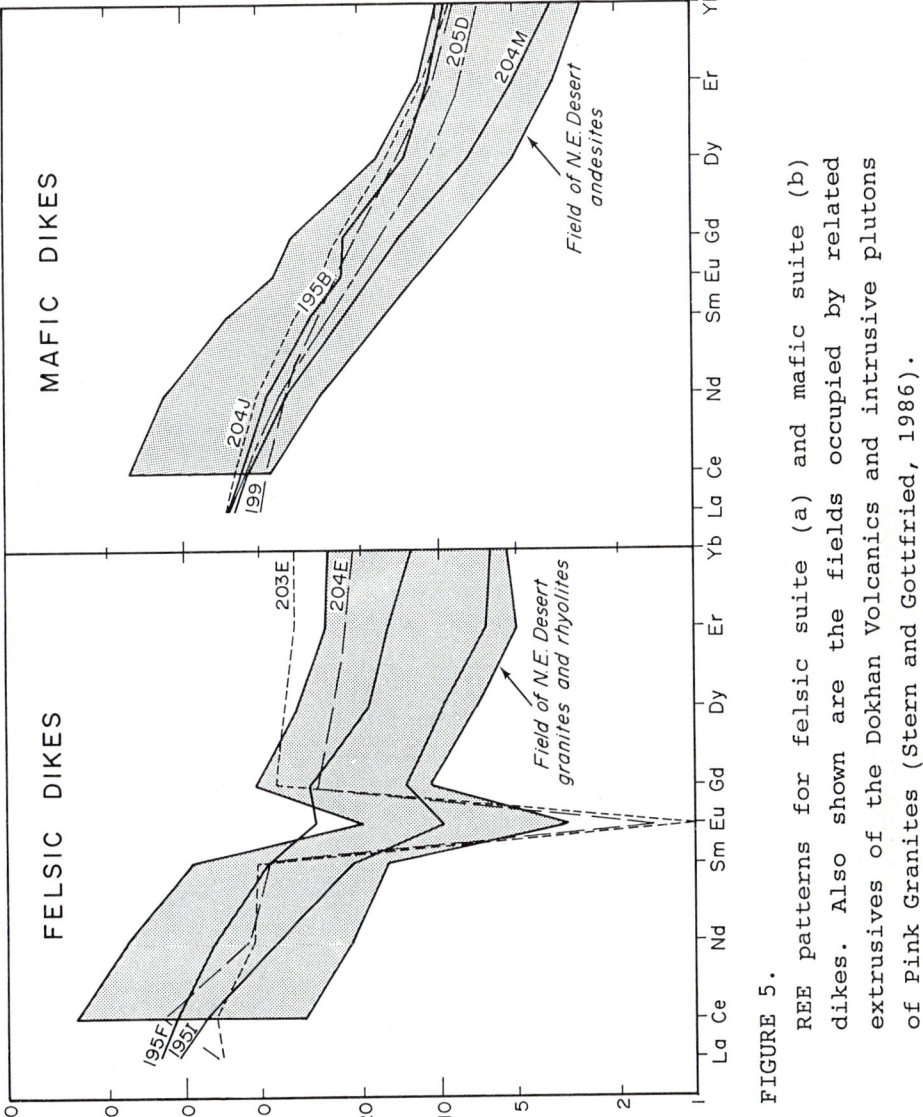

FIGURE 5.   REE patterns for felsic suite (a) and mafic suite (b) dikes. Also shown are the fields occupied by related extrusives of the Dokhan Volcanics and intrusive plutons of Pink Granites (Stern and Gottfried, 1986).

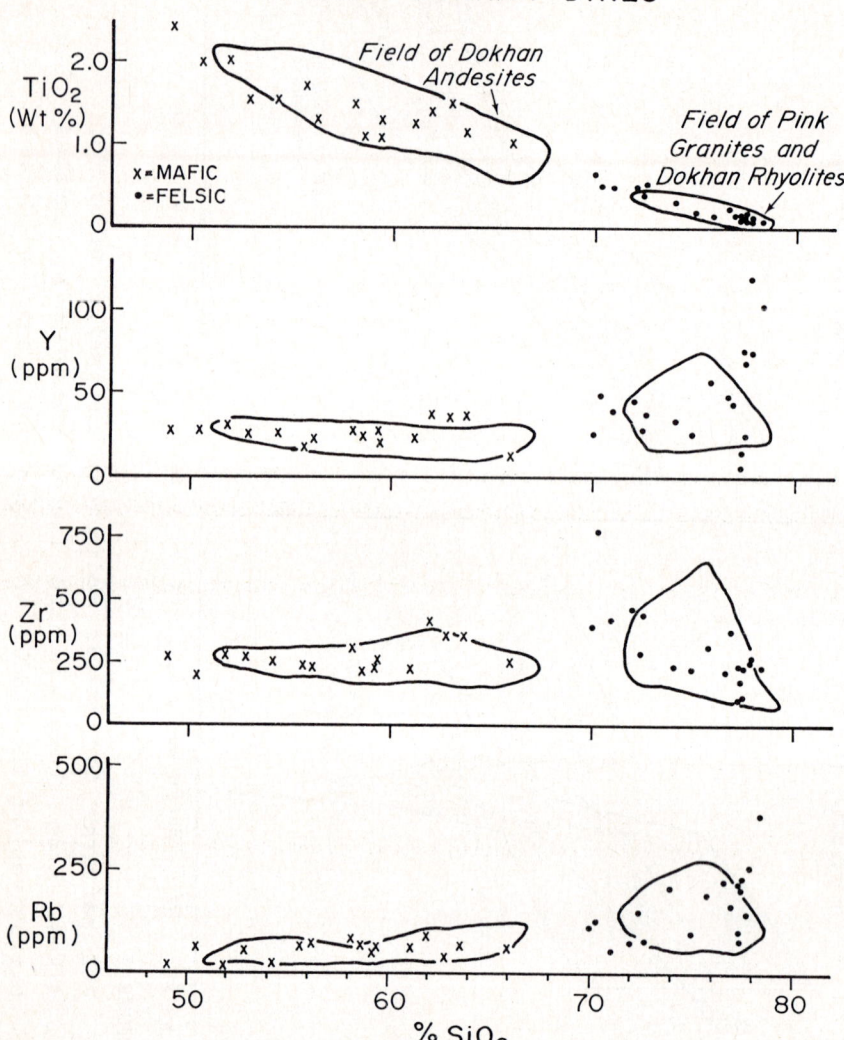

FIGURE 6.

Variation diagrams for the mafic and felsic suite dikes.
Also shown are the fields occupied by the Dokhan Volcanics
and Pink Granites (Stern and Gottfried, 1986).

similar isotopic compositions, REE patterns, K/Rb (basalt = 425 ± 94; andesite = 320 ± 74), K/Ba (basalts = 24 ± 6; andesites = 28 ± 9), and overall enrichment in LIL and HFSC incompatible elements.

On the basis of REE modelling, Stern and Gottfried (1986) argued that the mafic members of the North Eastern Desert bimodal suite were formed either by 10 % batch melting of LREE-enriched garnet lherzolite or by 10 - 20 % batch melting of eclogite. The similarity of the mafic dike bulk geochemistry and REE patterns to those of the mafic samples studied by Stern and Gottfried (1986) indicates that these conclusions hold for the mafic dikes as well. The fact that alkali basalts with high Cr, Ni, and MgO are more common in the dikes strongly indicates that these were formed by melting of LIL- and LREE-enriched garnet lherzolite in the upper mantle (e.g., Kay and Gast, 1973). The andesitic dike rocks could either have been formed by lower degrees of hydrous melting of similar garnet lherzolite (e.g., Mysen and Boettcher, 1975) or by melting of eclogite (e.g., Apted, 1981).

b) The Felsic Dikes

The felsic dike rocks are chemically indistinguishable from the Pink Granites of the Eastern Desert. They contain 70 - 78 % SiO_2 and 3.2. - 5.9 % K_2O, have K_2O/Na_2O = .65 - 1.34 ($\overline{x}$ = 1.08 ± .26), and are predominantly metaluminous. They contain the high $Fe_2O_3^*/MgO$ characteristic of post-tectonic suites (Anderson and Cullers, 1978). If found in a plutonic body, these compositions would correspond to "alkali granites" (Rogers and Greenberg, 1981) or "A-type" granites (Collings et al., 1982), terms that refer to the distinctive compositional characteristics observed for intrusive rocks of post-tectonic or extensional environments. Greenberg (1981) subdivided the Pink Granite plutons of the Central Eastern Desert into three groups. Group I granites are the most evolved with ~ 75 % SiO_2, 4.4 % K_2O, .07 % TiO_2, relatively low Ba, Sr, and Zr, and relatively high Rb, Y, and Nb. Group III granites are the least evolved, with 70 % SiO_2, 3.6 % K_2O, .5 % TiO_2, relatively high Ba, Sr, and Zr and relatively low Rb, Y, and Nb. Group II granites are compositionally intermediate but are more similar to Group I than Group III. The 19 felsic dike samples listed in Table 2 encompass the full range of chemical variations observed on a regional scale by Greenberg (1981). The low silica felsic dikes (70 - 73 % SiO_2) are similar to Group III granites and contain relatively high concentrations of TiO_2 (.3 - .6 %), $Fe_2O_3^*$ (1.8 - 3.3 %), P_2O_5 (.07 - .18 %), alkaline earths (MgO = .4 - 1.0 %; CaO = .9 - 2.2 %; Sr = 90 - 400 ppm; Ba = 500 - 1300 ppm), and Zr (240 - 780 ppm). These contain relatively little Rb (55 - 210 ppm), Y (25 - 50 ppm) and Nb (10 - 25 ppm). The high silica felsic dikes (77 - 78 % SiO_2) are similar to Group I granites. Compared to the low silica felsic dikes, these contain little TiO_2 (.08 - .18 %), $Fe_2O_3^*$ (.7 - 1.4 %), P_2O_5 (.04 - .1 %), alkaline earths (MgO = .01 - .1 %; CaO = .4 - .6 %; Sr = 5 - 120 ppm; Ba = 20 - 70 ppm), and Zr (190 - 380 ppm) and more Rb (125 - 380 ppm), Y (30 - 120 ppm) and Nb (20 - 100 ppm).

The relationship between the low- and high-silica members of the felsic dikes suite is problematic. The decreases in Ti, Fe, P, Mg, Ca, Sr, Ba, and Zr with increasing silica are consistent with fractional crystallization of amphibole or biotite, apatite, plagioclase, K-feldspar, and zircon from a low silica parent. However, the very large increases in Y (2 - 4x; Fig. 6) cannot be explained without calling on extreme fractionation. This should have been accompanied by large scale removal of biotite and K-feldspar, a prediction that is inconsistent with the observed 2-fold increase in Rb (Fig. 6). A similar conclusion can be drawn from the felsic dike REE patterns (Fig. 5a). It is very difficult to argue on this basis that crystal-liquid fractionation was the most important factor in the evolution of low-silica to high-silica felsic melts. One is instead forced to call on more exotic (and non-quantifiable) models of magmatic element fractionation such as thermogravitational diffusion (Hildreth, 1981) or halide- or carbonate-complexing and transport (Harris and Marriner, 1980; Drysdall et al., 1984).

c) Petrogenetic Relationship between Mafic and Felsic Dikes

Two different hypotheses have been advocated for the generation of the late Precambrian "A-type" granites of Afro-Arabia: anatexis of the lower crust and fractional crystallization of mafic magma. Since the felsic dike rocks are chemically indistinguishable from the "A-type" granite plutons, the same general types of models are applicable. One group of scientists argue that the late Precambrian accretion of arc systems to form the Afro-Arabian Shield resulted in a thickening of the juvenile crust which led to deep crustal anatexis (e.g. Greenberg, 1981). This was expanded on by Drysdall et al., (1984) who, following the model of Collings et al. (1982), argued that the "A-type" granites of Afro-Arabia were generated by partial melting, under high-temperature, vapor-absent conditions, of a crust from which anatectic granodiorite melts had previously been extracted. This model faces the following objections:

1. How to produce such a large volume of K-rich melts from depleted lower crust;
2. Flat HREE patterns indicate that the melts did not equilibrate with garnet-bearing rocks expected for the lower crust (Stern and Gottfried, 1986);
3. The felsic magmas are undersaturated in P_2O_5 and all potential anatectic source rocks would produce saturated melts (Stern and Gottfried, 1986); and
4. Extensive melting of depleted lower crust cannot produce the observed enrichments of HFSC without also increasing Ca and Al (Harris et al. 1986)

A modification of this model that overcomes objections 1 and 2 calls for anatexis of previously unmelted amphibolite-facies crust to generate the felsic magmas (Stern and Gottfried, 1986). In the case of the felsic dikes, this model would call on the intrusion of basic dikes into the juvenile crust to supply the heat necessary for melting.

At the other extreme are models that call on the mafic and felsic members of the bimodal suite to have evolved, the latter from the former, by magmatic fractionation. One possibility is that the two members are immiscible liquids which separated during ascent through the crust. While this model is attractive in terms of the field relations (i.e., composite dikes with mafic margins and a felsic core; Voegeli, 1985) and in explaining the compositional gap between the mafic and felsic suites, this model is not favored because the HFSC are not strongly partitioned into the mafic relative to the felsic rocks, as would be predicted from experimental studies (Watson, 1976; Ryerson and Hess, 1978). A second possibility is that the felsic magmas evolved from the mafic magmas by crystal-liquid fractionation. Stern and Gottfried (1986) developed a major-element fractionation model for the generation of North Eastern Desert felsic melts (75 % SiO_2; 4.8 % K_2O) by 67 % fractional crystallization of Dokhan Andesite (59 % SiO_2, 2.0 % K_2O). Testing of the model using the REE data did not support the model, because a match between the observed and predicted REE patterns could only be accomplished using partition coefficients appropriate for mafic magmas. A similar model can be proposed for the relationship between the mafic and felsic dikes and similar objections raised as well.

SIGNIFICANCE OF NORTH EASTERN DESERT DIKES FOR UNDERSTANDING LATE PRECAMBRIAN BIOMODAL IGNEOUS ACTIVITY

The previous discussion demonstrates that we still have an incomplete understanding of the petrogenesis of the felsic dike rocks and, by inference, the petrogenesis of all the "A-type" granites in northern Afro-Arabia. The present data set coupled with field observations does allow us to draw a number of significant conclusions regarding the relative timing of mafic and felsic magmatism, relation of dikes and plutons and lava, significance for magmatic evolution, and tectonic geochemistry. These items will be considered in turn in the following discussion.

a) Timing of Bimodal Igneous Activity

The Qattar-Dokhan dikes were intruded at 589 ± 8 Ma. This is indistinguishable from the ages of the Dokhan Volcanics (592 ± 13 Ma) but is slightly older than the felsic plutons of the immediate area (579 - 583 Ma; Stern and Hedge, 1985). Were the felsic plutons emplaced 10 - 15 Ma after the mafic volcanism ceased or do these differences reflect differences in cooling times? Age relationships among the felsic and mafic dikes indicate these were emplaced during a similar time interval. Where cross-cutting relationships can be observed, mafic dikes cutting felsic dikes are as common as felsic intruding mafic. In other instances, composite dikes consisting of felsic cores and mafic margins indicate the two magma types simultaneously exploited the same conduit (Voegeli, 1985). On this basis, we conclude that the bimodal igneous activity in at least this part of northern Afro-Arabia occurred concurrently, and that the observed difference in radiometric ages largely manifest the slower cooling of the larger plutons relative to the dikes.

b) Relationship of Dikes to Plutons and Lavas

The relationship of the dikes to the larger plutons and volcanic fields is important because understanding this may allow us to better understand the relationship between extensional tectonics and magmatism. The relationship between the mafic dikes and the Dokhan Volcanics is clear - the former represent magmatic conduits for the latter. The relationship between the felsic dikes and the Pink Granites, however, is more equivocal. Because both the felsic dikes and the plutons occupy similar crustal levels, it is not obvious whether the felsic dikes fed into or issued from the granite plutons. Field observations, however, favor the first possibility. The

following observations are pertinent: First, the felsic dikes
are more common below the granites than above. On the south
side of G. Dokhan is a thick sill-like body of granite. This
was emplaced between the overlying Dokhan Volcanics and the
underlying Hammamat Group. Felsic dikes are common in the
Hammamat and much less common in the Dokhan Volcanics. This
relationship indicates that granite sills were emplaced higher
in the crust and thus likely represent magmatic reservoirs
where melts issuing from the felsic dikes were impounded.
Secondly, rhyolite flows and ignimbrites in this area are very
rare. If the felsic dikes had largely continued to the surface
we would expect to find substantial volumes of felsic
volcanics. The fact that such lithologies are not common
strongly suggests that the felsic dikes did not generally vent
to the surface. A likely place for these melts to have been
trapped is in the sub-Dokhan felsic magma chambers previously
discussed. We conclude that the felsic dikes represent shallow
feeders for the epizonal sill-like granite bodies of the
Qattar-Dokhan area.

In the model summarized in Figure 7, the deposition of the
Hammamat Group sediments has just ceased (Fig. 7A). Following
this, Dokhan Volcanics were erupted conformably on the Hammamt.
These lava flows are fed by the numerous NE-SW trending dikes
that exploited tension-related crustal fissures. Felsic dikes
were also emplaced at this time. These magmas, however, did not
erupt but ponded as tack-shaped bodies within the crust
(Fig. 7B). Bimodal igneous activity continued with the
progressive thickening of the supracrustal basalt-andesite
volcanic field while the shallow felsic magma reservoirs
thickened and spread laterally. Continued extension was
accompanied in the upper crust by the emplacement of new dikes
and the widening of older ones (Fig. 7C). This process
continued until a volcanic field ~ 1 km thick is constructed
over slowly cooling magma cushions, 10's of km in diameter but
a km or less thick (Fig. 7D). Igneous activity ceased in the
Qattar-Dokhan area by ~ 580 Ma; the present erosion level has
been established since that time (Fig. 7E).

d) Tectonic Geochemistry

Previous investigators have noted the enriched composition of the mafic to intermediate rocks associated with latest Precambrian magmatic activity throughout Afro-Arabia but a consensus has not been reached regarding the tectonic setting in which these lavas were erupted. The Dokhan Volcanics have previously been interpreted as products of Andean-type magmatism (Basta et al., 1980) as were analogous volcanics in the Kid Group, Sinai (Furnes et al., 1985) and "Series A" volcanics in the Arabian Shield (Roobol et al., 1983). These interpretations were first questioned by Ressetar and Monrad (1983) who noted that the Dokhan lavas had elevated concentrations of incompatible trace elements with respect to igneous rocks typical of active continental margins. Stern and Gottfried (1986) re-interpreted the Dokhan Volcanics and Pink Granites as extrusive and intrusive components of a bimodal suite associated with crustal extension. Independently, as a result of detailed field studies of "Series A" units in Arabia, Agar (1986) discounted the convergent-margin tectonic interpretation of Roobol et al., (1983). He argued instead that much of the "Series A" volcanic activity occurred in pull-apart grabens of the Najd strike-slip orogen.

Much of the disagreement stems from uncertainties in interpreting the geochemistry of especially the mafic igneous rocks. In general, all of the latest Precambrian igneous products in Afro-Arabia (Dokhan/mafic dikes in NE Egypt; Kid Group in Sinai; Jibalah, Shammar, Murdama, Furyh, and Hadiyah Groups in northern Saudi Arabia) share compositions that on a chemical basis alone could be interpreted as either related to intra-plate rifting or a mature convergent margin (e.g., Ressetar and Monrad, 1983). That is, these are largely calc-alkaline, show moderate enrichments of the incompatible elements, and are rarely silica-undersaturated. Because, until recently, proposed tectonic hypotheses for the evolution of Afro-Arabia were models of arc-accretion, it was appropriate to try to fit even late magmatic activity into that conceptual framework (e.g., Bakor et al., 1976; Fleck et al., 1980). More recent interpretations of late crustal evolution in Afro-Arabia have emphasized the role of extensional tectonics and Najd strike-slip faulting (Stern et al., 1984; Stern, 1985; Burke and Sengor, 1986; Agar, 1986). Such recognition requires a re-interpretation of the tectonic signature of related igneous rocks.

Tectono-magmatic interpretations are greatly simplified for the case of the dike swarms observed here. The field exposures give unequivocal evidence of a regional NW-SE extensional tectonic regime. The composition of the mafic suite volcanics is best explained in a within-plate magmatic setting (Fig. 8). These rocks nevertheless present a problem in having high Zr/Nb (17 - 40) and Sr/Nd (19 - 39) considered otherwise diagnostic of convergent zone magmatism (Gill, 1980; DePaolo and Johnson, 1979). Barberi et al. (1982) have pointed out that systematic compositional variations can be observed between continental rifts that produce large and small volumes of melts. Low volcanicity rifts (e.g., Rhinegraben, Baikal, West African) erupt a wide range of mafic lavas but with relatively little felsic material. Instead, these rifts produce strongly alkalic and undersaturated lavas with strong LREE-enrichments ($(La/Yb)_n$ ~ 10 - 50). High volcanicity rifts (e.g., Afar, East African Rift) produce abundant felsic magmas and the mafic melts are less commonly undersaturated; much less LREE-enrichment is observed ($(La/Yb)_n$ ~ 4 - 6). Barberi et al. (1982) suggest that the low- and high-volcanicity rifts are associated with rifts undergoing relatively little and great extension, respectively.

The evolution of the Jemez Volcanic Field in the Rio Grande Rift is illuminating in the regard. Although this volcanic center began at ~ 16 Ma with the eruption of olivine tholeiites and high-silica rhyolites, by 10 Ma magmatism was dominated by andesites that are remarkably similar to the Dokhan andesites (Keres Group andesites; 60 - 63 % SiO_2; .8 - 1.1 % TiO_2;

2.5 - 3.2 % K_2O; $(La/Yb)_n$ = 12 - 15). Gardner et al. (1986) argue that these andesites comprise nearly half of the Jemez Volcano. Rhyolitic domes and extrusives are common throughout the volcanic stratigraphy, becoming the dominant unit with the eruption of the 1.45 - 1.12 Ma Bandelier Tuff (Self et al. 1986). These rhyolites are compositonally very similar to the Egyptian felsics (68 - 78 % SiO_2; .04 - .44 % TiO_2; 3.2 - 5.4 % K_2O; $(La/Yb)_n$ = 3 - 8; Gardner et al., 1986; Self et al. 1986). Nearly all the Jemez volcanics with more than 53 % SiO_2 are subalkaline (Gardner et al., 1986). If the tectonic setting of these rocks were not known a priori, their geochemical signatures would very likely have been interpreted as indicating their formation in an Andean-type convergent margin. These results indicate that it may not be appropriate to expect all continental rifts to share common compositional characteristics but that the rate or degree of extension may be as important as the process itself in determining magma compositions. Large degrees of compositional overlap thus may be expected between lavas erupted in highly magmatic, rapidly distending rifts and Andean-type convergent margins.

CONCLUSIONS

The ~ 590 Ma dike swarms of the North Eastern Desert of Egypt are critical for our understanding of late Precambrian crustal evolution in northern Afro-Arabia because they represent unequivocal evidence of NW-SE directed crustal extension accompanied by widespread bimodal igneous activity. The mafic dikes are composed of medium- to high-K andesite and alkali basalts and represent hypabyssal conduits for surficial eruptions of the Dokhan Volcanics. The felsic dikes are rhyolites that represent feeders for epizonal tabular granite plutons. Relationships between mafic and felsic dikes indicate that mafic and felsic magmas co-existed in time as well as in space.

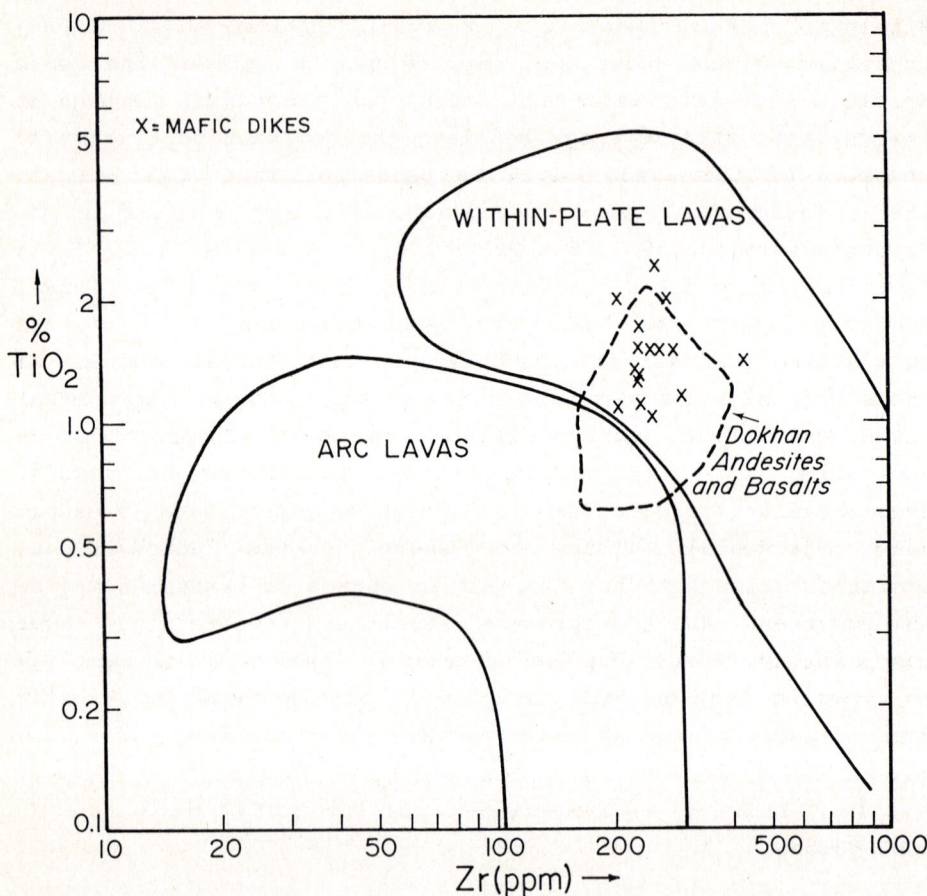

FIGURE 8.

Log-log plot of TiO_2 vs. Zr showing fields occupied by arc vs. within-plate lavas (Pearce, 1980). Samples of the mafic suite are plotted and the field occupied by Dokhan basalts and andesites (Stern and Gottfried, 1986) is also outlined.

Basaltic melts were generated by melting in the upper mantle. Andesitic melts were produced either by lower degrees of melting of hydrous upper mantle or eclogite. Processes responsible for the generation of the felsic melts remain enigmatic; processes of either mid-crustal anatexis or fractional crystallization of mafic melts remain as plausible, if not yet convincing, hypotheses.

The question of how the late Precambrian "A-type" granitic dikes and plutons were generated persists as the outstanding petrologic problem of Afro-Arabia. Resolution of this problem will require a two-pronged attack. First, advocates of either fractional crystallization or crustal anatexis must strive to create and test quantifiable petrogenetic models. The fact that especially the anatectic model has yet to be formally defined and tested should be rectified in the near future. Second, field and laboratory studies should be developed that focus on small areas where the two processes can be documented or discounted. The deep crust of Afro-Arabia should be studied for the purpose of determining whether or not this could have yielded the late Precambrian "A-type" felsic magmas. Such studies should be carried out on exposures of exhumed deep crustal terranes or on suites of xenoliths brought up with Cenozoic basalts. Studies should also be initiated to study plutonic bodies which might record fractionation of mafic to felsic magmas. Such layered composite intrusions as Kadabora (Greenberg, 1981) in Egypt, Timna (Beyth, in press) in Israel and Martaban (Douch et al., 1986) in Arabia are likely candidates for such detailed investigation.

ACKNOWLEDGEMENT

This work was made possible through the collaborative efforts of U.S. and Egyptian scientists under the auspices of NSF grant INT-7801469 to Prof. W. H. Kanes. Logistical assistance and scientific advice from Prof. E.M. El Shazly, Hafez Aziz, and Dr. Ahmed Abdulla Abdel-Meguid was critical to the project and

is gratefully acknowledged, as is the assistance of D. Voegeli
in the laboratory. The laboratory investigations were supported
by grants from NSF (EAR8205802) and the Petroleum Research Fund
of the American Chemical Society to RJS. We expecially
appreciate thorough reviews of the manuscript given by
Drs. T. Reischmann and D. G. Hadley. This is U.T.D. Programs in
Geosciences contribution # 511.

REFERENCES

Agar, R.A., 1986. The Bani Ghayy Group: Sedimentation and
    volcanism in pull-apart grabens of the Najd strike-slip
    orogen, Saudi Arabian Shield. Precambrian Res. 31,
    259-274.
Anderson, J.L. and R.L. Cullers, 1978. Geochemistry and
    evolution of the Wolf River batholith, a late Precambrian
    Rapikivi massif in north Wisconsin, U.S.A. Precambrian
    Res. 7, 2825-324.
Apted, M.J., 1981. Rare earth element systematics of hydrous
    liquids from partial melting of basaltic eclogite: a re-
    evaluation. Earth Planet. Sci. Lett. 52, 172-182.
Bakor, A.R., I.G. Gass and C.R. Neary, 1976. Jabal al Wask,
    northwest Saudi Arabia: an Eocambrian back-arc ophiolite.
    Earth Planet. Sci. Lett. 30, 1-9.
Barberi, F., R. Santacroce und J. Varet, 1982. Chemical aspects
    of rift magmatism. In: Palmason, G. (ed.). Continental and
    Oceanic Rifts. Geodynamics Series 8, Am. Geophys. Union.
    Washington, DC, 223-258.
Basta, E.Z., H. Kotb and M.F. Awadallah, 1980. Petrochemical
    and geochemical characteristics of the Dokhan Formation at
    the type locality, Jabal Dokhan, Eastern Desert, Egypt.
    In: Al-Shanti, A.M.S. (ed.). Evolution and Mineralization
    of the Arabian-Nubian Shield. I.A.G. Bull. 3. Pergamon
    Press, NY, 122-140.
Bentor, Y.K., 1985. The crustal evolution of the Arabian-Nubian
    Massif with special reference to the Sinai Peninsula.
    Precambrian Res. 28, 1-74.
Beyth, M. (in press). The Precambrian rocks of Timna Valley,
    southern Israel. Precambrian Res.
Bielski, M., E. Jager and G. Steinitz, 1979. The geochronology
    of Iqna Granite (Wadi Kid Pluton), Southern Sinai.
    Contrib. Mineral petrol. 70, 159-165.
Burke, K. and C. Sengor, 1986. Tectonic escape in the evolution
    of the continental crust. In: Barazangi, M. and Brown, L.
    (eds.). Reflection Seismology: The Continental Crust.
    Geodynamics Series 14. Am. Geophys. Union. Washington, DC,
    41-53.
Cavanagh, B.J., 1979. Rb/Sr geochronology of some Pre-Nubian
    igneous complexes of Central and North East Sudan.
    Ph.D. thesis, U. Leeds, 239 p.

Collins, W.J., S.D. Beams, A. J. R. White and B.W. Chappell, 1982. Nature and origin of A-type granites with particular reference to southeastern Australia. Contrib. Mineral. Petrol. 81, 189-200.

Depaolo, D.J. and R.W. Johnson, 1979. Magma genesis in the New Britain island-arc: constraints from Nd and Sr isotopes and trace-element patterns. Contrib. Mineral Petrol. 70, 367-379.

Douch, C.J., H. Al-Hazmi and A. Aidrous, 1986. Martabah gabbro-monzonite complex, Hijaz region, Kingdom of Saudi Arabia; petrography and structure. J. Afr. Earth Sci. 4, 135-138.

Drysdall, A.R., N.J. Jackson, C.R. Ramsay, C.J. Douch and D. Hackett, 1984. Rare element mineralization related to Precambrian alkali granites in the Arabian Shield. Econ. Geol. 79, 1366-1377.

El-Gaby, S., 1975. Petrochemistry and geochemistry of some granites from Egypt. N. Jb. Miner. Abh. 124, 147-189.

El-Ramly, M.F., 1972. A new geological map for the basement rocks in the eastern and southwestern deserts of Egypt. Annals Geol. Surv. Egypt 2, 1-18.

Fleck, R.J., W.R. Greenwood, D.G. Hadley, R.E. Anderson and D.L. Schmidt, 1980. Rubidium-strontium geochronology and plate-tectonic evolution of the southern part of the Arabian Shield. USGS Prof. Paper 1131, 38 p.

Fullagar, P.D., 1980. Pan-African age granites of northeastern Africa: new or reworked sialic materials? In: Salem, J.J. and Busrewil, M.I. (eds.). Geology of Libya. Second Symposium on the Geology of Libya 3. Academic Press, 1051-1058.

Furnes, H., A.E. Shimron and D. Roberts, 1985. Geochemistry of Pan-African volcanic arc sequences in southeastern Sinai Peninsula and plate tectonic implications. Precambrian Res. 29, 359-382.

Gardner, J.N., F. Goff, S. Garcia and R.C. Hagan, 1986. Stratigraphic relations and lithlogic variations in the Jemez Volcanic Field, New Mexico. J. Geophys. Res. 91, 1763-1778.

Ghanem, M., A.A. Dardir, M.H. Francis, A.A. Zalata and K.M. Abu Zeid, 1973. Basement rocks in eastern desert of Egypt north of latitude $25^\circ 40'N$. Annals Geol. Surv. Egypt 3, 33-38.

Ghobrial, M.G. and M. Lotfi, 1967. The geology of Gebel Gattar and Gebel Dokhan areas. Geol. Surv. Egypt, Paper No. 40, 6 p.

Gill, J., 1981. Orogenic Andesites and Plate Tectonics. Springer-Verlag, New York, 390 p.

Greenberg, J.K., 1981. Chracteristics and origin of Egyptian younger granites: summary. Geol. Soc. Am. Bull. I, 92, 224-232.

Grothaus, B., D. Eppler and R. Ehrlich, 1979. Depositional environment and structural implications of the Hammamat Formation, Egypt. Annals Geol. Surv. Egypt 9, 564-590.

Halpern, M. and N. Tristan, 1981. Geochronology of the Arabian-Nubian Shield in southern Israel and eastern Sinai. J. Geol. 89, 639-648.

Harris, N.B.W. and G.F. Marriner, 1980. Geochemistry and petrogenesis of a peralkaline granite complex from the Midian Mountains, Saudi Arabia. Lithos 13, 325-337.

Harris, N.B.W., F.M.H. Marzouki and S. Ali, 1986. The Jabel Sayid complex, Arabian Shield: geochemical constraints on the origin of peralkaline and related granites. J. geol. Soc. Lond. 143, 287-295.

Hildreth, W., 1981. Gradients in silicic magma chambers: Implications for lithospheric magmatism. J. Geophys. Res. 86, 10153-10192.

Jackson, N.J., J.N. Walsh and E. Pegram, 1984. Geology, geochemistry and petrogenesis of late Precambrian granitoids in the central Hijaz region of the Arabian Shield. Contrib. Mineral. Petrol. 87, 205-219.

Jarrar, G., A. Baumann and H. Wachendorf, 1983. Age determinations in the Precambrian basement of the Wadi Araba area, southwest Jordan. Earth Planet. Sci. Lett. 63, 292-304.

Kay, R.W. and P.W. Gast, 1973. The rare earth content and origin of alkali-rich basalts. J. Geol. 81, 653-682.

Massey, K.W., 1984. Rubidium-strontium geochronology and petrography of the Hammamat Formation in the northeastern desert of Egypt. MS thesis, U. Texas at Dallas, 75 p.

Mysen, B.O. and A.L. Boettcher, 1975. Melting of a hydrous mantle: II. Geochemistry of crystals and liquids formed by anatexis of mantle peridotite at high pressures and high temperatures as a function of controlled activities of water, hydrogen, and carbon dioxide. J. Petrol. 16, 549-593

Pearce, J.A., 1980. Geochemical evidence for the genesis and eruptive setting of lavas from Tethyan ophiolites. Proceedings of the International Ophiolite Symposium. Nicosia, Cyprus, 261-272.

Ressetar, R. and J.R. Monrad, 1983. Chemical composition and tectonic setting of the Dokhan Volcanic formation, eastern desert, Egypt. J. Afr. Earth Sci. 1, 103-112.

Ries, A.C., R.M. Shackleton, R.H. Graham and W.R. Fitches, 1983. Pan-African structures, ophiolites and melange in the eastern desert of Egypt: a traverse at 26°N. J. geol. Soc. Lond. 140, 75-95.

Rogers, J.J.W. and J.K. Greenberg, 1981. Trace elements in continental margin magmatism: Part III. Alkali granites and their relationship to cratonization. Geol. Soc. Am. Bull. 92, 6-9.

Roobol, M.J., C.R. Ramsay, N.J. Jackson and D.P.F. Darbyshire, 1983. Late Proterozoic lavas of the Central Arabian Shield - evolution of an ancient volcanic arc system. J. geol. Soc. Lond. 140, 185-202.

Ryerson, F.J. and P.C. Hess, 1978. Implications of liquid-liquid distribution coefficient to mineral-liquid partitioning. Geochim. Cosmochim. Acta 42, 921-932.

Self, S., F. Goff, J.N. Gardner, J.V. Wright and W.M Kite, 1986. Explosive rhyolitic volcanism in the Jemez Mountains: Vent locations, caldera development and relation to regional structure. J. Geophys. Res. 91, 1779-1798.

Stern, R.J., 1985. The Najd fault system of Saudi Arabia and Egypt: a late Precambrian rift-related transform system? Tectonics 4, 497-511.

Stern, R.J. and D. Gottfried, 1986. Petrogenesis of a late Precambrian (575-600 Ma) bimodal suite in northeast Africa. Contrib. Mineral. Petrol 92, 492-501.

Stern, R.J., D. Gottfried and C.E. Hedge, 1984. Late Precambrian rifting and crustal evolution in the northeastern desert of Egypt. Geology 12, 168-172.

Stern, R.J. and C.E. Hedge, 1985. Geochronologic and isotopic constraints on late Precambrian crustal evolution in the eastern desert of Egypt. Am. J. Sci. 285, 97-127.

Stern, R.J. and W.I. Manton, 1987. Age of Feiran basement rocks, Sinai: implications for late Precambrian crustal evolution in northern Afro-Arabia. J. geol. Soc. Lond. 144, 569-575.

Stoeser, D.B., 1986. Distribution and tectonic setting of plutonic rocks of the Arabian shield. J. African Earth Sci. 4, 21-46.

Streckeisen, A., 1976. To each plutonic rock its proper name. Earth Sci. Rev. 47, 1-33.

Stuckless, J.S., C.E. Hedge, D.B. Wenner and I.T. Nkomo, 1985. Isotopic studies of post-orogenic granites from the northeastern Arabian Shield, Kingdom of Saudi Arabia. U.S. Geol. Surv. Open File Report OF-85-726, 40 p.

Sultan, M., R. Batiza and N.C. Sturchio, 1986. The origin of small-scale geochemical and mineralogic variations in a granite intrusion. Contrib. Mineral. Petrol 93, 513-523.

Vail, J.F., 1970. Tectonic control of dikes and related irruptive rocks in eastern Africa. In: Clifford, T.N. and Gass, I.G. (eds.). African Magmatism and Tectonics. Hafner, Darien, Conn., 337-354.

Voegeli, D.A., 1985. The origin of composite dike rocks from the north eastern desert and Sinai, Egypt. MS thesis, U. Texas at Dallas, 165 p.

Watson, E.B., 1976. Two-liquid partition coefficients: experimental data and geochemical implications. Contrib. Mineral. Petrol 56, 119-134.

Windley, B.F., 1977. The Evolving Continents. J. Wiley and Sons, New York, 385 p.

Ophiolites and the Evolution of the Arabian-Nubian Shield

The Pan-African Belt of northeast Africa/Arabia contains the oldest ophiolites known as yet. Ophiolitic fragments are widespread and sometimes concentrated in linear belts, which separate domains of mainly calc-alkaline igneous rocks (the products of ophiolite subduction) and perhaps mark the sites of ancient ocean basins. Therefore, the following papers by Shackleton (Chapter 7), Vail (Chapter 8), Stoeser and Stacey (Chapter 9) and Church (Chapter 10) are all in a way concerned with Pan-African ophiolites and related rocks and discuss their tectonic significance and evolution. Relative to the preceding chapters they cover a wider region of northeast Africa and Arabia and the more general aspects of Late Proterozoic orogeny.

Shackleton presents a variety of possible structural settings of obducted ophiolites. Vail reviews the regional distribution of Pan-African island-arc and ophiolite complexes in northeast Africa between older granitoid basement towards west, the Nile craton, and east. (NB: other authors use East Sahara craton as a synonym for Nile craton, see Introduction). Importance and advantages of geochronology and isotope geology for the assessment of the Pan-African evolution in Arabia, the distinction of different terranes and their oceanic or continental affinity, are demonstrated by Stoeser and Stacey. Church's paper is again more concentrated on the structural aspects of terrane accretion in northeast Africa and critically reviews the tectonic setting of ophiolite generation and the modes of ophiolite obduction.

Important contributions to Pan-African tectonics in northeast Africa are also found in the following chapter by Pohl (11) on metallogeny and in El-Gaby et al.'s paper (Chapter 2) on the Pan-African of Egypt. The present state of the earth's crust in Egypt, generated in Pan-African times, is discussed in the chapter (12) on geophysics by Makris et al.

Chapter 7

Contrasting Structural Relationships of Proterozoic Ophiolites in Northeast and Eastern Africa

R. M. Shackleton
Department of Earth Sciences, The Open University, Walton Hall, Milton Keynes,
MK7 6AA, England

Keywords: Proterozoic, Pan-African, NE Africa, ophiolite, obduction, ophiolitic mélange, thrusting, gravity tectonics

Abstract: Of the many recognised models of occurrence of ophiolites, which include obducted slabs, thrust slices, fault-bounded blocks, serpentinite thrust-mélanges, olistostromic and tectonic ophiolitic mélanges, ejected serpentinite lenses and serpentinite diapirs, most are found in the Upper Proterozoic terranes of Arabia and NE and E Africa. Many of the occurrences are far from sutures: these must be identified on structural and other evidence. Ophiolite slabs obducted onto less dense continental crust tend to separate into units which subside, so that they are preserved in synforms. The Ingessena ophiolite in Sudan is an example.

INTRODUCTION

Numerous ophiolite occurrences are now recognised in the Upper
Proterozoic terranes of northeast and eastern Africa (Fig. 1).
Most are dismembered and incomplete. All are detached masses
with tectonic contacts, but the ways in which they have become
detached and transported vary widely, resulting in a variety of
structural relationships. These are examined here.

OBDUCTED OPHIOLITE SLABS

Initial detachment from the parent oceanic crust, and
obduction, is generally assumed to imply a destructive plate
margin. The most distinctive hallmark of primary obduction is
the occurrence of a metamorphic sole of ocean floor basalts and
sediments beneath the obducted ophiolite, as seen, for example,
under the Late Cretaceous Semail ophiolite in Oman. Such a
metamorphic sole implies that the detachment surface cuts deep
enough into the oceanic crust to reach hot mantle which was
then carried over the rocks which formed the metamorphic sole.
No such metamorphic sole has yet been recognised under any of
the Upper Proterozoic ophiolites of northeast or eastern
Africa. In the absence of such a sole, recognition of pristine
obducted ophiolite is more subjective. A complete and not
dismembered ophiolite sequence, on a suture identified by
criteria discussed below, would suggest an obducted slab. The
Sol Hamed ophiolite in northeast Sudan appears to be an example
(Fitches et al., 1983).

OPHIOLITE SYNFORMS - RESULT OF GRAVITATIONAL INSTABILITY

A characteristic feature of large obducted ophiolite slabs is
the tendency of the dense ophiolite to sink unequally into the
underlying lighter crustal rocks, forming an array of
subcircular synforms. Such a tendency seems to be displayed
(Fig. 2) in the Semail ophiolite of Oman (Rothery, 1982) and

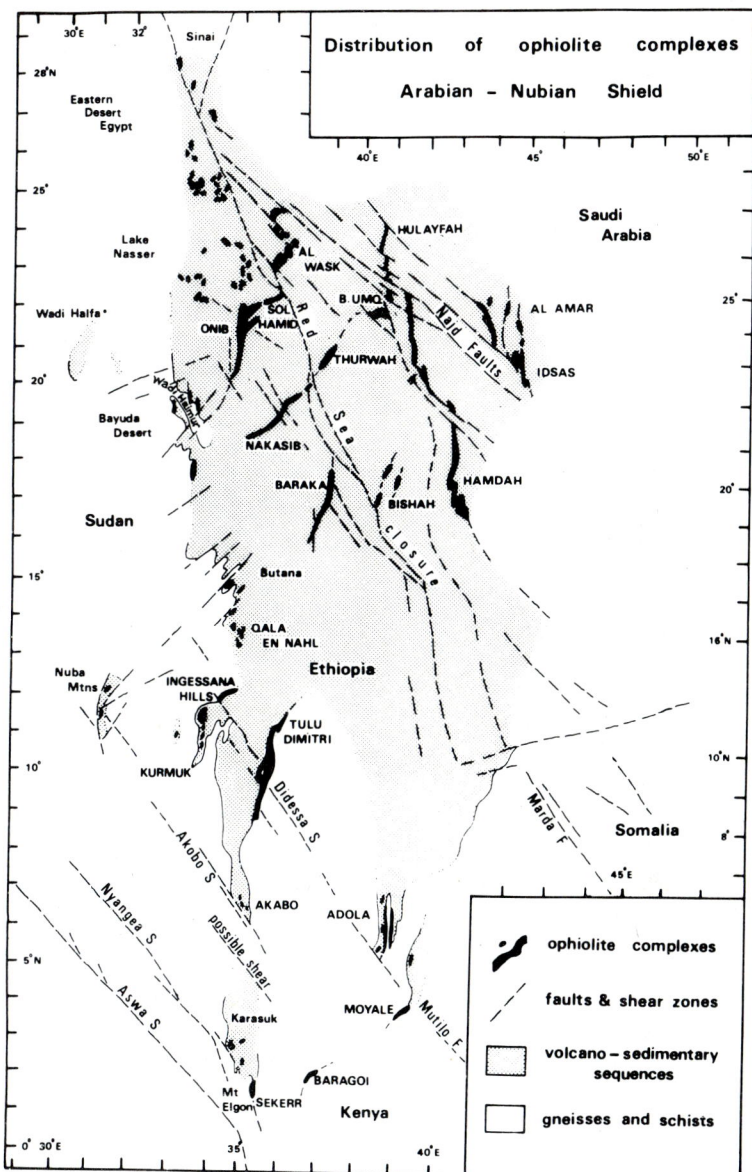

FIGURE 1.

Ophiolite occurrences in northeast and eastern Africa, based on Vail (1983).

OUTCROP EXTENT AND STRUCTURAL UNITS
OF THE OMAN OPHIOLITE

BLOCK NAMES

1	Khawr Fakkan	8	Sarami
2	Aswad	9	Muqniyat
3	Sumeini	10	Haylayn
4	Fizh	11	Rustaq
5	West Jizi	12	Bahla
6	Salahi	13	Ibra
7	Wuqbah	14	Muscat

MINOR OCCURENCES

Sh Sham
 A Ahin
 G Ghuzayn
AA Al Arid
 Si Sint
 H Haliw

FIGURE 2.

Subsidence synforms in the Semail Ophiolite, Oman (after Rothery, 1982).

more spectacularly by the Morais, Bragança and Cabo Ortegal synforms of Portugal and northwest Spain (Ries and Shackleton, 1971). The mean diameter of the Oman synforms is of the order of 30 km. More eroded examples would be smaller. These synforms are not associated with similarly shaped antiforms, as they would be if they were due to crossfolding. There are indications in Oman that underlying rocks have been ejected upwards through the narrow zones between the synforms. The pattern seems to be a large-scale analogue of sedimentary load casts, the result of instability when a sheet of dense material is emplaced over less dense soft or ductile material underneath. I interpret the Ingessena ophiolite complex in Sudan (Fig. 3) as an example of such a density-driven synform. Like other such complexes, including those in Portugal and northwest Spain, the Ingessena complex has been interpreted as a mantle diapir, but it displays clear though steep inward dips of the external contact and chromitite sheets in the mantle sequence and layering, which in the few places where I could determine its polarity, faced inwards in the gabbro cumulates; the sequence, from mantle harzburgite through a dunite-rich zone, to gabbro cumulates, with a small central area of sheeted dykes, confirms the synformal structure, which is clearly incompatible with a diapiric origin. Such synformal masses must necessarily be allochthonous. They are probably erosional remnants of larger sheets. Their position does not mark the outcrop of a suture between two plates. This may be a hundred km or more distant.

THRUST SLICES OF OPHIOLITES

Thrust slices and imbrications of ophiolites are common. Typical examples are seen in the Sekerr area of western Kenya (Vearncombe, 1983) and in the Baragoi area in the central part of the Mozambique belt in Kenya (Baker, 1963). A characteristic feature of ophiolite thrust slices is their lenticular shape. Usually both mafic and ultramafic components of the ophiolite are involved. A similar lenticular shape of the ophiolite

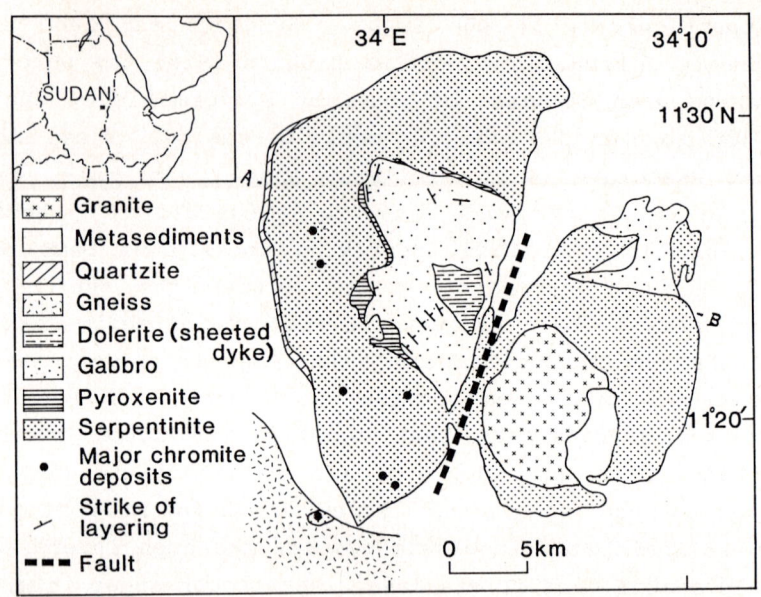

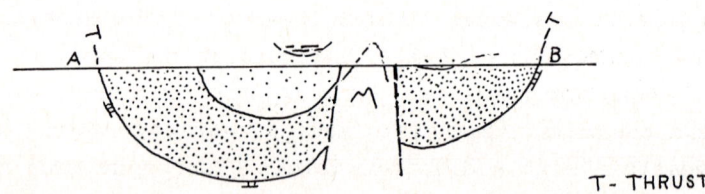

T- THRUST

FIGURE 3.

Ingessena ophiolite, Sudan (after Price 1984) and sketch
section.

masses is seen in the Wadi Haimur area of Egypt (El-Shazly et al., 1975) although it is not impossible that these lenticles represent flattened masses in a highly deformed olistostromic ophiolitic mélange.

FAULT BOUNDED OPHIOLITES

Fault-bounded blocks of ophiolite yield little information of the primary emplacement mechanism. An example is the Qala en Nahl mass in Sudan. This is limited by steep faults, between which the steeply-dipping layering in the ophiolite strikes obliquely to the faults.

SERPENTINITE MÉLANGES

Another structural type of ophiolite occurrence is in serpentinite mélange such as occurs immediately below the Semail ophiolite in Oman. Large masses of ophiolite, together with some other rocks, are distributed through the serpentinite matrix. This is a thrust-mélange. An example is the mélange in the Al Amar suture zone in the eastern part of the Saudi Arabian shield (Al-Shanti and Gass, 1983).

OPHIOLITIC MÉLANGE

Ophiolitic mélanges (Gansser, 1974), in which the ophiolitic masses are embedded in a sedimentary matrix are widely distributed in the upper Proterozoic of Egypt (El Sharkawy and El Bayoumi, 1979; Church, 1979; Shackleton et., 1980; Ries et al., 1983). The sedimentary matrix is characteristically unstratified, in sharp contrast to the normally stratified associated sediments. Blocks and masses of a variety of rocks are randomly associated, with much less order than in a thrust mélange where trails of the same rock horizon can often be traced as pseudo-stratigraphy for considerable distances. When

least deformed the blocks are angular, sometimes with re-
entrant angles, in contrast to the lenticular shapes in thrust
mélanges. Such mélanges are interpreted as olistostromes due to
submarine gravity sliding and are probably derived from the
thrust front of the obducted ophiolite. The upper Proterozoic
ophiolite mélange in the Eastern Desert of Egypt is so
interpreted although it has also been involved in thrusting and
mylonitization (Ries et al., 1983; Sturchio et al., 1983).

EJECTION AND DIAPIRISM OF OPHIOLITE

Serpentinite is both exceedingly slippery and less dense than
average crustal rocks, so it is to be expected that the upper
mantle component of an ophiolite complex, if serpentinitised,
as is common near the basal thrust of an obducted sheet, will
be liable to migrate, either in response to a pressure
gradient, by which lenses are ejected sideways or upwards, or
diapirically like a salt diapir. Neither ejected serpentinite
lenses nor serpentinite diapirs have yet been recognised in the
upper Proterozoic terrains of northeast or eastern Africa, but
they probably exist [see Greiling et al., 1987].

This review of the structural relationships of the northeast
African ophiolites shows that after obduction, masses of
ophiolite are commonly dismembered, displaced and dispersed,
both by gravity, at surface, and by deformation in the crust.

OPHIOLITES AND SUTURES

The implication, as has been pointed out previously (Church,
1979) is that ophiolites cannot be used, as they were in the
early stages of the application of plate tectonic models to the
Precambrian, to map out sutures by drawing lines on a map to
connect ophiolite occurrences. Ophiolites from a suture can be,
and often are, spread over a zone a hundred or more km wide.

Recognition of sutures may be based on the continuity over a long distance of ophiolites, as along the Indus-Tsangpo suture in Tibet or the Nabitah suture in Saudi Arabia; the occurrence of ophiolites in zones of intense deformation, which may be separated, as in Saudi Arabia, by less deformed calc-alkaline volcanic arc terrains (Stoeser and Camp, 1985); by evidence of opposing provenance from different plates, of contemporaneous sediments; geological evidence of misfit of opposing plates or microplates; or by palaeomagnetic misfit. Correlations of sutures across the upper Proterozoic domains of northeast and eastern Africa must be based not only on these criteria but needs to be supported by geochronological evidence of contemporaneity. It must be emphasised that a suture, since it marks a plate boundary, cannot just stop. It must continue until replaced by another type of plate boundary, either a transform fault, or a constructive margin, the different types of boundary together encircling the plates.

There is also the need to determine, especially from geochemical data, the tectonic environment, whether active mid-ocean spreading axis, narrow or wide ocean, forearc or back arc basin, of the various ophiolites. Geochemical data have been used to identify some of the Arabian and northeast African Proterozoic ophiolites as of back arc origin. The non-recognition, so far, of forearc prisms may suggest the absence of wide-ocean subduction zones.

The dip and age sequence of these upper Proterozoic ophiolitic sutures is still uncertain. An easterly or southeasterly dip is indicated for the Al Amar suture, the Yanbu-Sol Hamed suture and the Sekerr suture in West Kenya; in Saudi Arabia the evidence suggests that the sutures become younger eastwards (Stoeser and Camp, 1985).

GENERAL CONCLUSIONS

The general conclusion from a study of these upper Proterozoic ophiolites must be that the region evolved by accretion and collision of a series of plates and microplates. Palaeomagnetic evidence (McWilliams, 1981) appears to indicate that the collisions involved the eventual closure of a wide ocean, because there is a drastic difference between the apparent polar wander paths of East and West Gondwana, until the Ordovician when they coincide.

REFERENCES

Al-Shanti, A.M.S. and I.G. Gass, 1983. The Upper Proterozoic ophiolite mélange zones of the easternmost Arabian Shield. J. geol. Soc. London 140, 867-876.
Baker, B.H., 1963. Geology of the Baragoi Area. Geol. Surv. Kenya. Rep. 53, 74pp.
Church, W.R., 1979. Discussion on 'Granitic and metamorphic rocks of the Taif area, western Saudi Arabia'. Geol. Soc. Am. Bull. 90, 893-896.
El-Sharkawy, M.A. and R.M. El-Bayoumi, 1979. The ophiolites of Wadi Gadir area, Eastern Desert of Egypt, Ann. Geol. Surv. Egypt, 9, 125-135.
El-Shazly, E.M., F.A. Bassyouni and M.L. Abdel Khalek, 1975. Geology of the Greater Abu Swayel Area, Eastern Desert, Egypt. J. Geol. 19, 1-41.
Fitches, W.R., R.H. Graham, I.M. Hussein, A.C. Ries, R.M. Shackleton and R.C. Price, 1983. The Late Proterozoic ophiolite of Sol Hamed, NE Sudan. Precambrian Research 19, 385-41.
Gansser, A., 1974. The ophiolitic mélange - a worldwide problem on Tethyan examples. Eclogae Geol. Helv. 67, 479-509.
McWilliams, M.O., 1981. Palaeomagnetism and Precambrian tectonic evolution of Gondwana, In: Developments in Precambrian Geology: Precambrian plate tectonics. (Ed) A. Kröner, Elsevier.
Price, R.C., 1984. Late Precambrian mafic-ultramafic complexes in northeast Africa, unpublished Ph.D. Thesis, Open Universitiy.
Ries, A.C. and R.M. Shackleton, 1971. Catazonal complexes in Northwest Spain and North Portugal, Remnants of a Hercynian Thrust Plate. Nature, 234, 65-68 and 79.
Ries, A.C., R.M. Shackleton, R.H. Graham and W.R. Fitches, 1983. Pan-African structures, ophiolites and mélange in the Eastern Desert of Egypt: a traverse at 26°N. J. Geol. Soc. London, 140, 75-95.
Rothery, D.A., 1982. The evolution of the Wuqbah Block and the application of remote sensing to the Oman Mountains. Unpublished Ph.D. thesis, Open University.

Shackleton, R.M., A.C. Ries, R.H. Graham and W.R. Fitches, 1980. Late Precambrian ophiolitic mélange in the Eastern Desert of Egypt. Nature, 285, 472-474.

Stoeser, D.B. and V.E. CAMP, 1985. Pan-African microplate accretion of the Arabian Shield. Geol. Soc. Am. Bull. 96, 817-826.

Sturchio, N.C., M. Sultan and R. Batiza, 1983. Geology and origin of the Meatiq Dome, Egypt: A Precambrian metamorphic core complex? Geology, 11, 72-76.

Vail, J.R., 1983. Pan-African crustal accretion in north-east Africa. J. Afr. Earth Sci. 1, 285-294.

Vearncombe, J.R., 1983. A dismembered ophiolite from the Mozambique Belt, West Pokot, Kenya. J. Afr. Earth Sci. 1, 133-143.

[Greiling, R.O., A. Kröner, M.F. El-Ramly, A.A. Rashwan, 1987. In: S. El-Gaby and R.O. Greiling (eds.) The Pan-African Belt of NE Africa and Adjacent Areas. Earth Evol. Sci., Vieweg, Wiesbaden.]

Chapter 8

Tectonics and Evolution of the Proterozoic Basement of Northeastern Africa

J. R. Vail
Department of Geology, Portsmouth Polytechnic, Burnaby Building, Burnaby Road,
Portsmouth PO1 3QL, U.K.

Keywords: Proterozoic, Pan-African, NE Africa, Arabia, volcano-sedimentary-ophiolite assemblage, tectonic structure

Abstract: Two distinct lithologic-tectonic units underlie the Phanerozoic cover of northeast Africa between Egypt and Kenya and western Arabia between Sinai and Somalia. They are a granitic gneissose basement of granulite to amphibolite metamorphic facies overlain by interfolded supracrustal metasedimentary shallow shelf deposits. These rocks are in structural contact with dominantly greenschist facies volcano-sedimentary island-arc type assemblages with which are associated generally linear belts of fragmented ophiolite complexes intruded by syn- to late tectonic calc-alkaline granitoid masses and post-tectonic granitic plutons. Many of the ophiolite belts mark suture zones which form tectonic boundaries to microplates. Nine oceanic terranes are recognized, the intraplate boundaries of which trend broadly northeastwards, whereas to west and east gneissic-metasedimentary continental terranes form N-S striking bounding margins. To the south, in northern Kenya the volcanic rocks abut against Mozambique Belt gneisses.

The geology of three areas in the Sudan along the western
margin of the volcano-sedimentary assemblage is described; all
share similarities of continental gneisses with overlying shelf
metasediments against ophiolitic sutures and island-arc or
plate margin volcanic rocks. The oceanic terranes formed
between about 950 and 650 Ma by crustal accretion, whereas the
continental gneisses although isotopically overprinted and
metamorphosed around 650 - 450 Ma (the Pan-African) are
probably Early to Middle Proterozoic. The present structure is
thought to have evolved through plate collision. Less disturbed
Archaean cratonic areas lie further west in the Tanganyika
Shield and Uweinat inlier of the Nile Craton and possibly also
to the east in southern Somalia.

REGIONAL CONSIDERATIONS

It has been only in the last decade that the geology of the
Precambrian basement of northeast Africa and adjoining Arabian
Peninsula has been properly studied. Following extensive
mapping in Saudi Arabia an appreciable insight into the geology
has now been obtained and this has enabled the less well mapped
regions of Egypt, Sudan, Ethiopia and Somalia to be better
understood.

This paper whilst reviewing the results of some of this work
attempts to put the geology in perspective in relation to the
rest of NE Africa, and to show the nature of the major
geological units now recognized in three specific areas, and
how the tectonic evolution of the entire region developed.

The basement of NE Africa and Arabia has been referred to as
the Arabian-Nubian Shield (Al-Shanti, 1979) and it is
convenient to consider the area in these terms. However, whilst
the definition of the Arabian Shield is facilitated by the
limiting exposure of basement rocks beneath the extensive cover
of undeformed Palaeozoic strata in Sinai, Saudi Arabia and
Yemen, the margins to the Nubian Shield are much less precise.
Precambrian basement is exposed in Egypt, the Sudan and
Ethiopia along the border of the Red Sea in the Eastern Desert,
Red Sea Hills, and Eritrean parts of those territories.
However, west of the Nile Valley there are many scattered
basement inliers because the Mesozoic and Cainozoic sedimentary
and volcanic cover is not everywhere present. In addition,
basement exposures are extensive in the southern Sudan and
Ethiopia, and in northern Uganda, Kenya and Somalia and there
is no clearly defined limit to the Shield. It is thus timely to
reconsider the concept and geographical constraints of the
Arabian-Nubian Shield.

It also has been generally accepted that the basement rocks of
NE Africa and Arabia are a continuation of those seen further
south in East Africa, where they comprise the Mozambique Belt

(Holmes, 1951). It has been shown that the latter belt is but a part of a far more extensive band of metamorphism, tectonism, and plutonic activity recognized as the Pan-African tectonic and thermal event by Kennedy (1964, 1965). In addition, these events characteristically provide single sample Rb-Sr and K-Ar isotopic ages in the order of 650 - 450 Ma, and thus define the Pan-African as a time of late Proterozoic tectogenesis.

NE Africa and Arabia also are areas of intense geological activity, and extensive isotopic studies (summarized in Cahen et al., 1984) have shown these events to have been active during the Pan-African, and possibly to extend back as early as 1000 Ma.

There is one main difference, however, between the geology and crustal evolution of NE Africa and that of the Mozambique Belt. In the present state of knowledge and interpretation of the geology of the two areas, the Mozambique Belt comprises continental para- and orthogneisses in amphibolite to granulite grade of regional metamorphism, whereas most of the Arabian-Nubian Shield comprises oceanic margin, island-arc type volcanism and associated syn- to post-tectonic intrusions, for the most part in the greenschist facies of metamorphism. The genesis of these two contrasting types are totally different, and thereby raises the question as to whether they can both be considered as belonging to the Pan-African without dividing the latter into two totally dissimilar terranes (Hepworth, 1979; Almond, 1984). The resolution of this will exercise the considerations of geologists for some time to come, for not only is there a conceptual model to resolve, but because the metallogenic potential of the two regions is quite distinct, and since the mineralization is controlled by quite different factors in each area (Pohl, 1984; Vail, 1985b) there are also direct economic consequences.

GEOLOGICAL FRAMEWORK

There are two major lithological associations in NE Africa
which represent two contrasting tectonic environments: gneisses
and metasedimentary rocks, and volcanic-sedimentary assemblages
with associated ophiolites and intrusive granitoids. The
distinction between the two is not always obvious, due to
locally more intense metamorphism, tectonism, or disruption by
igneous activity. Also small blocks of one group are to be
found within the other group and their identification,
structural relationship, and genetic and tectonic emplacement
is not always unambiguous. In an area stretching from the
Mediterranean coast 3000 km to the equator and extending across
the Red Sea for a distance of over 1500 km there is inevitably
variation in the quality and intensity of geological
information, and in the coverage of analyses and detailed work.
Nevertheless, the concept of plate-tectonic mechanisms and
continent-ocean margin collision environments provides a
suitable model in which to consider the regional geology.
Because the events are related to late Proterozoic or earlier
activity, it is convenient to consider the palinspastic
reconstruction of the region prior to the Tertiary opening of
the Red Sea and to treat most of the Palaeozoic and Cainozoic
activity as unrelated to the Pan-African events under review
(Fig. 1).

Gneiss-Metasedimentary Unit

Quartzo-feldspathic gneisses, migmatites and paragneissic
metasediments generally in the granulite to amphibolite grade
of metamorphism crop out between southern Egypt (Schandelmeier
et al., 1983; Huth et al., 1984) and northern Uganda-Kenya
approximately along the Nile valley and westwards from $34^{O}E$
longitude (Vail, 1978). Similar gneisses make up most of the
basement exposures of northern Somalia (Warden and Horkel,
1984) and in the Bur region (Daniels, 1965) and they are
extensively developed in southern Ethiopia (Kazmin et al.,

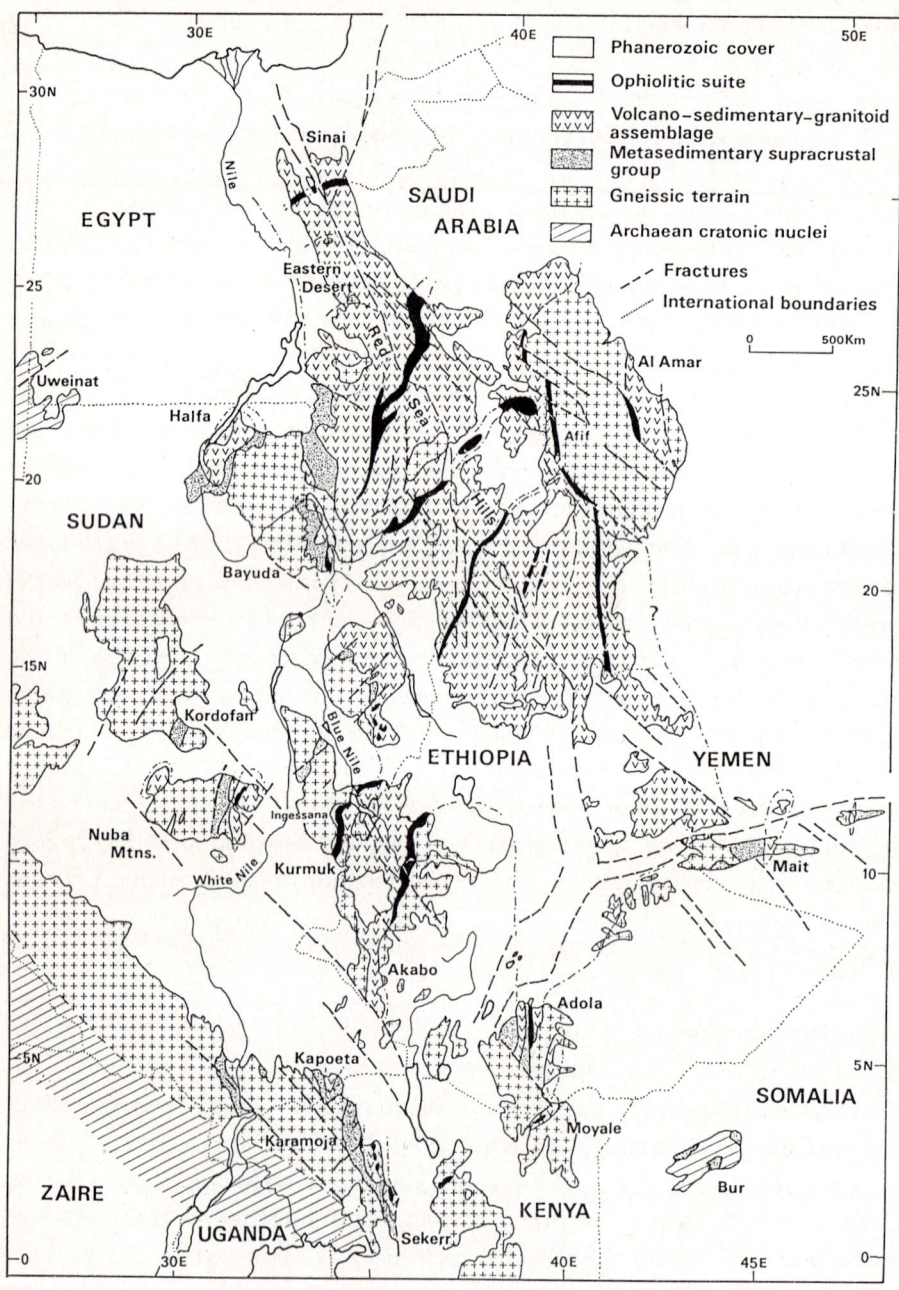

FIGURE 1.

Lithological-tectonic units in the palinspastic Proterozoic basement of NE Africa and Arabia.

1978; Davidson et al., 1973, 1976). In the latter areas they have been considered to constitute the north extension of the East Africa Mozambique Mobile Belt.

At least some of the rocks in the Afif region of the eastern Arabian Shield have been identified (Stacey and Hedge, 1984,; Stoeser et al., 1984) as ancient gneisses which although having suffered later remobilization and subsequent tectonic events are comparable with the continental gneisses further south and west.

In places, overlying the gneisses and infolded within them are extensive metasedimentary rocks comprising quartz-mica psammites and quartzites, garnet-mica pelites and graphitic schists, pink or white marbles and calc-silicates, and irregular amphibolite bodies. They are in places probably unconformable over the gneisses, elesewhere their relationships are unclear due to subsequent tectonism. Generally they are in amphibolite to granulite facies. These metasediments are well developed in the Sudan in the Halfa District (Griffiths, 1985; Curis and Griffiths, 1985), in the eastern Bayuda Desert (Vail, 1971; Meinhold, 1979; Dawoud, 1980), eastern Blue Nile Province (Abdel Magid, 1983; Abdel Rahman, 1983), Nuba Mountains inlier of Kordofan Province (El Ageed and El Rabaa, 1981; Sadig and Vail, 1986) and in the southern Sudan (Civetta et al., 1980; Hunting, 1979).

In western Kenya and eastern Uganda the Cherangani metasediments (Miller, 1956; McCall, 1964) and Karasuk metasediments (Almond, 1969) also are of this type, whereas the Middle Complex of the Adola zone in southern Ethiopia (Kazmin, 1973; Kazmin et al., 1978), the Dinsor metasediments in the Bur region of Somalia (D'Amico et al., 1981), and the Gebile psammites of north Somalia are comparable (Warden and Horkel, 1984; Warden and Daniels, 1984).

Small inliers of amphibolite facies metasedimentary and gneissic rocks occur within the lower grade volcano-sedimentary-ophiolite greenschist assemblages. They form exotic terranes within the oceanic rocks and can be variously considered to be the basement floor, thrust klippe, locally areas of more intense metamorphism or shearing of the lower grade rocks, or tectonically detached microcontinental blocks. Such inliers in the Eastern Desert of Egypt form the Hafafit and Mitiq domes (Greiling et al., 1984; El Ramly et al., 1984; Sturchio et al., 1984) and the high grade rocks near Abu Swayel and Wadi Haimur (Hashad et al., 1972; El Ramly, 1972) and elsewhere. In the Red Sea Hills of the Sudan paragneissic-metasedimentary blocks make up the Sasa Plains (Ruxton, 1956; Almond, 1980), southern Red Sea Hills and J. Eyob (El Samani, 1985), and the Qeissan block in the Blue Nile Valley (Toum, 1985). Isolated patches also occur in Ethiopia and Saudi Arabia along the closure line of the Red Sea and possibly in the Khamis Mushayet area and elsewhere in the Arabian Shield (Kazmin, 1973; Johnson, 1983; Johnson et al., 1984).

Volcano-Sedimentary-Ophiolite Assemblages

The Sinai peninsula, most of the Eastern Desert of Egypt, the Red Sea Hills physiographic province of eastern Sudan, most of the basement of Ethiopia and the western Arabian Shield and Yemen basement are underlain by a heterogeneous accumulation of oceanic island-arc, and plate margin Andean type volcanic and associated pyroclastic, volcanogenic and shallow water shales, silts and limestones. The volcanic rocks range in composition from basaltic and andesitic to rhyolitic, and are dominantly calc-alkaline. The grade of metamorphism is characteristically greenschist to lower amphibolite facies, which together with their lithology contrasts strongly with the amphibolite-granulite gneiss-metasedimentary unit.

In addition, mafic and ultramafic linear masses are closely associated with the volcano-sedimentary sequences. Serpentinized pyroxenites and peridotites, layered gabbros, sheeted dyke complexes, pillow lavas and rare siliceous bands and plagiogranites all point to an ophiolite suite (Bakor et al., 1976; Gass, 1977; Vail, 1983; Hussein et al., 1984; etc.). However, these rocks are usually fragmented, tectonically disrupted, disturbed by plutonic emplacements, and covered by younger deposits. Nevertheless, they tend to form long, narrow bands within the metavolcanic assemblages, some of which have been traced for several hundreds of kilometres (Frisch and Al-Shanti, 1977). In contrast, in the Eastern Desert of Egypt these ophiolite components are contained in an extensive tectonic mélange and the suite is completely dismembered and their original distribution and stratigraphic character has been totally disrupted (El Sharkawy and El Bayoumi, 1979; Ries et al., 1983). Ophiolite mélanges are also developed elsewhere, but not to the extent they are in Egypt.

In keeping with the oceanic, plate margin environment of the volcano-sedimentary-ophiolite assemblage, the entire area in which these units occur has been subjected to intense plutonic activity. This takes the form of syn-tectonic batholithic intrusions of gabbros and diorites, granodiorites, trondjhemites, tonalites, adamellites and granites. There is also the development of late and post-tectonic plutonic emplacements, which are characteristically bimodal calc-alkalic gabbro-granite complexes giving way to alkaline granites and syenites. The plutons show evidence of high level emplacement and ring complex structures are commonly developed (Stoeser and Elliot, 1980; Stoeser, 1986; Vail, 1983; Vail and Kuron, 1978). Later in the same area anorogenic granite, syenite, and rare foid syenite ring complexes and plutons constitute an extensive alkali province extending from northern Uganda to southern Egypt (Vail, 1985a).

Apart fromt the main outcrop for the volcano-sedimentary-ophiolite and associated granitoids, similar sequences are to be found as isolated detached outliers within the continental gneiss terranes. The most notable of these blocks are near Wadi Ḥalfa in northern Sudan (Vail et al., 1973; Griffiths, 1985), the eastern Bayuda desert along the Nile river (Vail, 1971;

El Rabaa, 1975; Ries et al., 1985), the Abu Zabed and eastern Nuba Mountains block of central Sudan (El Ageed and El Rabaa, 1981; Saig and Vail, 1986), the Moyale outlier in northern Kenya-southern Ethiopia (Vail, 1983), and possible outliers in the Mait greenstones of north-eastern Somalia (Warden and Horkel, 1984), south Yemen, and the Al Amar-Idsas region of the easternmost Arabian Shield (Al-Shanti and Gass, 1983).

AGE OF EVENTS

Geochronological data are unevenly distributed across the Arabian-Nubian Shield and adjacent areas and many individual units have not yet been adequately dated, particularly in parts of Egypt, the Sudan, Ethiopia and Somalia. Nevertheless, from a regional viewpoint there are now sufficient age data to enable a broad pattern to be recognized across most of the region. It is also apparent that the entire area suffered a metamorphic and/or magmatic event during the Pan-African period (650-450 Ma) and thus older events were isotopically overprinted and partially obliterated. In addition, magmatism and tectonism continued for some time and this protracted activity has made it difficult to identify the earliest phases, which in many places have not been isotopically analysed.

The oldest ages to have been obtained are all from the gneiss-metasedimentary units. In the Uweinat basement inlier of SE Libya, SW Egypt and NW Sudan, good evidence from Rb-Sr isochron and K-Ar determinations indicates the oldest granulite facies metamorphism to be as old as 2900 Ma, with retrogressive metamorphism around 2632 Ma and a late granitization event around 1800 Ma (Klerkx and Deutsch, 1977; Klerkx, 1980, Cahen et al., 1984). Similarly, Archaean and early Proterozoic ages have been obtained from northern Uganda where granite gneisses yielded 2536 Ma, with cooling ages reflecting a tectono-metamorphic event around 1850 ± 40 Ma (Leggo, 1974; Cahen et al., 1984). In the southern Sudan K-Ar ages up to 1906 ± 50 Ma have been measured in the basement gneisses west of the White

Nile (Civetta et al., 1980), and a Sm-Nd model age of 1950 Ma
has been determined for the Turbo migmatites of western Kenya
(Harris et al., 1984).

Farther north, the gneisses in Kordofan and Bayuda desert have
yielded Sm-Nd model ages of 1550 Ma and in southern Egypt
1800 Ma (Harris et al., 1984) and between 2322 Ma and 1403 Ma
for granitoid rocks in the northern Sudan and southern Egypt
(Schandelmeier et al., 1987). Zircons, of possible detrital
origin, from the Hafafit gneisses gave a U-Pb concordia age of
1770 Ma (Abdel-Monem and Hurley, 1979) and zircons from
granitic pebbles ages up to 2300 Ma (Dixon, 1981). It is
reasonable, therefore to suppose that an early Precambrian
terrane in which granitic rocks were emplaced around 1800 Ma,
exists close to the Eastern Desert. A later metamorphic event
at about 1185 Ma has also been detected affecting
metasedimentary rocks in the Abu Swayel and Wadi Haimur area of
southern Egypt (El Shazly et al., 1973; El Manharawy, 1977,
Cahen et al., 1984).

There have been no systematic geochronological studies on the
gneisses of northern Kenya, Somalia or Yemen, although Dolginov
and Said (1976) considered the basement to be Archaean in
Somalia, and Warden and Horkel (1984) suggested there may be
affinities in the Bur inlier to the Nyanzian (Archaean) of
Kenya.

The gneisses of the Afif region of Saudi Arabia have yielded
model Pb ages indicative of ancient continental crust 1830 to
2067 Ma old (Stacey and Hedge, 1984; Stoeser et al., 1984;
Stacey and Stoeser, 1984) and the granodioritic gneiss at Jabal
Khida yielded Rb-Sr, U-Th-Pb and Sm-Nd data indicating an age
of 1628 Ma subsequently remobilized at 660 Ma (Stacey and
Hedge, 1984).

It is thus apparent that although the age of gneissic rocks has
not been established specifically in all the outcrop areas,
nevertheless, there is regional evidence for ancient crust

2550 Ma or older in the west. Granite emplacement and metamorphism took place around 1800 ± 150 Ma. The subsequent tectonic events which followed from around 900 Ma down to about 550 Ma affected these older rocks.

Due to metamorphism and igneous intrusions the direct dating of the metavolcanic and associated metasedimentary and ophiolitic rocks has proved difficult, and interpretation of the results has led to disagreement. The plutonic events, however, provide time limits and a chronological sequence has been built up for much of the region, particularly Saudi Arabia (Jackson and Ramsay, 1980; Roobol et al., 1983; Stoeser, 1986). From these data it is clear that the earliest volcanic events, related to oceanic magmatism, began shortly before 900 Ma (Marzouki et al., 1982) and volcanism continued to about 570 Ma (Jackson and Ramsay, 1980). The latter authors recognized three sequences in the Central Arabian Shield: Sequence C older than 900 Ma located mainly in the west, Sequence B 950 - 650 Ma consisting of several volcano-sedimentary complexes deposited in elongated basins adjacent to island-arcs across much of the Shield, and sequence A 650 - 570 Ma occurring mainly in the eastern parts of the central Shield. Roobol et al. (1983) believed that the evolution of the volcanic rocks was by superimposition of sequences, and it is unequivocal that renewed igneous activity occurred across much of the Shield at more than one time.

A different concept of the relative ages was proposed in different areas by different authors. In the Sudan, Embleton et al. (1984) envisaged structurally bounded blocks to contain similarly derived volcano-sedimentary sequences, and suggested their age might progress from block to block. Since their analyses were mainly from the central Red Sea Hills which they considered to have been volcanically active about 730 Ma (Vail et al., 1984) they were unable to set limits to the activity across the entire Nubian Shield; subsequently Ries et al. (1985) dated metavolcanic rocks in the eastern Bayuda Desert at 800 ± 83 Ma.

In Saudi Arabia, Stoeser and Camp (1985) also proposed a model in which the age of volcanism and magmatism was envisaged as being coeval in structurally separated blocks or microplates, with some later renewal of activity across the blocks. The oldest arc related magmatism was located in the south-east of the Arabian Shield erupting from about 950 to 800 Ma ago, this is adjacent to a block in which the volcanic rocks have been dated between about 800 and 715 Ma. Farther northwards most of the magmatic rocks are younger than 680 Ma, and activity may have continued down to about 580 Ma, as was the case in northern Egypt (Stern and Hedge, 1985).

As already indicated, post-tectonic plutonism continued across the Shield. In Saudi Arabia intercratonic peraluminous and peralkaline alkali-feldspar granites and minor syenites were emplaced from about 620 to 510 Ma (Stoeser, 1986), whereas across the Sudan and southern Egypt anorogenic plutonism continued down through 465 - 360 Ma, 251 - 221, and 169 - 132 Ma (Vail, 1985a).

WESTERN MARGIN OF THE VOLCANO-SEDIMENTARY-OPHIOLITE SEQUENCES

The nature of the contact between the oceanic volcanogenic sequences and associated plutonic emplacements in the east and the continental gneisses and metasedimentary deposits in the west is pertinent to understanding the evolution of north-eastern Africa.

Exposures are poor, and systematic geological mapping not yet complete, but three areas in particular indicate the nature of this margin. They are the eastern Bayuda Desert of northern Sudan, the Ingessana-Kurmuk region of Blue Nile Province of eastern Sudan, and the Sekerr-Karamoja-Kapoeta region of East Africa and the south-eastern Sudan.

1. Eastern Bayuda Desert

The Bayuda Desert of the northern Sudan lies within the big
bend of the Nile river north and west of Atbara (Fig. 2). The
basement is exposed beneath Cainozoic and Mesozoic sandstones
and volcanic rocks, and is intruded by younger granite ring-
complexes. Early work by Vail (1971) and El Rabaa (1975, 1976)
recognized three main units, subsequently confirmed and
extended by Dawoud (1980). These are a western granitic gneiss
unit, the Grey Gneiss of Vail (1979); a central quartzo-
feldspathic psammite, pelite, marble and amphibolite unit, the
Metasedimentary Group; and an eastern greenschist facies
metavolcanic sequence with minor limestones. Intrusive into all
three are syntectonic batholithic granitic masses, most of
which are foliated and which grade in places through migmatites
to quartzo-feldspathic gneisses. In addition, El Rabaa (1975,
1976) recognized a strip of metagabbro on the west bank of the
Nile between Atbara and Berber (Fig. 2), and Dawoud (1980) and
Ries et al. (1985) identified ultramafic lenses along the
contact of the volcanics around Wadi El Koro south of Abu
Hamed, and also a weakly deformed purple molasse, the Amaki
Series, north of Shereik, on the edge of the volcanic
assemblage. The contacts between the units are highly deformed,
displaced by thrust faulting and shear zones, and intruded by
granites, and although it is believed the metasediments
unconformably overlie the Grey Gneisses, it is difficult to
prove so. However, the metamorphic and structural histories of
the three groups differ, becoming less complex with time, and
the geochronological evidence (Harris et al., 1984; Ries et
al., 1985) indicates the Grey Gneisses to be considerably older
than the Metasedimentary Group, which might be older than
898 ± 51 Ma (granitic/aplitic intrusive material) and about
1050 Ma (Nd model age). The basic to dacitic volcanics at Wadi
El Koro yielded a Rb-Sr isochron age of 800 ± 83 Ma, which Ries
et al. (1985) suggested to be the age of extrusion. The latter
authors also believed the contact beneath the northern
volcanics represents a tectonic suture zone, which can be
envisaged as extending along the contact, southwards to the

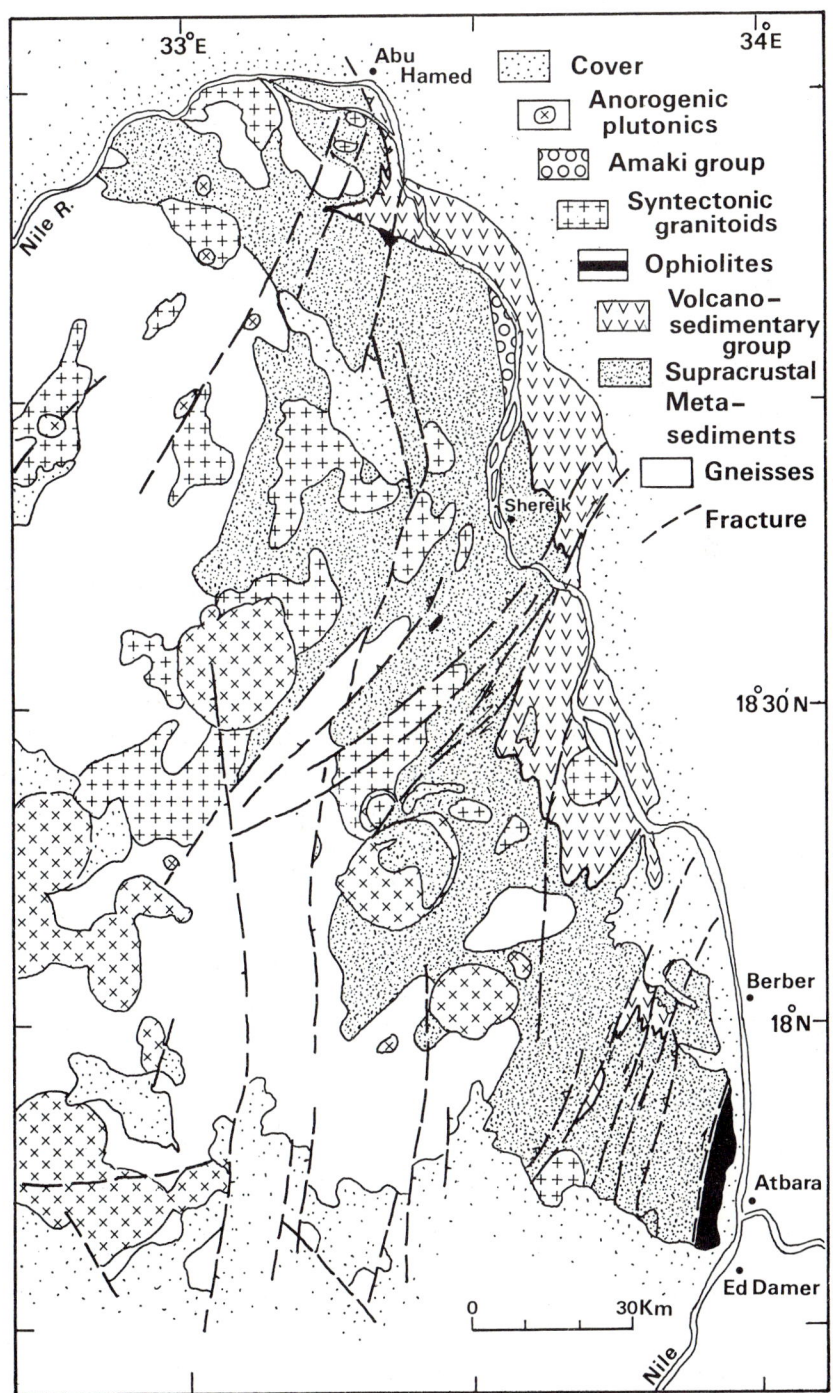

FIGURE 2.

Geological sketch map of the eastern Bayuda Desert, Sudan.

deformed basic rocks near Atbara. Geochemistry and regional distribution of the metavolcanic assemblage indicates they represent an island-arc, adjacent to Metasedimentary Group shallow water shelf deposits resting upon the continental foreland; subsequently collision-induced metamorphism and subduction-related calc-alkaline granitic emplacements occurred in a broad band along the contact zone.

The above interpretation does not support the evolutionary model proposed by Barth and Meinhold (1979) and Meinhold (1979, 1983), who provided a detailed geological map of the Bayuda Desert (scale 1 : 250 000, 1981), in which the various lithologies were grouped into somewhat different units. They believed the volcano-sedimentary greenschist assemblages to be retrograded Metasedimentary rocks (which they named the Kurmut and Absol Series of the Bayuda Formation). Most of the Grey Gneiss and parts of the Metasedimentary Group they combined as the Rahaba Series, of the Bayuda Formation, and in addition they recognized an old basment, the Abu Harik Series, in the Berber-Atbara region to be unconformably overlain by the younger sediments. The isotopic, geochemical and regional tectonic work reported earlier does not so far substantiate this view and further studies will be needed to resolve inconsistencies between the two models and to decide between the conflicting interpretations.

2. Ingessana-Kurmuk

The Blue Nile Province of Sudan and Wollega Province of western Ethiopia are adjacent in the vicinity of the Blue Nile (Abbai) river valley. In this region the contact between the volcano-sedimentary-ophiolite sequence and the metasediments-gneisses is exposed in a generally north-south band of country between Damazien and Kurmuk (Fig. 3).

Most of the geological work here has been concentrated on the Ingessana Hills complex because of the presence of exploitable

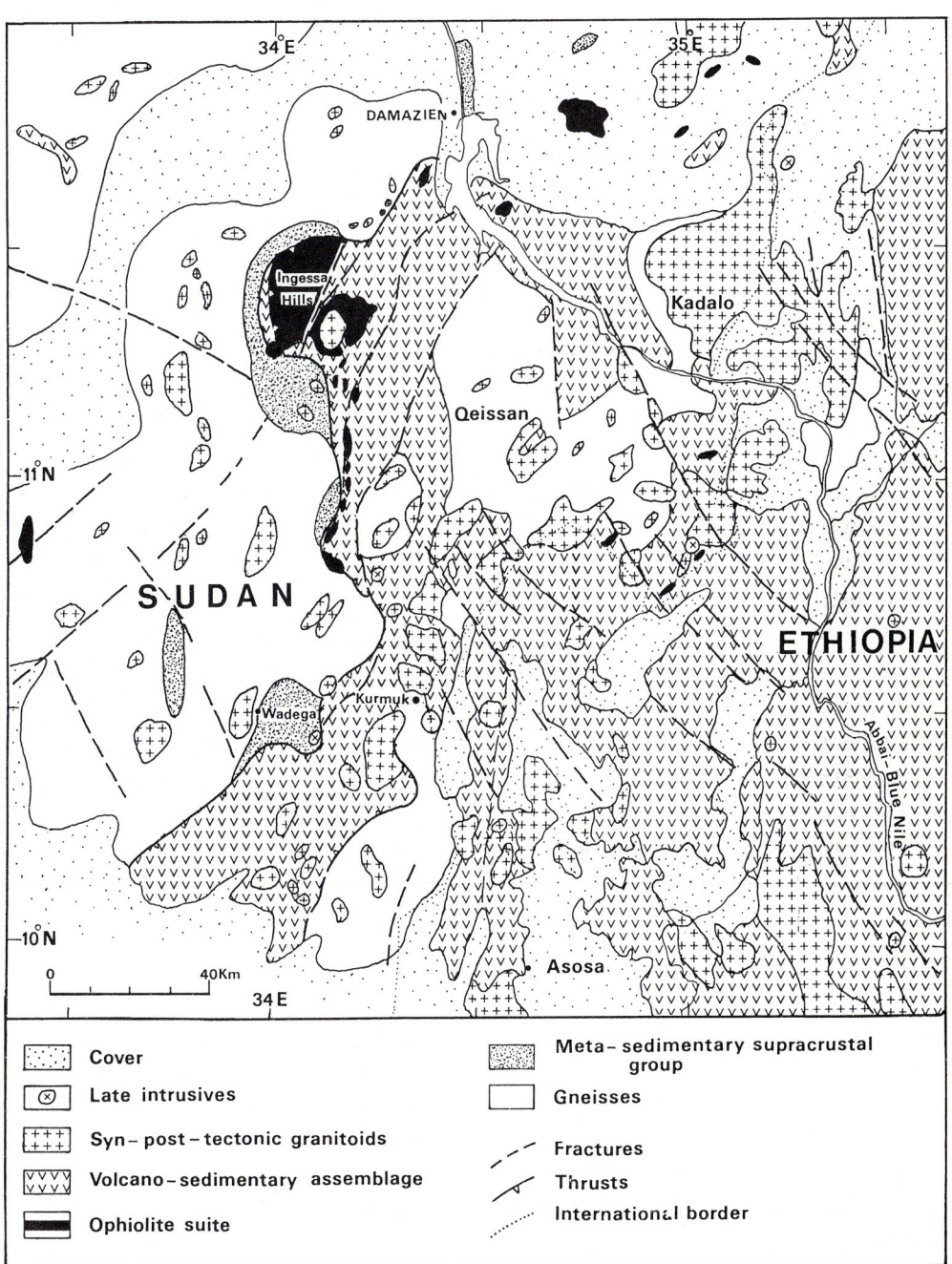

FIGURE 3.

Geological sketch map of the Ingessana-Kurmuk region of
Sudan and parts of Wollega, Ethiopia.

lenses of chromite (Kabesh, 1961; Hunting Geology and
Geophysics, 1969; Shaddad, 1974; Babiker, 1977; Karrar, 1980).
Recent work (Abdel Magid, 1983; Abdel Rahman, 1983; Toum, 1985)
has distinguished a number of units, comparable to those
occurring elsewhere along the contact, and recognized the mafic
masses to be remnant ophiolite suites within island-arc and
continent margin volcanic sequences, and the thrust faulted,
mylonitized, and tectonically disrupted nature of the belt.

In the west, migmatitic high grade grey orthogneisses and minor
amphibolites (the Selak Formation) are highly folded, and
intruded by synorogenic granitoids (Fig. 3). These continental
rocks are probably unconformably overlain by a supracrustal
metasedimentary and paragneissic succession, containing marble
bands, now in amphibolite to granulite facies, the Gonak
Formation of the Tin Group, and which are infolded with the
underlying gneisses in which small patches of sillimanite
schists are present. Above these two units, generally in thrust
contact is the Uffat Group comprising a metavolcanic unit
(Kurmuk Formation) of basic and intermediate island-arc type
lavas, interbanded with a sedimentary sequence (Marafa
Formation) of pyroclastics, volcanoclastics and greywackes,
silts and slates. Associated with these volcano-sedimentary
sequences, now in the greenschist facies of metamorphism, are
lenses of mafic and ultramafic rocks, the Ingessana Formation.
They include serpentinized pyroxenites and lherzolites,
gabbros, chromite lenses, rare sheeted dyke complexes and
plagiogranites. Disruption by thrusting is widespread and
tectonic mélanges have developed, particularly south of
Ingessana in the J. Danderu massif. Small plutons and
batholiths of diorites, granodiorites, and granites intrude the
Uffat Group rocks close to the gneiss contact. Further east
synorogenic calc-alkaline batholiths cover large areas
(Fig. 3). Post-tectonic granitic plutons, such as the Bau
Granite in the Ingessana complex, represent a late magmatic
phase, dated at a minimum of 550 Ma by K-Ar (Vail and Rex,
1971). Small anorogenic syenite ring complexes and plugs,
thought to be Mesozoic in age (Vail, 1985a) indicate within-

plate magmatism, prior to peneplanation and cover by Cretaceous and younger sedimentary and volcanic rocks.

Although geochronological data are insufficient, the geology of the area is comparable to other parts of the basement in NE Africa, and can be explained satisfactorily by the continent-oceanic collision model, with possible subduction zones marked by ophiolite remnants, although intense dislocation probably has displaced large slabs in regionally imbricate thrust stacks over the sediment-draped continental shelf. Outliers of the volcano-ophiolite assemblage rest upon gneisses to the west, and exotic blocks of gneiss-metasediment terrane, particularly in the Kadalo-Qeissan area north and south of the Blue Nile, may represent allochthonous basement, or displaced microcontinental fragments.

3. Sekerr-Karamoja-Kapoeta

The strip of basement rocks west of the Gregory Rift in the Pokot region of Kenya, and along the international frontier in the Karamoja region of Uganda extends into the Kapoeta-Nagishot district of the south-eastern Sudan (Fig. 4). The geology of this extensive area is here reinterpreted on the basis of the earlier work of the Kenyan Geological Survey (for references see Vearncombe, 1983), the Ugandan Geological Survey Karamoja Map (1966), and unpublished reconnaissance work in the Sudan by Belgium Technical Aid, and follows recent interpretations of the geology by Shackleton (1977, 1986), Vearncombe (1983a, 1983b) and Vail (1983).

Like the Bayuda and Ingessana areas, this region comprises three major lithological units, and also the eastern edge of the Archaean Tanganyika craton south of Mt. Elgon. In addition, the intensity of metamorphism and degree of thrust faulting is perhaps greater than in the north, thus making it more difficult to identify the major rock units.

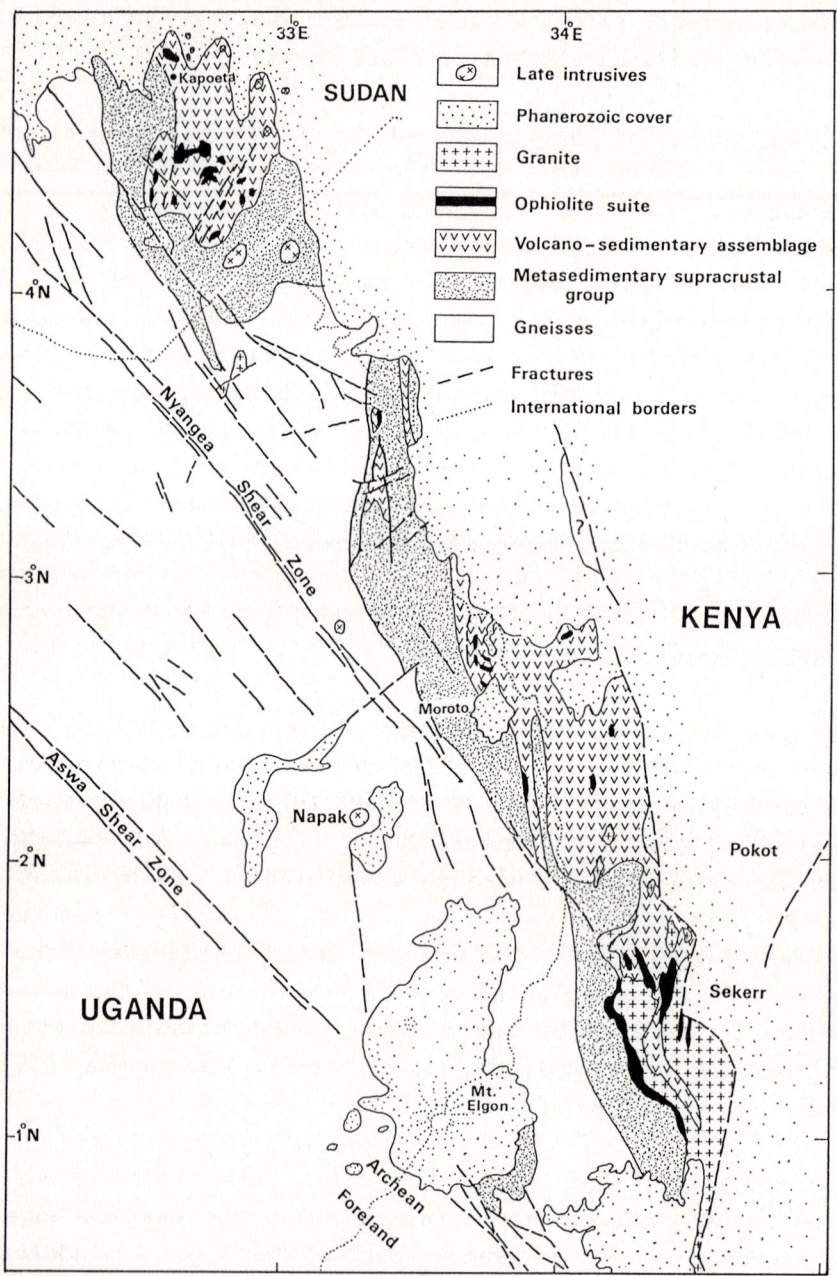

FIGURE 4.

Geological sketch map of the Kapoeta-Karamoja-Sekerr region of north-eastern Africa.

Most of the northern Uganda basement comprises gneisses and hornblende schists, up to granulite grade, of migmatitic character which although indicative in places of Archaean or Early Proterozoic origin, are now strongly overprinted with Pan-African ages (Almond, 1969; Leggo, 1974; Cahen et al., 1984). Overlying these rocks, in part in thrust contact, is an extensive metasedimentary unit made up of quartzites, quartzo-feldspathic psammites, mica and graphitic schists and crystalline limestones. In Uganda they are referred to as the Karasuk group, in western Kenya as the Cherangani metasediments. Sanders (1963) interpreted them as miogeosynclinal deposits along the margin of a stable continent to the west.

Resting upon the metasediments, in a complicated imbricate zone in the Sekerr area (Vearncombe, 1983a), and with some strong thrusting in Karamoja district, but with less well understood relationships at the northern end of this belt around Kapoeta, are a sequence of andesitic metavolcanic rocks, pillow lavas, gabbros with preserved layering, hornblende schists, serpentinites with podiform chromite, basic dykes, perhaps representing a sheeted dyke complex, marble lenses, and narrow bands of quartzites considered to have been original chert layers; psammites and mica schists were probably tuffs, pyroclastics and turbiditic sediments, the whole making up an island-arc, ophiolite succession. In SE Sudan gabbro and pyroxenite fragments are tightly infolded with marbles, hornblende schists and gneissic metasediments and may also represent dismembered island-arc and ophiolite suite rocks, although this remains speculative. Unlike the volcano-sedimentary-ophiolite terranes farther north synorogenic batholithic intrusive masses are rare. In West Pokot a flat granite, migmatitic in places, intrudes this ophiolite-island-arc assemblage; in composition it is granodioritic in part. In north Karamoja diorites occur; in SE Sudan post-tectonic granite plutons and anorogenic alkaline ring-complexes are also present. Late Tertiary to Recent volcanism and rift faulting affected much of the area.

Geochronological data are limited in this region. Late
pegmatites from Kenailmet indicate a U-Pb age of 597 Ma and
K-Ar age of 624 ± 40 Ma, while an earlier amphibolite facies
metamorphic event probably affected the metasedimentary unit
around 840 Ma (Cahen et al., 1984). Amphibolite from Sekerr
yielded Rb-Sr and model Sm-Nd ages of c. 1000 Ma, (Harris et
al., 1984). These authors also reported Sm-Nd model ages of up
to 1950 Ma for Marich Granite, which provided a Rb-Sr
regression calculation of 593 ± 49 Ma (MSWD 51), whereas
metasediments gave model Sm-Nd ages of 1150 Ma and Rb-Sr ages
of 584 Ma.

Vearncombe (1983a) considered the dismembered ophiolite at
Sekerr to mark a N-S suture zone, which possibly extends
northwards via Moroto through the Akabo region of SW Ethiopia
(Fig. 1) to the Kurmuk-Ingessana region of the eastern Sudan
(Vail, 1983; Vail in press). The edge of the Mozambique mobile
belt lies further west passing beneath Mt. Elgon.

TECTONIC STRUCTURE OF NE AFRICA AND ARABIA

The recognition that the basement of western Arabia and northeastern Africa comprises oceanic volcano-sedimentary sequences
in an island-arc setting with associated belts of ophiolite
complexes has led to a reexamination of how the different
regions relate to one another and whether a pattern of ocean
floor fragments can be established within the now
metamorphosed, intruded and disturbed rocks. The initial
suggestions (Bakor et al., 1976; Greenwood, et al., 1976; Gass,
1977; Shackleton, 1977) that a number of ophiolite belts
separating volcanic areas extended across both Arabia and
Africa stimulated a more specific review of the region by Vail
(1983), in which over twenty ophiolite complexes were
identified. At the same time detailed work on the Saudi Arabian
Shield led to the recognition that the Proterozoic tectonic
processes were comparable to Mesozoic and modern plate

interactions, in which interplate collisions along tectonic boundaries were the primary feature and that it was thus possible to recognize individual plates. An environment of microplate collisions would explain the juxtaposition of dissimilar blocks, whose margins could be considered to be ophiolite draped suture zones, or collision structures marked by transcurrent faults, mylonite belts, thrust sheets, or imbricate klippen. Significant contributions to this microplate concept and the naming of the various units were published by Camp (1984), Johnson et al. (1984), Stoeser and Camp (1985), Embleton et al. (1984) and Vail (1985b, 1985c, in press).

A pattern of tectonic terranes and orogenic sutures has now been established (Fig. 1). At least nine volcano-sedimentary plates are recognized, some of which have been subdivided further, and an additional half dozen detached blocks occur, some with contained ophiolites, as in the Nuba Mountains of central Sudan (Hirdes and Brinkmann, 1985), surrounded by continental gneisses and metasediments.

The main internal features of the oceanic plates are their rather similar geology, their disruption by syn- and post-orogenic granitoids which tend to be aligned in belts parallel to the ophiolite sutures, and the broad parallelism of their bounding interplate suture-shear margins. The latter trend NE-SW and intersect the continental-oceanic boundaries obliquely, whereas the continental margins trend generally N-S and are roughly parallel. Subdivision of the continental terranes has only recently been attempted (Schandelmeier et al., 1983; Huth et al., 1984; Bernau et al., 1987). In the far west, Archaean nuclei form the Tanganyika and Nile Cratons, and between these and the metavolcanic assemblages the extensive regions of gneisses occur within the oceanic terranes, and many of these differ from the gneisses on the cratons. The southern termination of the meta-volcano-sedimentary terranes is in two prongs, and it is generally felt that the rocks in the three separated basement areas although superficially very similar are in fact different, and thus should constitute distinctive gneissic terranes as well. This is a problem for future research.

The gneisses, whilst certainly paragneissic in part, are generally distinguishable from the supracrustal Metasedimentary groups. The latter appear to be shallow water, miogeosynclinal deposits resting upon the gneissic foreland and because they occur in a restricted band all along the nearly 2000 km of contact from southern Egypt to Uganda they are thought to be closely related with the evolution of the oceanic terranes, either as fill to early continental marginal rifting or as shallow shelf deposits, and thus represent a sedimentary prism along the African plate margin (Vail, 1983). Their relationship with other paragneissic and metasedimentary units between Jebel Uweinat and the Nile Valley near Halfa (Bernau et al., 1987) or in western Kordofan and west of the Nuba Mountains (Vail, 1978) (see Fig. 1) remains to be determined.

CONCLUSIONS

The Arabian-Nubian Shield, representing an imprecisely defined extent of Precambrian basement outcrops beneath undisturbed Phanerozoic deposits, is now known to comprise two major lithological-tectonic units. These are an oceanic volcano-sedimentary-ophiolite-granitoid calc-alkaline assemblage and a continental gneiss-metasedimentary group. The contact between the two is abrupt and usually marked by ophiolite-bearing sutures or thrust faulting. It would be useful therefore to distinguish the oceanic terrane and continental terrane provinces and to define the limits of the Arabian-Nubian Shield. The continental areas are of two recognizable types; farthest from the Shield are ancient gneisses and schists of the Archaean cratons, and these stable areas are separated from the oceanic terranes by gneiss belts. In Kenya these gneisses comprise the Mozambique Belt, and since they continue without recognizable difference northwards into Somalia, and possibly Yemen and Saudi Arabia, and through northern Uganda into the western and central Sudan, understandably it has been inferred that the Mozambique Belt extends through these territories as well. However, there are indications from these northern areas that the gneisses are remobilized older crust, perhaps Early Proterozoic or even Archaean, whereas in Mozambique, Tanzania and Kenya no gneisses older than about 1100 Ma have been found, other than in the Lurio belt (1311 Ma) near Nampula and along the Archaean cratonic borders of the Mozambique Belt (Cahen et al., 1984). There is therefore a problem to determine the real nature of the latter belt in the type area, and hence its tectonic evolution and thus the reason for the widespread Pan-African tectono-thermal event affecting the whole of eastern Africa. It could be that the clue to the solution lies in the presence of the oceanic material so extensively preserved in the Arabian-Nubian Shield. The evolution of that area envisages collision tectonics; in the south, continent-to-continent contact may now be the position (Shackleton, 1986), with one of the continental plates overriding the other, and perhaps generating sufficient orogenesis to account for the Pan-African event.

The establishment and identification of tectonic terranes (Fig. 1) is not only of scientific interest in the evolution of Africa, for it can be shown that the various micro-plates have associated characteristic economic mineral deposits (Pohl, 1984; Vail 1985b; Vail in press) and a proper understanding of the tectonic environment can aid the exploration and exploitation of the mineral wealth of the region.

ACKNOWLEDGEMENTS

Jackie Duggua and John Davidson are thanked for secretarial and cartographic assistance in the preparation of this paper. Heinz Schandelmeier is thanked for additional information.

REFERENCES

Abdel Magid, A.E.M., 1983. The geology of the metamorphic terrain south-west of the Ingessana mafic-ultramafic complex, E. Sudan. M. Phil. thesis, Porthsmouth Polytechnic, UK (unpubl.).

Abdel-Monem, A.A. and P.M. Hurley 1979. U-Pb dating of zircons from psammitic gneisses, Wadi Abu Rosheid-Wadi Sikait area Egypt. In: S.A. Tahoun (editor) Evolution and Mineralization of the Arabian-Nubian Shield. Vol. 2, 165-170. Pergamon Press, Oxford.

Abdel Rahman, E.M., 1983. The geology of mafic-ultramafic masses and adjacent rocks south of the Ingessana igneous complex, Blue Nile Province, E. Sudan. M. Phil. thesis, Portsmouth Polytechnic, UK (unpubl.).

Almond, D.C., 1969. Structure and metamorphism of the Basement Complex of north-east Uganda. Overseas Geol. Miner. Resour., London, 10, 146-163.

Almond, D.C., 1980. Precambrian events at Sabaloka, near Khartoum, and their significance in the chronology of the Basement complex of north-east Africa. Precambrian Res., 13, 43-62.

Almond, D.C., 1984. The concept of "Pan-African Episode" and "Mozambique Belt" in relation to the geology of East and North East Africa. Bull. Fac. Earth Sci., King Abdulaziz Univ. Jeddah, 6, 71-88.

Al-Shanti, A.M.S., 1979. Evolution and Mineralization of the Arabian-Nubian Shield. Pergamon Press, Oxford. Bull. Inst. Applied Geol., King Abdulaziz Univ., Jeddah, 3.

Al-Shanti, A.M.S. and I.G. Gass, 1983. The Upper Proterozoic ophiolite melange zones of the easternmost Arabian Shield. J. geol. Soc. London, 140, 867-876.

Babiker, I.M., 1977. Aspects of the ore geology of Sudan - with particular reference to the chromitiferous ultrabasic rocks of Jebal El Ingessana, eastern Sudan. Ph. D. thesis, Cardiff Univ., UK (unpubl.).

Bakor, A.R., I.G. Gass and C.R. Neary, 1976. Jabal al Wask, N.W. Saudi Arabia: an Eocambrian back-arc ophiolite. Earth Planet. Sci. Lett., 30, 1-9.

Barth, H. and K.-D. Meinhold, 1979. Mineral prospecting in the Bayuda Desert. Bundesanstalt Geowiss. Rohstoffe, Hannover. Part 1, 2 vols. (unpubl. report).

Barth, H. and K.-D. Meinhold, 1981: Geological map of the Bayuda Desert, Sudan. Scale 1 : 250 000. Bundesanstalt Geowiss. Rohstoffe, Hannover.

Bernau, R., D.P.F. Darbyshire, G. Franz, U. Harms, A. Huth, N. Mansour, P. Pasteels and H. Schandelmeier, 1987. Petrology, geochemistry and structural development of the Bir Safsaf-Aswan uplift, southern Egypt. J. African Earth Sci. 6, 79-90.

Cahen, L., N.J. Snelling, J. Delhal and J.R. Vail, 1984. The geochronology and evolution of Africa. Clarendon Press, Oxford.

Camp, V.E., 1984. Island arcs and their role in the evolution of the western Arabian Shield. Geol. Soc. Am. Bull. 95, 913-921.

Civetta, L., D. de Vivo, G. Giunta, F. Ippolito, A. Lima, G. Orsi, V. Perrone and A. Zuppetta, 1980. Geological and structural outlines of the southern Sudan. Atti Conv. Lincei No. 47, Geo-dynamic evolution of the Afro-Arabian Rift System, Rome 1979, 175-183.

Curtis, P.A.S. and P. Griffiths, 1985. Geological mapping and mineral exploration in northern Sudan using satellite remote sensing. 13th Colloq. African Geol., St. Andrews. Abstracts, 258-259.

D'Amico, C., H.A. Ibrahim and F.P. Sassi, 1981. Outline of the Somalian basement. Geol. Rdsch., 70, 882-896.

Daniels, J.L., 1965. A photogeological interpretation of the Bur region, Somali Republic. Overseas Geol. Miner. Resour., London, 9, 427-436.

Davidson, A., J.M. Moore and J.C. Davies, 1973. Preliminary report on the geology and geochemistry of parts of Sidamo, Gemu Gofa, and Kefa provinces, Ethiopia. Rept. Omo River Project 1. Ministry of Mines, Addis Ababa.

Davidson, A., J.M. Moore, and J.C. Davies, 1976. Preliminary report on the geology and geochemistry of parts of Gemu Gofa, Kefa and Ilubabor provinces. Rept. S.W. Ethiopia Omo River Project 2. Ministry of Mines, Addis Ababa.

Dawoud, A.S., 1980. Structural and metamorphic evolution of the area south-west of Abu Hamed, Nile Province, Sudan. Ph. D. Thesis, Univ. Khartoum (unpubl.).

Dixon, T.H., 1981. Age and chemical characteristics of some pre Pan-African rocks in the Egyptian Shield. Precambrian Res., 14, 119-133.

Dolginov, E.A. and M.A. Said, 1976. Precambrian structures of the Northern Province of the Somali Democratic Republic. In: H. Tsegaye (editor) African Geology, Addis Ababa 1973, p. 77, Geol. Soc. Africa.

El Ageed, A.I. and S.M. El Rabaa, 1981. The geology and structural evolution of the northeastern Nuba Mountains, southern Kordofan Province, Sudan. Bull. Geol. Miner. Resour. Dept. Sudan, 32, 1-50.

El Manharawy, M.S., 1977. Geochronological investigations of some Basement rocks in the central Eastern Desert, Egypt between Lat. 25° and 26° North. Ph. D. thesis, Univ. Cairo (unpubl.).

El Rabaa, S.M., 1975. Geological setting and structural control of Rubatab pegmatites, Northern Province, Sudan. Rec. Res. Univ. Khartoum, 1 (1972), 4-6.

El Rabaa, S.M., 1976. Structural and metamorphic evolution of west Berber District, Sudan with special reference to the structural control of mica-bearing pegmatites. In: H. Tsegaye (editor) Proceedings 2nd Conference on African Geology, Addis Ababa 1973, pp. 81-96. Geol. Soc. Africa.

El-Ramly, M.F., 1972. A new geological map for the basement rocks in the Eastern and south-western deserts of Egypt. Scale 1 : 1 000 000. Ann. Geol. Surv. Egypt 2, 1-18.

El-Ramly, M.F., R. Greiling, A. Kröner and A.A.A. Rshwan, 1984. On the tectonic evolution of the Wadi Hafafit area and environs, Eastern Desert of Egypt. Bull. Fac. Earth Sci., King Abdulaziz Univ. Jeddah, 6, 113-126.

El Samani, Y., 1985. Sediments hosting massive sulphide deposits from the Proterozoic of the Red Sea Hills, Sudan. 13th Colloq. African Geol., St. Andrews. Abstracts, 156-157.

El Sharkawy, M.A. and R.m. El Bayoumi, 1979. The ophiolites of Wadi Ghadir area, Eastern Desert, Egypt. Ann. Geol. Surv. Egypt, 9, 125-135.

El Shazly, E.M., A.H. Hashad, T.A. Sayyah and E.A. Bassyuni, 1973. Geochronology of Abu Swayel area, south Eastern Desert. Egyptian J. Geol., 17, 1-18.

Embleton, J.C.B., D.J. Hughes, P.M. Klemenic, S. Poole and J.R. Vail, 1984. A new approach to the stratigraphy and tectonic evolution of the Red Sea Hills, Sudan. Bull. Fac. Earth Sci., King Abdulaziz Univ. Jeddah, 6, 101-112.

Frisch, W. and A. Al Shanti, 1977. Ophiolite belts and the collision of island arcs in the Arabian Shield. Tectonophysics, 43, 293-306.

Gass, I.G., 1977. The evolution of the Pan African crystalline basement in N.E. Africa and Arabia. J. geol. Soc. London, 134, 129-138.

Greenwood, W.A., D.G. Hadley, R.E. Anderson, R.J. Fleck and D.L. Schmidt, 1976. Late Proterozoic cratonization in southwestern Saudi Arabia. Philos. Trans. R. Soc. London, A280, 517-527.

Greiling, R., A. Kröner and M.F. El Ramly, 1984. Structural interference patterns and their origin in the Pan-African basement of the south-eastern Desert of Egypt. In: A. Kröner and R. Greiling (editors) Precambrian tectonics illustrated. Schweizerbart, Stuttgart, 401-412.

Griffiths, P., 1985. The role of remote sensing geology in mineral exploration in northern Sudan. EEC Workshop: Remote Sensing in Mineral Exploration, Brussels 1985.

Harris, N.B.W., C.J. Hawkesworth and A.C. Ries, 1984. Crustal evolution in north-east and east Africa from model Nd ages. Nature, London, 309, 773-776.

Hashad, A.H., T.A. Sayyah, S.B. El-Kholy and A. Youssef, 1972. Rb/Sr isotopic age determinations of some basement Egyptian granites. Egyptian J. Geol., 16, 269-281.

Hepworth, J.V., 1979. Does the Mozambique orogenic belt continue into Saudi Arabia? In: S.A. Tahoun (editor) Evolution and Mineralization of the Arabian-Nubian Shield. Vol. 1, pp. 39-51, Pergamon Press, Oxford.

Hirdes, W. and K. Brinkmann, 1985. The Kabus and Balula serpentinite and metagabbro complexes - a dismembered Proterozoic ophiolite in the north-eastern Nuba Mountains, Sudan. Geol. Jahrbuch, B58, 3-43.

Holmes, A. 1951. The sequence of pre-Cambrian orogenic belts in south and central Africa. 18th Inter. Geol. Cong. London (1948), 14, 254-269.

Hunting Geology and Geophysics Ltd., 1969. Photogeological survey of the Eastern Area: mineral survey in three selected areas in the Republic of Sudan. Report to United Nations Development Programme, Khartoum (unpubl.).

Hunting Geology and Geophysics Ltd., 1979. Geological Map, mineral exploration of the Juba area, south Sudan. Scale 1 : 250 000. Southern Regional Government, Democratic Republic of Sudan (unpubl.).

Hussein, I.M., A. Kröner and St. Durr, 1984. Wadi Onib - a dismembered Pan-African ophiolite in the Red Sea Hills of Sudan. Bull. Fac. Earth Sci., King Abdulaziz Univ. Jeddah, 6, 319-327.

Huth, A., G. Franz and H. Schandelmeier, 1984. Magmatic and metamorphic rocks of NW Sudan: a reconnaissance survey. Berliner Geowiss. Abh., A50, 7-21.

Jackson, N.J. and C.R. Ramsay, 1980. Time-space relationship of Upper Precambrian volcanic and sedimentary units in the central Arabian Shield. J. geol. Soc. London, 137, 617-628.

Johnson, P., 1983. A preliminary lithofacies map of the Saudi Arabian Shield. Scale 1 : 1 000 000. Deputy Ministry Miner. Resour. Jeddah. Tech. Rec. RF-TR-03-2.

Johnson, P., E. Scheibner and E.A. Smith, 1984. Mineral ccurrence base map of the Saudi Arabian Shield. Scale 1 : 1 000 000.

Kabesh, M.L., 1961. The geology and economic minerals and rocks of the Ingessana Hills. Bull. Geol. Surv. Sudan, 11, 1-61.

Karrar, Y.H., 1980. The geology of the chromite deposits of the ultramafic rocks of the Ingessana Hills (Sudan). PhD thesis, Patrice Lumumba People's Friendship Univ., Moscow (unpubl.).

Kazmin, V., 1973. Geological map of Ethiopia. Scale 1 : 2 000 000. Geol. Surv. Ethiopia, Addis Ababa.

Kazmin, V., A. Shifferaw and T. Balcha, 1978. The Ethiopian basement: stratigraphy and possible manner of evolution. Geol. Rdsch., 67, (2), 531-546.

Kennedy, W.Q., 1964. The structural differentiation of Africa in the Pan-African (± 500 million years) tectonic episode. Eighth Annu. rep. Res. Inst. African Geol., Univ. Leeds, 48-49.

Kennedy, W.Q., 1965. The influence of basement structure on the evolution of the coastal (Mesozoic and Tertiary) basins of Africa. In: Salt basins around Africa. Institute of Petroleum, London. pp. 7-16.

Klerkx, J., 1980. Age and metamorphic evolution of the Basement Complex around Jabal al 'Awaynat. In: M.J. Salem and M.T. Busrewil (editors) The geology of Libya. Academic Press, Oxford. pp. 901-906.

Klerkx, J. and S. Deutsch, 1977. Resultats préliminaires obtenus par la méthode Rb/Sr sur l'âge des formations Précambrienes de la région d'Uweinat (Libye). Rapp. annu. (1976) Mus. R. afr. Centr., Tervuren (Belg.), Dépt. Géol. Min., 83-94.

Leggo, P.J., 1974. A geochronological study of the basement complex of Uganda. J. geol. Soc. London, 130, 263-277.

Marzouki, F.M.H., N.J. Jackson, C.R. Ramsay and D.P.F. Darbyshire, 1982. Composition, age and origin of two Proterozoic diorite-tonalite complexes in the Arabian Shield. Precambrian Res., 19, 31-50.

McCall, G.J.H., 1964. Geology of the Sekerr area. Rep. Geol. Surv. Kenya, 65.

Meinhold, K.-D., 1979. The Precambrian Basement Complex of the Bayuda Desert, northern Sudan. Géogr. phys. Géol. dyn., 21, 395-401.

Meinhold, K.D., 1983. Summary of the regional and economic geology of the Bayuda Desert, Sudan. Bull. Geol. Miner. Resour. Dept. Sudan, 33, 1-45.

Miller, J.M., 1956. The geology of the Kitale-Cherangani hills area. Rep. Geol. Surv. Kenya, 35.

Nassief, M.O., R. Macdonald and I.G. Gass, 1984. The Jebel Thurwah Upper Proterozoic ophiolite complex, western Saudi Arabia. J. geol. Soc. London, 141, 537-546.

Pohl, W., 1984. Large-scale metallogenetic features of the Pan-African in East Africa, Nubia and Arabia. Bull. Fac. Earth Sci., King Abdulaziz Univ. Jeddah, 6, 591-602.

Ries, A.C., A.S. Dawoud and R.M. Shackleton, 1985. Geochronology, geochemistry and tectonics of the NE Bayuda Desert, N. Sudan: implications for the western margin of the late Proterozoic fold belt in NE Africa. Precambrian Res., 30, 43-62.

Ries, A.C., R.M. Shackleton, R.H. Graham and W.R. Fitches, 1983. Pan-African structures, ophiolites and mélanges in the Eastern Desert of Egypt: a traverse at 26°N. J. geol. Soc. London, 140, 75-95.

Roobol, J., C.R. Ramsay, N.J. Jackson and D.P.F. Darbyshire, 1983. Late Proterozoic lavas of the Central Arabian Shield - evolution of an ancient volcanic arc system. J. geol. Soc. London, 140, 185-202.

Ruxton, B.P., 1956. Major rock groups of the northern Red Sea Hills, Sudan. Geol. Mag., 93, 314-330.

Sadiq, A.A. and J.R. Vail, 1986. Geology and regional gravity traverses of the Nuba Mountains, Kordofan Province, Sudan. J. African Earth Sci., 5, (4), 329-338.

Sanders, L.D., 1963. Geology of the Eldoret area. Rept. Geol. Surv. Kenya, 64.

Schandelmeier, H., D.P.F. Darbyshire, U. Harms and A. Richter, 1987. The East Sahara Craton: evidence for pre-Pan-African crust in NE Africa west of the Nile. In: S. El Gaby and R.O. Greiling (eds.) The Pan-African Belt of NE Africa and Adjacent Areas. Vieweg.

Schandelmeier, H., A. Richter, and G. Franz, 1983. Outline of the geology of magmatic and metamorphic units between Gebel Uweinat and Bir Safsaf (SW Egypt/NW Sudan). J. African Earth Sci., 1, 275-283.

Shackleton, R.M., 1977. Possible late-Precambrian ophiolites in Africa and Brazil. 20th Annu. rep. Res. Inst. African Geol., Univ. Leeds, 3-7.

Shackleton, R.M., 1986. Precambrian collision tectonics in Africa. In: M.P. Coward and A.C. Ries (editors) Collision tectonics. Geol. Soc. London Special publ. 19, pp. 329-349.

Shaddad, M.Z., 1974. Geology and chromite mineralization of the Ingessana Hills area, Democratic Republic of Sudan. Diss. Kandidat Geol. Miner. Sci., Friendship Univ., Moscow (unpubl.).

Stacey, J.S. and C.E. Hedge, 1984. Geochronologic and isotopic evidence for early Proterozoic continental crust in the eastern Arabian Shield. Geology, 12, 310-313.

Stacey, J.S. and D.B. Stoeser, 1984. Distribution of oceanic and continental leads in the Arabian-Nubian Shield. Contrib. Miner. Petrol., 84, 91-105.

Stern, R.J. and C.E. Hedge, 1985. Geochronologic and isotopic constraints of late Precambrian crustal evolution in the Eastern Desert of Egypt. Am. J. Sci., 285, 97-127.

Stoeser, D.B., 1986. Distribution and tectonic setting of plutonic rocks of the Arabian Shield. J. African Earth Sci., 4, 21-46.

Stoeser, D.B. and V.E. Camp, 1985. Pan-African microplate accretion of the Arabian Shield. Bull. geol. Soc. Am., 96, 817-826.

Stoeser, D.B. and J.E. Elliot, 1980. Post-orogenic peralkaline and calc-alkaline granites and associated mineralization of the Arabian Shield, Kingdom of Saudi Arabia. In: S.A. Tahoun (editor) Evolution and Mineralization of the Arabian-Nubian Shield. Vol. 4, pp. 1-23. Pergamon Press, Oxford.

Stoeser, D.B., J.S. Stacey, W.R. Greenwood and L.B. Fischer, 1984. U/Pb zircon geochronology of the southern portion of the Nabitah mobile belt and Pan-African continental collision in the Saudi Arabian Shield. Saudi Arabia Deputy Ministry Miner. Resour. Tech. Rec., USGS-TR-04-05.

Sturchio, N., M. Sultan, P. Sylvester, R. Batiza, C. Hedge, E.M. El Shazly and A. Abdel-Maguid, 1984. Geology, age, and origin of the Meatiq Dome: implications for the Precambrian stratigraphy and tectonic evolution of the Eastern Desert of Egypt. Bull. Fac. Earth Sci., King Abdulaziz Univ. Jeddah, 6, 127-143.

Toum, I.M., 1985. The geology of the Qeissan area, Blue Nile Province, S.E. Sudan. M. Phil. thesis, Portsmouth Polytechnic, UK (unpubl.).

Vail, J.R., 1971. Geological reconnaissance in the Berber District, Northern Province, Sudan. Bull. Geol. Surv. Sudan, 18, 1-76.

Vail, J.R., 1978. Outline of the geology and mineral deposits of the Democratic Republic of the Sudan and adjacent areas. Overseas Geol. Miner. Resour., London, 49.

Vail, J.R., 1979. Outline of geology and mineralization of the Nubian Shield east of the Nile valley, Sudan. In: S.A. Tahoun (editor) Evolution and Mineralization of the Arabian-Nubian Shield, Vol. 1, pp. 97-107. Pergamon Press, Oxford.

Vail, J.R., 1983. Pan-African crustal accretion in north-east Africa. J. African Earth Sci., 1, 285-294.

Vail, J.R., 1985a. Alkaline ring complexes in Sudan. J. African Earth Sci., 3, (1), 51-59.

Vail, J.R., 1985b. Relationship between tectonic terrains and favourable metallogenic domains in the central Arabian-Nubian Shield. Trans. Inst. Min. Metall., Appl. Earth Sci., B94, 1-5.

Vail, J.R., 1985c. Pan-African (late Precambrian) tectonic terrains and the reconstruction of the Arabian-Nubian Shield. Geology, 13, 839-842.

Vail, J.R., in press. Late Proterozoic tectonic terranes in the Arabian-Nubian Shield and their characteristic mineralization. In: P. Bowden and J. Kinnaird (editors). African geology reviews. John Wiley, London.

Vail, J.R., D.C. Almond, D.J. Hughes, P.M. Klemenic, S. Poole, S.E.M. Nour and J.C.B. Embleton, 1984. Geology of the Wadi Oko-Khor Hayet area, Red Sea Hills, Sudan. Bull. Geol. Miner. Resour. Dept. Sudan, 34, 1-20.

Vail, J.R., A.S. Dawoud and F. Ahmed, 1973. Geology of the Third Cataract, Halfa District, Northern Province, Sudan. Bull. Geol. Miner. Resour. Dept. Sudan, 22, 1-34.

Vail, J.R. and J.L. Kuron, 1978. High level igneous emplacements in the Red Sea Hills, Sudan. Geol. Rdsch., 67, 521-530.

Vail, J.R. and D.C. Rex, 1971. Potassium-argon age measurements on pre-Nubian basement complex rocks from Sudan. Proc. geol. Soc. London, 1664, 205-214.

Vearncombe, J.R., 1983a. A dismembered ophiolite from the Mozambique Belt, West Pokot, Kenya. J. African Earth Sci., 1, 133-143.

Vearncombe, J.R., 1983b. A proposed continental margin in the Precambrian of western Kenya. Geol. Rdsch., 72, 663-670.

Warden, A.J. and J.L. Daniels, 1984. Evolution of the Precambrian of northern Somalia. Bull. Fac. Earth Sci., King Abdulaziz Univ. Jeddah, 6, 145-164.

Warden, A.J. and A.D. Horkel, 1984. The geological evolution of the NE-branch of the Mozambique Belt (Kenya, Somalia, Ethiopia). Mitt. Österr. geol. Ges. Vienna, 77, 161-184.

Chapter 9

Evolution, U-Pb Geochronology, and Isotope Geology of the Pan-African Nabitah Orogenic Belt of the Saudi Arabian Shield

Douglas B. Stoeser[1] / John S. Stacey[2]
[1] U.S. Geological Survey, MS 905, DFC, Box 25046, Denver, Colo., USA 80225
[2] U.S. Geological Survey, MS 937, 345 Middlefield Rd., Menlo Park, Calif., USA, 94025.

Keywords: Proterozoic, Pan-African, Arabia, U-Pb geochronology, Nabitah Belt, isotopic provinces, terranes, tectonic evolution

Abstract: U-Pb zircon age data were used to define a major episode of collisional orogenesis and tectonism within the eastern half of the Arabian Shield between 680 and 600 Ma. This orogenesis included two major orogenies, the Nabitah in the central Shield and Al Amar in the east. The Nabitah orogeny resulted from the collision of an ensimatic accreted island arc terrane in the west with the Afif composite terrane in the east at about 690 - 683 Ma to form the 1200 kilometer-long north-striking Nabitah suture. Deformation and crustal remobilization along the Nabitah suture, during the period 680 to 640 Ma, resulted in formation of the 100 to 200 kilometer-wide Nabitah orogenic belt. This belt is characterized by widespread early catazonal (680 - 650 Ma) synorogenic gneiss domes and complexes of leucocratic granodiorite, trondhjemite and monzogranite that are intruded by a bimodal mesozonal suite of late synorogenic (650 - 640 Ma) diorite, tonalite, and monzogranite. The early gneiss complexes are partly migmatitic and have metamorphic aureoles in the amphibolite to granulite facies. More than

10 kilometers of uplift occurred along the Nabitah orogenic belt and erosion of these highlands resulted in the deposition of the kilometer-thick Murdama group clastic sedimentary rocks. During the Al Amar orogeny (670 - 630 Ma), the Afif composite terrane collided with the Ar Rayn terrane to form the Al Amar suture zone. After orogenesis waned (640 - 600 Ma), the northwest-striking Najd left-lateral transcurrent fault system and postorogenic silicic magmatism developed throughout much of the Shield. The Al Amar orogeny and the Najd fault system are interpreted to be the result of collision and plate interaction between the Arabian Shield and a concealed continental plate to the east.

Feldspar Pb and Rb-Sr whole-rock isotopic data for the southern and central Saudi Arabian Shield reveal two isotopic provinces separated from each other by the Nabitah suture. Rocks from accreted island arc terranes west of the suture (950 - 720 Ma) contain ensimatic lead. Magmatic arc rocks from the Afif composite terrane (720 - 690 Ma) and Ar Rayn terrane >670 Ma) east of the Nabitah suture contain leads that are isotopically somewhat more evolved than those to the west. In addition, the south-central part of the Afif composite terrane is underlain by the early Proterozoic Khida basement, which contains evolved continental lead. The more evolved lead of the eastern arc terranes may be the result of mixing western arc ensimatic lead with evolved lead from subducted sediments containing continental material, or it may reflect differences in the lead isotope composition of the asthenosphere beneath the two isotopic provinces. Initial $^{87}Sr/^{86}Sr$ ratios for arc rocks from the eastern Shield (>0.7029) are slightly elevated relative to those of the west (<0.7027); this difference in initial Sr ratios is consistent with either model.

INTRODUCTION

In this paper, we review the evolution of the Nabitah orogenic belt of the central Arabian Shield (Figs. 1, 2) and relate it to a broad period of Pan-African age orogenesis that affected much of the Shield. In addition, we present U-Pb zircon and feldspar lead isotopic data for plutonic rocks within the Nabitah orogenic belt. Some of these data were presented in an internal report for the Saudi Arabian Deputy Ministry for Mineral Resources (Stoeser et al., 1984b) and some are presented here for the first time. We follow the petrographic nomenclature for plutonic rocks established by the International Union of Geological Sciences (Streckeisen, 1976).

GEOLOGIC SETTING

Modern concepts regarding evolution of the Proterozoic Arabian Shield were first presented in 1972 when Jackaman observed that volcanic rocks throughout the southern Shield were similar to those formed in modern island arcs. In addition, Brown (1972) and Brown and Coleman (1972) proposed that several belts of ultramafic rocks within the Shield might represent suture zones. The first tectonic syntheses of the Shield were by Brown (1972) and Schmidt et al. (1973), who recognized a widespread suite of synorogenic gneissic, migmatitic, and plutonic rocks and an episode of orogenesis at about 660 Ma.

In 1976 and 1977, key papers published by Al-Shanti and Mitchell (1976), Bakor et al. (1976), Greenwood et al. (1976), and Frisch and Al-Shanti (1977) presented the first plate tectonic interpretations for the Arabian Shield. In particular, Frisch and Al Shanti (1977) proposed that the Arabian Shield formed through a process of arc accretion similar to that presently occurring in the southwest Pacific. Since 1977, a considerable literature supporting and proposing island-arc models for the Shield has developed (e.g. Delfour, 1981; Gass, 1981; Duyverman, 1984; Camp, 1984; Jackson et al., 1984;

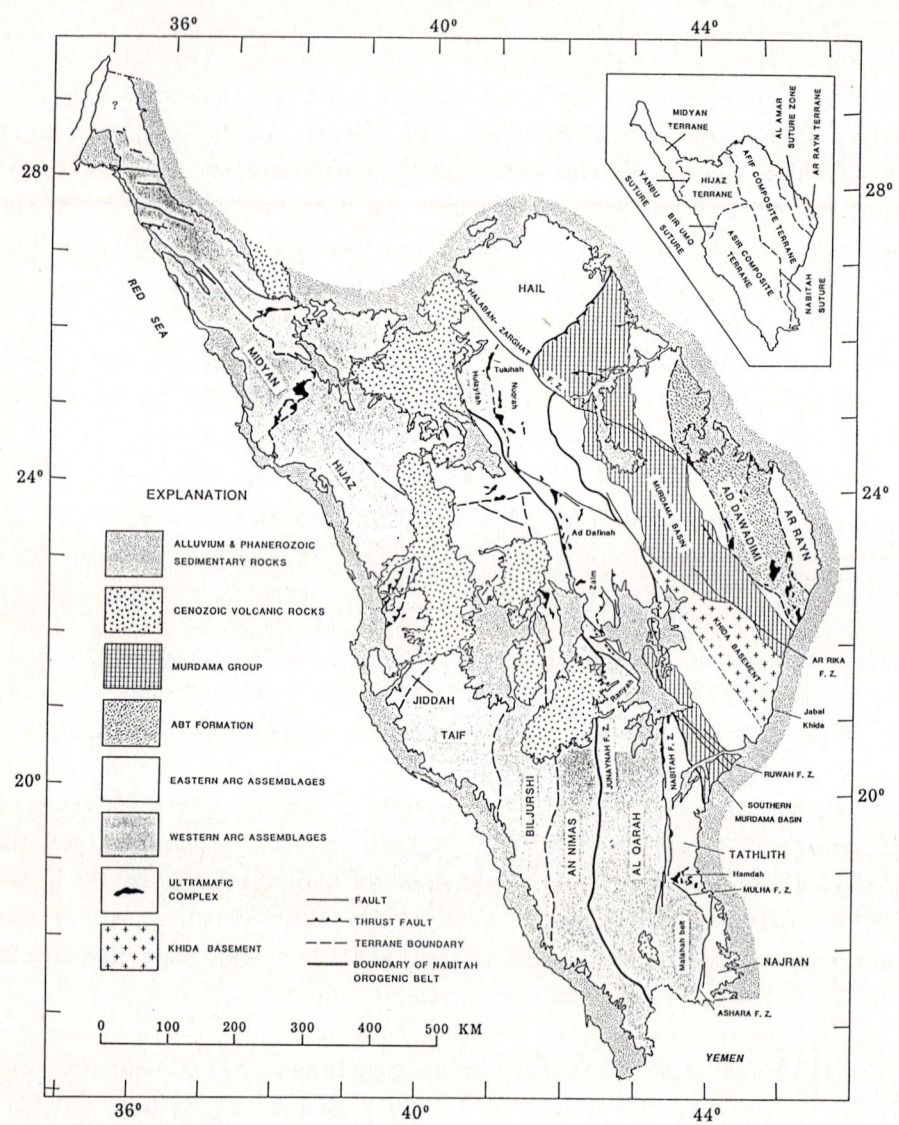

FIGURE 1.

Tectonic map of the Saudi Arabian Shield showing the location of terranes (large upper-case lettering), geologic features (small upper-case lettering), and geographic locations discussed in text (lower-case lettering). Inset map at the upper right shows sutures and major terranes. (F.Z., fault zone).

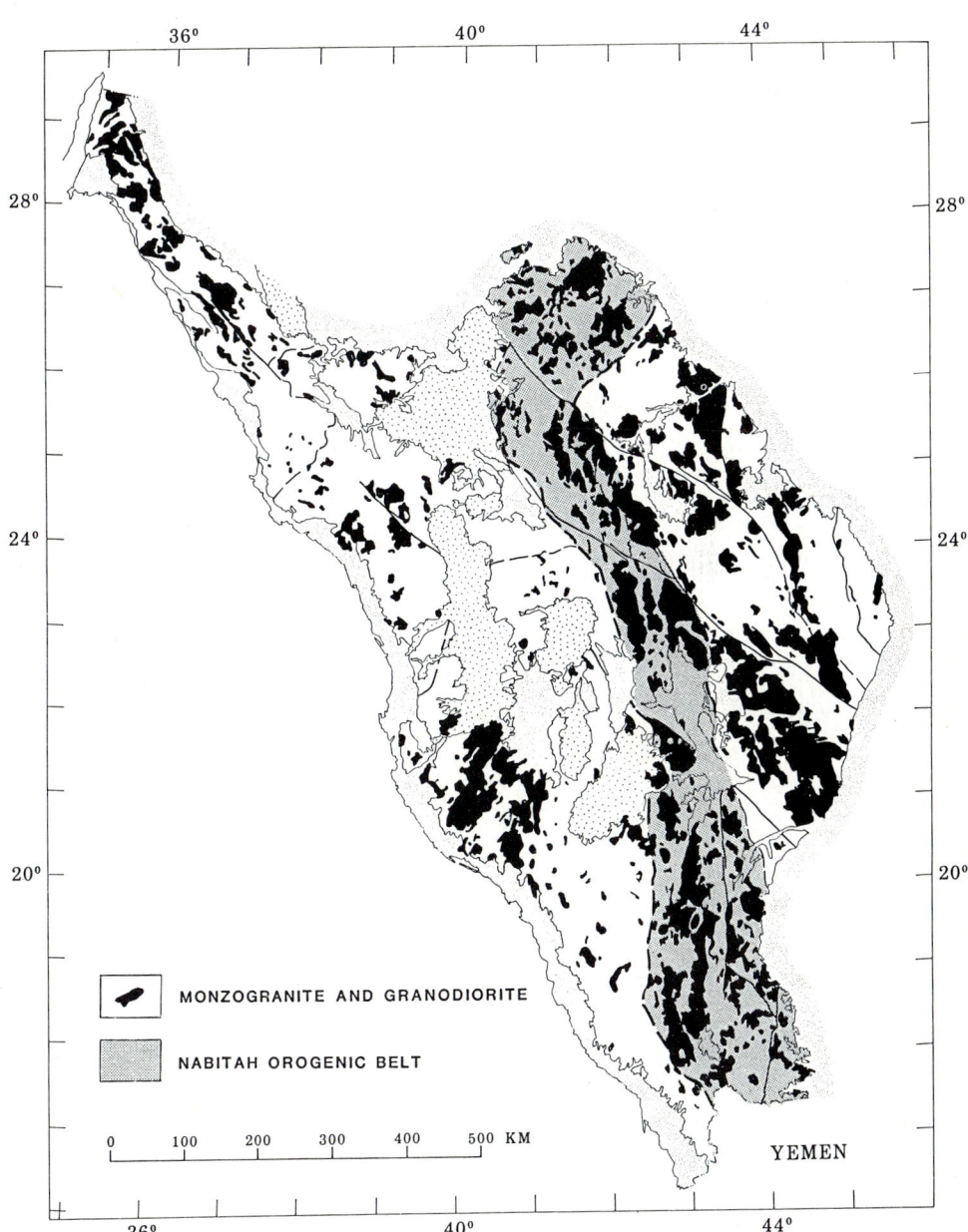

FIGURE 2.

Map of the Saudi Arabian Shield showing the distribution of monzogranite and granodiorite. Other units and boundaries same as Figure 1.

Jackson, 1986). Geochronologic studies established that magmatic arc rock assemblages of late Proterozoic age formed between approximately 950 and 670 Ma (e.g. Baubron et al., 1976; Cooper et al., 1979; Fleck et al., 1980; Bokari and Kramers, 1981; Marzouki et al., 1982; Calvez et al., 1984; Stacey et al., 1984). In our paper, the term "magmatic arc" refers to magmatic rock assemblages formed above subduction zones along convergent plate margins and thus includes both oceanic island and Andean type arcs.

Sillitoe first suggested in 1979 that the Arabian Shield may have formed by microplate or terrane accretion, but the first serious application of this concept was by Kröner (1983, 1985). Other microplate accretion models have been presented by Johnson and Vranas (1984), Stoeser et al. (1984b), Stoeser and Camp (1985), Vail (1985), and Pallister et al. (in press). For the definition of the term "terranes", we follow Coney et al. (1980), who refer to them as tectonostratigraphic domains which "are chracterized by internal homogeneity and continuity of stratigraphy, tectonic style and history". Although the implication is that terranes are allochthonous and have accreted together, Coney et al. "note that identification of a terrane is based primarily on its stratigraphy and need not carry any genetic or even plate tectonic implication. At the start of investigations, the identified terranes are simply considered as domains in the descriptive sense". A number of such tectonostratigraphic terranes have been recognized within the Arabian Shield (Fig. 1; Delfour, 1981; Greenwood, et al., 1982; Johnson and Vranas, 1984), but few detailed studies of these terranes or their boundaries are available. In addition, we use the term composite terrane for terranes which themselves are composed of two or more terranes. That is, composite terranes are single crustal units in the context of our discussion, but were sometime earlier in their history formed by the terrane accretion process and therefore are terrane complexes. Terrane boundaries are abrupt and usually consist of fault zones. If correctly defined terrane boundaries generally represent "sutures, now largely cryptic, and most have been reactivated by concurrent and post-collisonal large-scale strike slip movements" (Coney et al., 1980). Although there are many potential sutures within the Shield, only four of the major terrane boundaries are marked by belts of ultramafic rocks, and which are generally accepted as sutures (Bakor et al., 1976; Al-Shanti and Mitchell, 1976; Frisch and Al-Shanti, 1977; Camp, 1984). We, therefore, discriminate on Figure 1 between terrane boundaries of suspected terranes and those which are thought to definitely represent sutures.

Following Stoeser and Camp (1985), the Arabian Shield is divided into five principle terranes (Asir, Hijaz, Midyan, Afif, Ar Rayn) that are separated by four belts of ophiolitic and ultramafic complexes interpreted to mark sutures (Bir Umq, Yanbu, Nabitah, Al Amar; Fig. 1). The three western terranes, Asir, Hijaz and Midyan, are underlain by magmatic assemblages that formed in an ensimatic oceanic island arc environment. Because the Asir terrane consists of multiple magmatic arc terranes as well as possible oceanic basin and plateau segments (Greenwood et al. 1982; Reischmann et. al., 1984; Kröner, 1985), we refer to it herein as the Asir composite terrane (Fig. 1). The western arc terranes were sutured together along the Yanbu and Bir Umq suture zones prior to 680 Ma, and it is assumed that these terranes were a single crustal block when they collided with the Afif terrane to the east to form the Nabitah suture (Camp, 1984; Stoeser and Camp, 1985; Pallister et al., in press).

The Nabitah ultramafic belt is the longest of the four belts (Fig. 1), and is clearly defined for 1200 km (Frisch and Al-Shanti, 1977; Schmidt et al., 1979). Numerous ultramafic masses occur along the belt within a major fault zone, the Nabitah fault zone (Fig. 1). The name Nabitah is taken from Jabal Nabitah, a 5-kilometer-long serpentinite massif in the fault zone. The first detailed study of the Nabitah ultramafic belt was by Frisch and Al-Shanti (1977), who interpreted the ultramafic complexes as ophiolites and the belt (their Hulayfah-Hamdah belt) as an intra-arc suture that had been the

site of an eastward-directed subduction zone. Unlike the
ultramafic complexes of the Bir Umq and Yanbu sutures, those of
the Nabitah belt are generally not normal ophiolites, but
typically consist of highly deformed serpentinite (Le Metour et
al., 1983; Pallister et al., in press). Dismembered ultramafic
complexes which appear to be ophiolitic in character are only
reported at the north and south ends of the belt: the Tuluhah
ultramafic complex in the north (Le Metour et al., 1983) and
the Hamdah complex in the south (Al-Rehaili and Warden, 1980;
Worl and Elsass, 1980; Fig. 1).

Schmidt et al. (1979) also interpreted the Nabitah ultramafic
belt as a suture between older island arc crust to the west and
a younger arc to the east. The younger arc was presumed to be
related to a westward-directed subduction zone located along
the site of the Al Amar suture (Fig. 1). In addition, Schmidt
et al. showed that the Nabitah suture lies within a
1200-kilometer-long, 100 to 200-kilometer-wide orogenic belt
characterized by gneiss domes, migmatites, greenschist to
granulite facies metamorphism, intense westward directed
compressional deformation, and voluminous granodioritic to
granitic plutonism (Fig. 2). They interpreted the orogenic belt
as resulting from a collision between the younger arc and a
continental plate east of the Al Amar suture zone of which the
Ar Rayn terrane was the leading edge (Fig. 1). They referred to
this orogeny as the "culminative orogeny" and, based on a few
Rb-Sr whole rock ages, placed it at about 625 Ma. Schmidt and
Brown (1982) discussed the geochemistry of the magmatic rocks
and the age of the "culminant orogeny" was revised to
approximately 660 Ma.

Stoeser et al. (1984a,b) supported the collisional orogenic
interpretation of Schmidt et al. (1979) for the gneiss belt and
described widespread gneiss doming, metamorphism and plutonism
within the southernmost part of the belt. Stoeser et al.
(1984b) provided the first detailed geochronology for the
orogenic belt, which they termed the Nabitah mobile belt and
designated the corresponding orogeny as the Nabitah orogeny.

They proposed that the Nabitah suture has been the site of a westward-directed subduction zone that produced a magmatic arc of 760 - 720 Ma age along the western margin of the suture. Finally, they proposed that the Nabitah orogeny was the result of a collison between the arc terranes to the west and a continental microplate, the Afif terrane, to the east.

Although the existence of continental basement within the Shield had been proposed by a number of workers on the basis of mapping, the basement issue remained speculative until Stacey et al. (1980), Stacey and Stoeser (1983), and Stacey and Hedge (1984) presented isotopic evidence for early to middle Proterozoic crust in the eastern Shield. Detailed mapping of Agar (1985) and Theime (in press) and associated U-Pb isotopic work of Stacey and Agar (1985) identified 1800 Ma continental basement within the Afif terrane. In the present paper, we will refer to this older crust as the Khida basement after the discovery location at Jabal Khida by Stacey and Hedge (1984) (Fig. 1). In addition, Agar (1985) and Stacey and Agar (1985) proposed that an Andean type arc, the Siham arc, formed along the west central margin of the Afif terrane (Zalim region of Fig. 1) from 720 to 685 Ma as a result of eastward-directed subduction from the Nabitah zone prior to collision.

Because the Afif terrane contains several magmatic arc terranes as well as the Khida basement, we use the term Afif composite terrane in the present paper. In addition, we refer to the southern and northern Afif composite terrane with the dividing line being the Ar Rika fault zone, and to the south-central Afif composite terrane as being the region between the Ar Rika and Ruwah fault zones (Fig. 1).

The fourth major zone of ultramafic complexes is located in the Ad Dawadimi terrane of the eastern Shield (Fig. 1). The northwest-striking Ad Dawadimi terrane is cored by highly deformed clastic rocks of the schistose Abt formation and flanked on either side by fault zones containing ultramafic complexes, although some ultramafic rocks are also intercalated within the Abt metasediments (Al-Shanti and Mitchell, 1976; Al-Shanti and Gass, 1983). The Ar Rayn terrane to the east is underlain by island arc volcanic and sedimentary rocks and is intruded by a complex series of dioritic and tonalitic rocks (Coulomb et al., 1981; Le Bel and Laval, 1986). A number of workers have interpreted the rocks of the Ad Dawadimi terrane as representing an accretionary wedge that was deformed when the Afif composite terrane and Ar Rayn terranes collided to form the Al Amar or Al Amar-Idsas suture zone (Fig. 1; e.g. Al-Shanti and Mitchell, 1976; Stacey et al., 1984; Le Bel and Laval, 1986). It is not clear, however, whether the Abt sediments were deposited on the eastern margin of the Afif composite terrane or on the western margin of the Ar Rayn terrane and, therefore, either the eastern or western boundary of the Ad Dawadimi terrane is probably a suture. An additional complication is that thrusting during or subsequent to collision appears to have added considerably to the tectonic complexity of the region (e.g. Al-Shanti and Mitchell, 1976; Delfour, 1981; Stacey et al., 1984). Because of these uncertainties and following earlier work (Stacey et al., 1984; Stoeser and Camp, 1985), we have referred to the whole of the Ad Dawadimi terrane as a suture zone, but clearly further work is needed to define the stratigraphic and tectonic elements of the Ad Dawadimi "terrane".

ANALYTICAL METHODS

Chemistry and mass spectrometry of lead and uranium from zircons follow procedures of Krogh (1973). Laboratory blank levels for lead ranged from 0.3 to 2.0 ng. Common lead corrections for zircons used the composition of lead from feldspar in the same sample. Analytical precisions (95 % confidence level) are ± 1.2 % for concentrations of lead and uranium and ± 0.1 % for $^{207}Pb/^{206}Pb$ as determined by replicate analyses of a standard concordant zircon sample. Uncertainties for intercepts between regression lines and concordia are 95 % confidence limits (Ludwig, 1982). For regression lines involving only two or three data points, intercept errors were computed on the basis of the assigned analytic errors only and may, therefore, be underestimates (Ludwig, 1982). Correlation coefficients for $^{206}Pb/^{238}U$ and $^{207}Pb/^{235}U$ (determined mainly by the common lead content of the zircons and the precision of the mass spectrometer analyses) range from 0.99 to 0.60. Uranium decay constants used for age calculations are those of Jaffey et al. (1971). More nearly concordant data were achieved for some zircon fractions by abrasion in an air-driven cell developed by J. N. Aleinikoff and S. S. Goldich (Goldich and Fischer, 1986).

To minimize in situ radiogenic lead, feldspar concentrates were washed with 7N HNO_3 and 6N HCl, then leached in warm 5 % HF (Ludwig and Silver, 1977). Feldspars were dissolved and lead extracted as outlined by Stacey and Stoeser (1983). Microwave dissolution of feldspars was also used (Fischer, 1986).

U-Pb ZIRCON RESULTS

In the first zircon geochronology study of rocks from the Late Precambrian Arabian Shield, Cooper at al. (1979) showed that U-Pb systems in zircons were substantially undisturbed by later metamorphic events. Most igneous rocks of the Saudi Arabian Shield were emplaced between 900 and 570 Ma, but widespread thermal activity between 660 and 570 Ma disturbed K-Ar and Rb-Sr relations in many of the older units (Baubron et al. 1976; Fleck et al., 1976, 1980). The ability of zircons from the southern Arabian Shield to remain unaffected by that activity has proved of great value. Nevertheless, most zircon fractions measured had evidently lost lead in a recent event probably related to the Cenozoic Red Sea rift. On the basis of the weighted mean of the lower intercepts for the regression lines with concordia, a common lower intercept age of 15 ± 15 Ma was determined for all samples following the model of Cooper et al. (1979). Anchoring the lower ends of the regression lines in this manner improved the relative precision of the upper intercepts, such that differences in age of 15 Ma between samples about 800 Ma old could be resolved.

Figure 3A emphasizes the validity of the 15 ± 15 Ma lead loss model for zircons from the southern Saudi Arabian Shield. The U-Pb data are for twenty zircon fractions from six plutonic samples with zircon model ages from 657 to 667 Ma. Four of the samples are from Cooper et al. (1979) and two are from this study. The rocks range in composition from gabbro to monzogranite and are separated by distances of as much as 300 km. In Figure 3A, the regression line has an upper intercept with concordia corresponding to a mean age of 662 ± 5 Ma. The good fit of the line indicates that the only subsequent disturbance for all samples was recent lead loss shown by the lower intercept of 15 ± 20 Ma.

New U-Pb zircon data are reported in Table 1 and sample locations shown in Figure 4 for three suites of samples collected between 1978 and 1982. The first suite of rocks is from the southern part of the Al Qarah terrane (locs. 1-11) and the Najran terrane (locs. 19, 20). U-Pb zircon ages (without supporting data) and feldspar lead isotope ratios have been published for these samples by Stacey and Stoeser (1983). The second suite (locs. 12-17, 22-30) was selected to date rocks west of the Nabitah suture in the Ranyah region of the Al Qarah terrane (locs. 12-16) and to compare their lead compositions with those east of the suture in the south-central Afif composite terrane (locs. 25-27, 29, 30). The third suite is from a traverse across the central part of the Nabitah orogenic belt (locs. 18, 28, 31 32).

Zircon ages are summarized in Table 2 and tabulated for three different models. Model I ages correspond to concordia upper intercepts for regression lines through the data; Model II ages correspond to upper intercepts for regression lines through the data and an assumed lower intercept of 15 ± 15 Ma; and for model III an upper intercept age is computed for a line anchored to 15 ± 15 MA and drawn through the data point having

Table 1. U-Pb data for zircons from the samples of this study. [NM, nonmagnetic; M, magnetic; A, abraded]

Locality & sample Number (mesh size)	Sample weight (mg)	U (ppm)	Pb (ppm total)	Atomic ratios			Apparent ages (Ma)			Measured
				$\frac{206Pb}{238U}$	$\frac{207Pb}{235U}$	$\frac{207Pb}{206Pb}$	$\frac{206Pb}{238U}$	$\frac{207Pb}{235U}$	$\frac{207Pb}{206Pb}$	$\frac{206Pb}{204Pb}$
WEST OF NABITAH SUTURE										
Southern Al Qarah terrane										
1. 111552										
(+100)	13.59	296	34.3	0.11085	0.9736	0.06371	676	690	732	1090
(100-150)	15.83	304	34.4	0.11461	1.0071	0.05373	699	707	733	7814
(200-250)	24.42	324	36.1	0.11297	0.9908	0.06361	690	699	728	8506
(-325)	15.43	355	38.7	0.10932	0.9597	0.06367	699	683	731	4456
2. 112683										
(100-150)	16.59	114	12.7	0.11595	1.0178	0.06367	707	713	730	6189
(250-325)	14.50	137	15.4	0.11599	1.0174	0.06361	707	713	729	4665
(-325)	19.13	158	17.9	0.11727	1.0278	0.06357	715	718	727	8344
3. 112805										
(+100)	20.87	175	17.4	0.10263	0.8760	0.06199	630	639	674	2940
(150-200)	25.46	265	27.4	0.10917	0.9318	0.06199	668	669	674	5888
(-325)	22.35	280	25.1	0.09399	0.8053	0.06222	579	600	681	4883
4. 111555										
(100-150)	3.46	756	62.1	0.07813	0.6658	0.06180	485	518	667	518
(150-200)	4.11	759	65.1	0.08469	0.7214	0.06178	524	562	667	758
(-250)NM	21.48	859	70.9	0.08030	0.6835	0.06174	498	529	665	685
5. 128918										
(+100)NM	19.14	274	29.9	0.10665	0.9085	0.06178	653	656	666	2458
(-250)NM	7.70	444	46.9	0.10152	0.8648	0.06184	623	633	667	1616
6. 112806										
(200-250)	16.48	515	44.6	0.08741	0.7406	0.06153	540	563	658	1182
(250-325)	22.92	550	46.0	0.08868	0.7502	0.06136	548	568	652	2185
7. 111564										
(+200)	9.79	997	85.6	0.08575	0.7238	0.06122	530	553	647	1766
(200-250)	4.99	958	84.1	0.08921	0.7542	0.06132	551	571	650	3089
(-325)	8.67	865	77.8	0.09083	0.7687	0.06138	560	579	652	2948
8. 111541										
(+150)NM	11.47	193	19.2	0.09557	0.8046	0.06106	588	600	641	1216
(+150)M	25.01	192	22.7	0.09546	0.8049	0.06114	588	600	645	272
(150-200)NM	11.99	221	26.3	0.09765	0.8222	0.06107	600	610	642	1522
9. 128916										
(+100)NM	16.52	1550	167.0	0.10073	0.8481	0.06098	619	616	638	9371
(-325)	10.61	1375	144.0	0.0993	0.8439	0.06098	610	616	638	5967
10. 111554										
(+100)	10.72	3097	300.8	0.10048	0.8438	0.06091	617	621	636	9760
(-250-325)	11.05	1761	168.6	0.09713	0.8175	0.06104	598	607	641	4392
(-325)	14.91	1651	153.0	0.09323	0.7855	0.06111	575	589	643	3705
11. 111558(gray)										
All sizes	1.64	1098	92.6	0.08196	0.6883	0.06092	508	532	636	935
111559(pink)										
(+150)	2.03	1139	68.2	0.04664	0.3901	0.06066	294	334	627	210
(150-200)	6.12	1030	55.1	0.04396	0.3640	0.06052	277	315	605	275

Table 1 continued, page 2:

Northern Al Qarah terrane (Ranyah region)

12.	Z-113									
	(+200)	3.56	672	72.2	0.10711	0.9355	0.06335	656	671	720	3450
	(200-250)	1.81	712	77.9	0.10900	0.9523	0.06337	667	679	721	1617
	(-250)	4.82	719	69.3	0.09816	0.8545	0.06314	604	627	713	5154
13.	Z-109									
	(+100)	15.8	1500	128.8	0.08617	0.727	0.0619	533	555	646	7397
	(-325)NM	11.57	1606	143.0	0.09104	0.7670	0.06100	562	578	643	3182
	(-325)M	4.52	2176	165.0	0.07608	0.6400	0.06100	473	502	639	1600
14.	Z-114									
	(+100)NM	11.98	2287	233.1	0.09735	0.8165	0.06083	599	606	633	1158
	(-250)M	11.45	1196	105.1	0.08350	0.7005	0.06085	517	539	634	1403
15.	Z-110									
	(+100)NM	7.90	993	80.1	0.07593	0.6319	0.06035	472	497	616	1131
	(-250)NM	11.73	1979	152.2	0.07024	0.5836	0.06027	438	467	613	979
16.	Z-111									
	(+100)NM	3.3	508	57.7	0.08751	0.72716	0.06027	541	555	613	235
	(250-325)NM	5.13	248	27.7	0.09804	0.82564	0.06108	603	611	642	352
	(-325)M	6.06	646	56.6	0.079	0.6527	0.05992	490	510	600	517

Northern An Nimas terrane (?) (Ad Dafinah region)

18.	175600									
	(-325)NM	7.45	128	16.9	0.12770	1.1623	0.06600	775	783	806	1702
	(+150)NM	16.59	88	11.1	0.12440	1.1345	0.06613	756	770	810	1676
	(150-200)NM	14.87	113	15.7	0.13110	1.1975	0.06624	794	792	814	989

EAST OF NABITAH SUTURE

Najran terrane

19.	128917									
	(+100)NM	12.38	401	39.4	0.09336	0.7858	0.06105	575	589	641	3406
	(200-250)NM	18.95	469	48.4	0.09968	0.8387	0.06102	613	618	640	5789
	(-400)	9.06	408	42.6	0.10046	0.8456	0.06105	617	622	641	5737
20.	128905									
	(+100)NM	16.95	134	14.3	0.10106	0.8507	0.06105	621	625	641	1671
	(-325)NM	24.69	202	21.0	0.09920	0.8351	0.06106	610	616	641	2210

Tathlith terrane

22.	Z-100									
	(+100)NMA	10.64	1756	195.1	0.10250	0.8847	0.06256	629	644	695	2182
	(+100)NM	16.53	1657	180.3	0.10074	0.8706	0.06268	619	636	697	2315
	(-400)M	9.16	1633	156.6	0.09188	0.7936	0.06265	567	593	696	2426

South-central Afif composite terrane

25.	Z-107									
	(+100)NM	16.12	530	57.5	0.10098	0.8485	0.06094	620	624	637	2528
	(-200)M	5.87	445	41.0	0.08612	0.7252	0.06107	533	554	642	1547
	(-325)NM	10.49	446	45.6	0.09794	0.8215	0.06084	602	609	633	3778
26.	Z-102									
	(+100)NMA	4.89	1411	116.1	0.06595	0.5506	0.06056	412	445	623	306
	(+100)NM	14.56	1536	120.5	0.06129	0.5107	0.06043	383	419	619	262
	(-325)M	10.83	3979	245.8	0.04146	0.3461	0.06054	262	302	623	167
27.	Z-106									
	(+200)NM	9.77	1164	111.6	0.09029	0.7529	0.06048	557	570	621	1801
	(-325)NM	4.45	1214	124.1	0.09462	0.7886	0.06044	583	590	619	2149
	(-325)M	1.96	2002	191.4	0.08163	0.6791	0.06034	506	526	615	1217

Table 1 continued, page 3:

28.	175599										
	(+150)NM	14.81	888	80.1	0.08864	0.7380	0.06038	547	561	617	1854
	(150-200)NM	6.92	886	89.6	0.09924	0.8285	0.06055	610	613	623	2477
	(-250)M	8.33	2154	187.0	0.08234	0.6827	0.06010	510	528	608	1441
29.	Z-108										
	(+100)	7.44	4144	567.6	0.08549	0.7100	0.06022	529	545	612	135
	(100-150)	2.90	4342	638.6	0.08793	0.7306	0.06026	543	557	613	131
30.	Z-101										
	(+100)NM	20.52	131	12.5	0.09030	0.7424	0.05963	557	564	590	962
	(100-150)NM	8.49	139	13.7	0.09192	0.7569	0.05972	567	572	593	672
	(-250)M	12.53	344	29.0	0.07967	0.6546	0.05959	494	511	589	1821
Northern Afif composite terrane											
31.	175590										
	(+200)	5.26	844	67.7	0.07317	0.6455	0.06398	455	506	741	581
	(-250)M	1.42	1662	151.0	0.07977	0.6802	0.06184	495	527	668	754
	(-200+250)	1.02	1303	88.7	0.05967	0.5076	0.06169	373	417	663	435
	(-250)NM	6.95	706	69.4	0.09518	0.8579	0.06537	586	629	786	1897
32.	175589										
	(+150)NM	14.98	644	70.8	0.09808	0.8276	0.06120	603	612	646	1679
	(-325)NM	4.69	538	61.7	0.09856	0.8352	0.06126	606	615	648	885

the lowest $^{207}Pb/^{206}Pb$ age. Model III ages are only computed
for three samples that may contain inherited older zircon
(locs. 16, 31, 32) and are maximum ages because even the zircon
fraction having the least radiogenic lead may contain some
older zircon. Table 2 shows that, for most samples, the model
used has little effect on computed ages, and therefore, model
II ages are generally used in the text because they allow all
of the data to be treated uniformly.

Southern Al Qarah and Najran terranes

Model II ages of zircons from the southern Al Qarah terrane
(Table 2) fall into four district groups and the U-Pb data for
each group are plotted together on the concordia diagrams of
Figure 3. Older tonalite gneisses of the Wadi Tarib Batholith
(locs. 1, 2, Fig. 3B) have nearly concordant data with similar
ages of 727 ± 5 and 732 ± 4 Ma, respectively, and a group
regression age of 731 ± 3 Ma. Data for synorogenic rocks (locs.
3, 4, 5, Fig. 3C) have model II ages of 675 ± 4, 670 ± 8 and
667 ± 4 Ma, respectively, and a group regression line age of
671 ± 4 Ma. Data for two gneisses from the southern terrane
(locs. 6, 7, Fig. 3D) have individual model ages of 654 ± 3 and
657 ± 3 Ma and a group regression age of 654 ± 4 Ma. Late-
orogenic tonalite to monzogranite samples from the southern
Al Qarah and Najran terranes (locs. 8, 9, 19, 20, Fig. 3E) all
have model II ages between 641 and 643 Ma. In Figure 3E, these
data are grouped with that of the postorogenic Jabal Thairwah
monzogranite (loc. 10) whose model age is 641 ± 10 Ma, but
whose actual age should be slightly younger than that of the
late-orogenic plutons as is permitted by the 10 Ma error limit.
The group model II age of 641 ± 3 Ma is essentially
indistinguishable from the individual model II ages.

The postorogenic Tindahah monzogranite of the southern Al Qarah
terrane (loc. 11, Fig. 3F) had two pink and gray phases at the
zircon sample site. Zircons from the pink phase have very high
common lead contents that greatly reduce precision of the U/Pb

ratios. If the same age is assumed for both phases, all fractions together yield a model II age of 638 ± 18 Ma.

Northern Al Qarah terrane

In the Ranyah region of the Al Qarah terrane (loc. 12, Fig. 3G), the oldest unit is the Shaib Hadhaq tonalite gneiss (Greene, 1980) which has a model II age of 719 ± 9 Ma. The Shaib Hadhaq monzogranite (loc. 13, Fig. 3H) has a model II age of 646 ± 4 and represents gneiss doming during the Nabitah orogeny (Greene, 1980). Younger plutons in the region include the Jabal Suily granite, which has an age of 635 ± 4 Ma (loc. 14, Fig. 3I), and the Al Jizah monzogranite (loc. 15, Fig. 3J) which has an age of 620 ± 7 Ma.

Zircons from the Kwar Barahah alkali-feldspar granite of the Ranyah region (loc. 16, Fig. 3K) have a high common lead content with low $^{206}Pb/^{204}Pb$ values of 235 to 517 (Table 1). The model I upper intercept age is 653 ± 25 Ma, with a very elevated lower intercept of 185 ± 77 Ma. The 653 Ma age is too old for an unfoliated granite typical of the postorogenic suite, which suggests either its age is closer to the lower error limit or it inherited older zircons during emplacement which would account for the 600 - 642 Ma range in $^{207}Pb/^{206}Pb$ ages. We use the model II age for the zircon fraction (- 325 mesh) that has the lowest $^{207}Pb/^{206}Pb$ ratio and also the lowest common lead component to calculate a model III age of 604 ± 15 Ma (Table 2). Similar difficulties with the U-Pb method apparently due to inheritance have been encountered in other zircon studies of the eastern Shield (Stacey et al., 1984; Aleinikoff and Stoeser, in press).

South-central Afif composite terrane

The oldest zircons dated in Table 2 for the south-central Afif composite terrane are from the granodiorite sample Z-103 from

Table 2.   U-Pb zircon ages (Ma) and sample descriptions. See text for description of models.
           Uncertainties quoted are model dependent, and computed for 95% confidence levels (Ludwig, 1982).
           Model III ages are marked in the table.

Locality number, sample number, Lat. N; Long. E	No. of fractions	Regression through data: model I intercept ages		Regression through 15±15 Ma Model II	Sample description
		lower	upper		

WEST OF NABITAH SUTURE
(ASIR COMPOSITE TERRANE)

Southern Al Qarah terrane

1. 111552 $17°47.6'$; $43°29.3'$	4	30±120	732±4	732±4	Talhah tonalite gneiss
2. 112683 $18°01.9'$; $43°19.8'$	3	*	*	727±5	Suwaydah tonalite gneiss
3. 112805 $18°23.1'$; $43°27.6'$	3	-70±42	669±3	675±5	Wadi Arin tonalite
4. 111555 $18°40.3'$; $43°14.5'$	3	*	*	670±8	Mahda two-mica granite
5. 128918 $18°15.8'$; $42°38.7'$	2	*	*	667±4	Khamis Mushayt quartz diorite
6. 112806 $18°14.4'$; $43°17.0'$	2	*	*	654±3	Hijrat trondhjemite gneiss
7. 111564 $18°22.2'$; $42°48.3'$	3	*	*	657±3	Khamis Mushayt granodiorite gneiss
8. 111541 $18°19.3'$; $43°13.4'$	3			643±5	Al Ar tonalite
9. 128916 $18°24.2'$; $43°29.6'$	2	*	*	641±3	Wadi Makhdhul quartz diorite
10. 111554 $18°01.3'$; $43°38.6'$	3	-116±89	634±4	641±10	Jabal Thairwah biotite monzogranite
11. 111558/9 $18°12.7'$; $43°00.3'$	3	25±174	642±130	638±18	Tindahah biotite monzogranite

Northern Al Qarah terrane (Ranyah Region)

12. Z113 $21°13.7'$; $32°36.3'$	3	85±56	727±8	719±9	Shaib Hadhaq tonalite gneiss
13. Z-109 $21°28.7'$; $42°59.3'$	2	24±290	648±82	646±4	Shaib Hadhaq hornblende-biotite monzogranite
14. Z-114 $21°20.0'$; $42°30.0'$	2	-4±52	633±5	635±4	Jabal Suily biotite-hornblende mesosolvus alkali-feldspar granite
15. Z-110 $21°16.9'$; $42°52.6'$	2	-62±46	602±11	620±7	Al Jizah biotite monzogranite
16. Z-111 $21°11.6'$; $42°07.1'$	3	185±77	653±25	604±15 (III)	Kwar Barahah hornblende perthite alkali-feldspar granite
17. Z-112 $21°05.5'$; $42°42.5'$			No zircon obtained		Kaersutite-olivine-clinopyroxene gabbro

Northern An Nimas terrane (?) (Ad Dafinah Region)

18. 175600 $23°46.2'$; $41°31.7'$	3	*	*	811±4	Leucotonalite

Table 2 continued:

EAST OF NABITAH SUTURE

Najran Terrane

19.	128917	3	-4±75	640±5	641±3	Wadi Simlal quartz diorite
	18°04.3'; 44°05.6'					
20.	128905	3	*	*	642±3	A'ashiba quartz diorite gneiss
	17°36.8'; 44°29.7					
21.	128912	No zircon recovered.		-	-	Jabal al Gaharra two-mica, subsolvus, alkali-feldspar granite
	18°04.0'; 44°16.0'					

Tathlith Terrane

| 22. | Z-100 | 3 | -15±47 | 695±6 | 698±3 | Hornblende tonalite |
| | 21°00.0'; 43°32.0' | | | | | |

South-central Afif composite terrane

23.	Z-103	5	658±107	1632±208	-	Jabal Khida biotite granodiorite (data from Stacey & Hedge, 1983)
	21°19.0'; 44°50.3'					
24.	Z-104		No zircon recovered			Muhayil hornblende anorthosite
	21°20.5'; 44°51.3'					
25.	Z-107	3	-61±38	634±79	635±5	Mylonitic epidote-hornblende-biotite granodiorite
	21°32.1'; 43°49.6'					
26.	Z-102	3	-2±25	621±18	630±16	Al Bahah alkaline granite
	20°55.2'; 44°20.2'					
27.	Z-106	3	28±52	622±6	621±4	Hornblende-biotite monzogranite
	21°30.0'; 43°50.5'					
28.	175599	3	83±35	625±5	620±11	Hornblende-biotite granodiorite
	23°50.2'; 42°00.0'					
29.	Z-108	2	*	*	614±10	Huqban leucocratic biotite perthite alkali-feldspar granite
	21°45.0'; 43°53.5'					
30.	Z-101	3	24±53	593±6	592±4	Cataclastic hornblende-biotite monzogranite
	20°56.0'; 43°47.4'					

Northern Afif composite terrane

31.	175590	4	#	#	673±9 (III)	Biotite granodiorite
	23°39.4'; 42°43.6'					
32.	175589	2	*	*	647±7 (III)	Cataclastic biotite-hornblende tonalite
	23°35.7'; 42°54.5'					

* Insufficient separation between data points to justify regression line.
# High degree of scatter, probably because of inherited material.

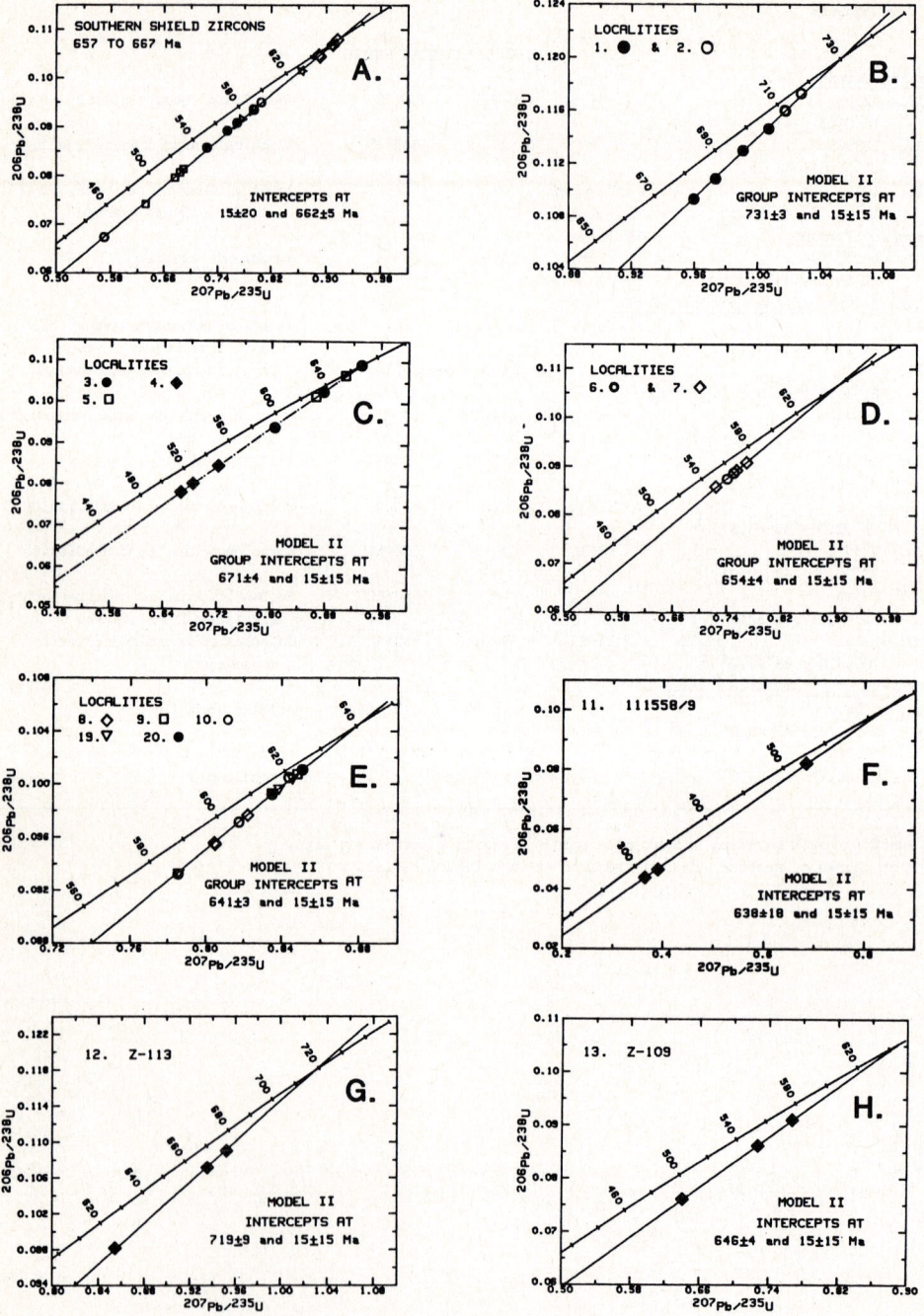

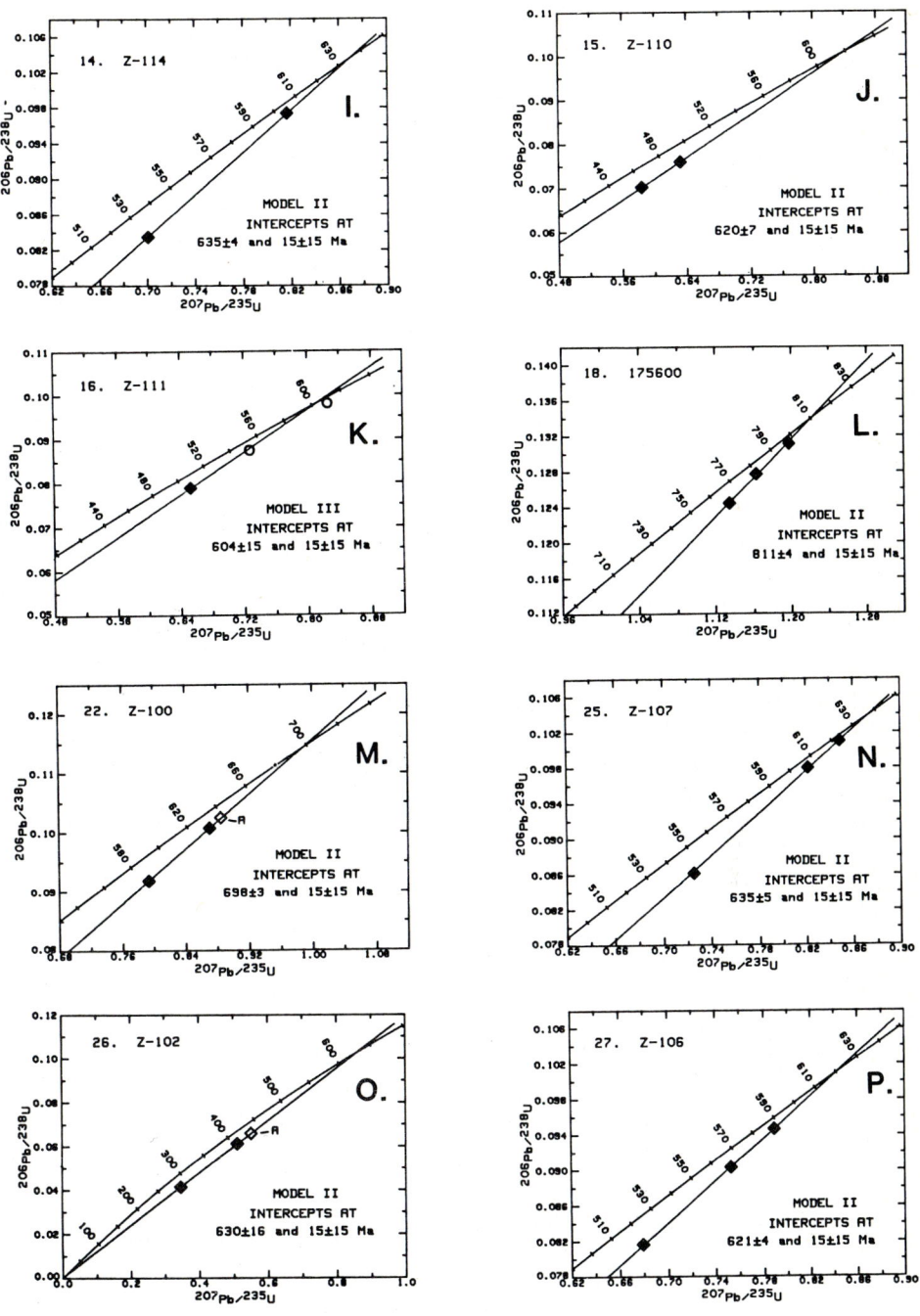

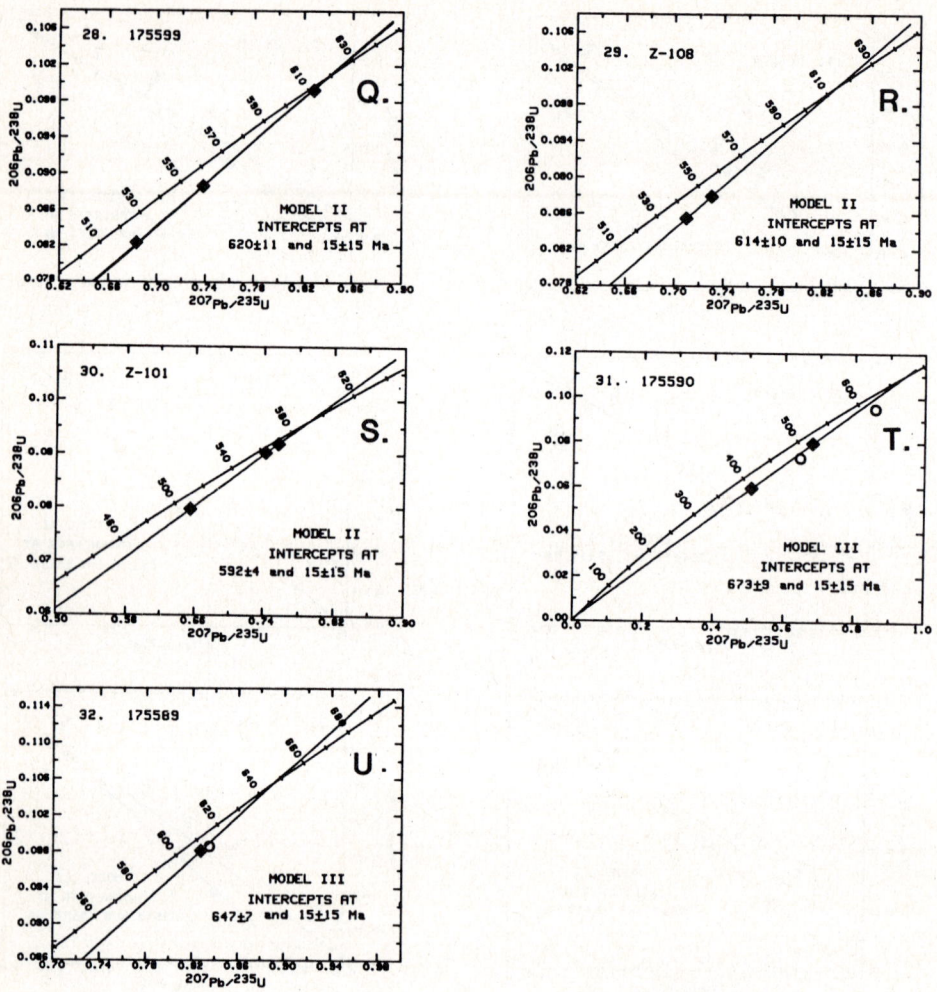

FIGURE 3.

    U-Pb concordia diagrams for zircon data presented in
Table 1. Sample locality numbers are presented in the
upper left corner of each diagram. Regression models are
discussed in the text. In Figure 3A, symbols indicate
different samples used for the regression (see text). Open
circles in Figures 3K, T, U indicate data not used for age
regression calculations. Open diamond symbols in Figures
3M, O indicate abraded zircon fraction (A).

Jabal Khida (loc. 23). Stacey and Hedge (1984) report upper and lower intercept ages for this sample of 1628 ± 200 Ma and 658 ± 107 Ma, respectively, and interpret the Z-103 data to indicate that either the Jabal Khida granodiorite represents 1628-Ma-old basement or it was emplaced at about 658 Ma and derived from 1628 Ma old basement. Theime (in press) indicates that the granodiorite was intruded into the Khida basement. We, therefore, assume that the lower intercept age for sample Z-103 of 658 ± 107 Ma represents the age of emplacement and that the granodiorite was derived directly from the Khida basement by melting and remobilization during the Nabitah orogeny.

A sample of the Muhayil anorthosite (loc. 24), from just immediately east of Jabal Khida, yielded no zircons, but feldspar lead data are presented in Table 3. Theime (in press) indicates that the Muhayil anorthosite intrudes the Khida granodiorite.

To the west of Jabal Khida four postorogenic granitic plutons were sampled. The oldest is the mylonitic granodiorite at locality 25 (Fig. 3N). The granodiorite has a model II age of 635 ± 5 Ma. Model II ages for younger postorogenic granites in the area (locs. 26, 27, 29, Table 2, Figs. 3O,P,R) are between 630 ± 16 and 614 ± 10 Ma.

A cataclastic monzogranite (loc. 30) was sampled at the suggestion of D.L. Schmidt, because it lies within and appears to be syntectonic with the Ruwah fault zone. Its model II age of 592 ± 4 Ma (Fig. 3S) probably represents a good estimate of the end of major Najd faulting in this region.

Three zircon fractions, including one abraded fraction, from a hornblende tonalite from the northern part of the Tathlith terrane (loc. 22, Fig. 3M) yield a good model II age of 698 ± 3 Ma.

Central Nabitah orogenic belt

A leucotonalite (loc. 18) from west of the Nabitah suture has a
model II age of 811 ± Ma and is the oldest sample taken by us
from west of the Nabitah suture. A granodiorite sample from the
south-central Afif composite terrane (loc. 28) has a model II
age of 620 ± 11 Ma (Fig. 3Q) and is contemporaneous with
granitic rocks farther south in the Zalm region (Fig. 1)
(Stacey and Agar, 1985). Zircon U-Pb data for a granodiorite
and a tonalite (loc. 31, 32) from the central Nabitah belt just
north of the Ar Rika fault zone in the northern Afif composite
terrane show scatter on concordia (Fig. 3T,U) that we attribute
to assimilation of older zircons during emplacement. The
presence of older material is also indicated by the very
elevated lead isotope ratios of the feldspars (Table 3). Model
III zircon ages are, therefore, preferred and fractions having
lowest $^{207}Pb/^{206}Pb$ ratios yield ages of 673 ± 9 Ma (loc. 31)
and 647 ± 7 Ma (loc. 32).

FELDSPAR LEAD RESULTS

Lead isotopic data have proved very useful in delineating
terranes within the Shield. Stacey et al. (1980) showed that
$^{208}Pb/^{204}Pb$ ratios of galenas from the eastern Shield are
distinctly higher than those in the western Shield, and they
postulated the existence of Early Proterozoic crust in the
eastern Shield. Stacey and Stoeser (1983) confirmed these
findings by analyzing lead from feldspars of plutonic rocks,
and concluded that the Nabitah suture is a fundamental line
within the Shield that separates terranes derived from late
Precambrian oceanic crust in the west from those in the east
that are somewhat more evolved and in some places have older
continental-type lead signatures (Fig. 5).

Table 3 presents new lead isotope data, as well as data from
Stacey and Stoeser (1983) and Stacey and Hedge (1984). These
data, as well as most other lead data for the Shield, are

Table 3. Pb isotope data for leached feldspars and two whole-rocks. [Al Qarah and
Najran terrane data from Stacey and Stoeser (1983); Z-103 feldspar data
from Stacey and Hedge (1984). Model isochron ages computed by method of
Stacey and Kramers (1985)]

Locality and sample number	206Pb 204Pb	207Pb 204Pb	208Pb 204Pb	Model Isochron Age (Ma) (M)	Zircon Model Age (Ma) (Z)	Difference (Z-M)	Pb group
WEST OF NABITAH SUTURE							
Southern Al Qarah terrane							
1. 111552	17.363	15.450	36.988	660	732	+72	I
3. 112805	17.512	15.458	36.982	560	675	+115	I
4. 111555*	17.556	15.461	37.009	530	670	+140	I
7. 111564*	17.589	15.467	37.014	520	657	+37	I
8. 111541*	17.388	15.490	37.003	620	643	+23	I
9. 128916	17.665	15.473	37.135	470	641	+171	I
10. 111554	17.643	15.498	37.159	540	641	+101	I
Northern Al Qarah terrane (Ranyah region)							
12. Z-113	18.084	15.525	37.354	253	719	+463	I
13. Z-109	17.622	15.468	37.125	494	646	+152	I
14. Z-114	17.940	15.503	37.365	318	635	+317	I
15. Z-110	17.728	15.479	37.238	434	620	+186	I
16. Z-111	18.623	15.536	37.848	-148	604	+752	I?
17. Z-112	17.670	15.483	37.219	487	-	-	I
Northern An Nimas terrane (?) (Ad Dafinah region)							
18. 175600	17.290	15.433	36.804	684	811	+127	I
EAST OF NABITAH SUTURE							
Najran terrane							
19. 128917	17.701	15.501	37.260	500	641	+141	I
20. 128905	17.459	15.481	37.252	645	642	-3	II
21. 128912	17.891	15.480	37.182	308	-	-	I
Tathlith terrane							
22. Z-100	18.437	15.549	38.552	29	698	+669	II?
South-central Afif composite terrane							
23. Z-103#	17.458	15.631	37.870	658	924	-264	III
Whole Rock	18.410	15.696	39.589	356	-	-	-
24. Z-104	16.747	15.570	36.800	1339	<658	-679	II
Whole Rock	17.198	15.598	37.430	1056	-	-	-
25. Z-107	17.545	15.484	37.331	587	635	+48	I-II
26. Z-102	17.650	15.579	37.655	690	630	-60	III
27. Z-106	17.540	15.513	37.360	647	621	-26	II
28. 175599	17.646	15.467	37.112	473	620	+147	I
29. Z-108	17.609	15.571	37.588	706	614	-89	III
30. Z-101	17.534	15.533	37.419	690	592	-98	II
Northern Afif composite terrane							
31. 175590	17.520	15.621	37.832	862	673	-189	II
32. 175589	19.352	15.719	39.125	-291	647	+938	II?

* Initial ratios from whole rock U-Th-Pb data of Stacey and Stoeser (1983).
# Data of Stacey and Hedge (1984).

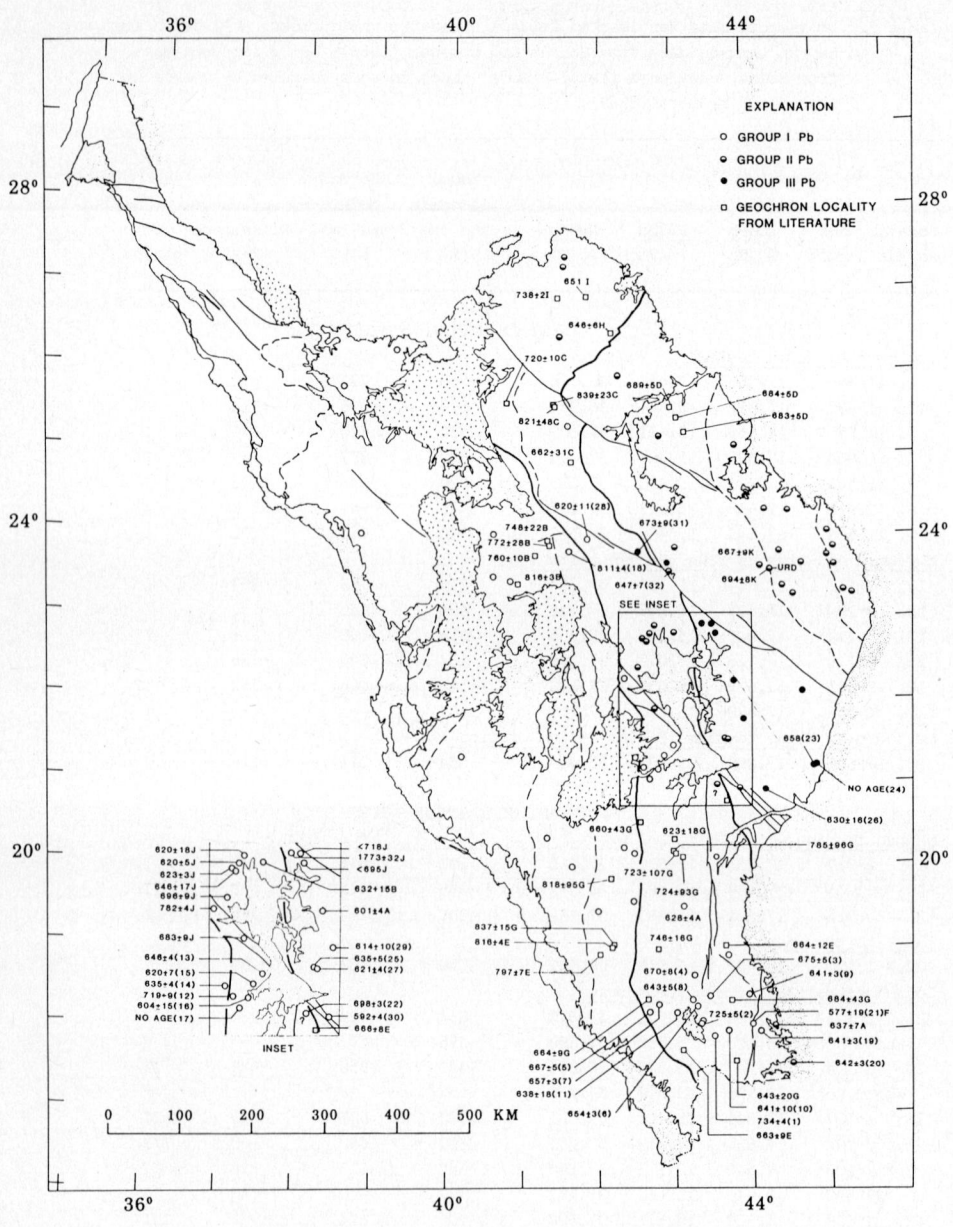

FIGURE 4.

Map of the Saudi Arabian Shield showing sample localities, other geochronology localities discussed in the text, and common Pb sample localities. The suffix to the ages indicates either a locality number from Tables 2 and 3 (in parentheses) or a literature source (A, Aleinikoff et al., in press; B, Calvez and Kemp, 1982; C, Calvez et al., 1984; D, Cole, 1986; E, Cooper et al., 1979; F, du Bray et al., in press; G, Fleck et al., 1980; H, Johnson and Williams, 1984; I, Quick and Doebrich, 1986; J, Stacey and Agar, 1985; K, Stacey et al., 1984). Common lead feldspar and galena localities are from Stacey et al. (1979), Stacey and Stoeser (1983), and Aleinikoff et al. (in press), and the present paper.

divided into three groups: group I from rocks of the western
arc terranes; group II from rocks east of the Nabitah suture,
except for group III from the Khida basement (Figs. 4, 5).
Groups I and II are the same as the groups I and II defined by
Stacey et al. (1980) and Stacey and Stoeser (1983).

Group I leads are ensimatic (Stacey and Stoeser, 1983) and
occur throughout the western arc terranes (Fig. 4). New data
for rocks having group I lead are from the Ranyah region of the
Al Qarah terrane (loc. 12 - 17) and the An Nimas terrane west
of Ad Dafinah (loc. 18; Fig. 1). The Ranyah data are mainly
from younger granites, and their leads are slightly more
radiogenic than those of the southern Al Qarah terrane (loc.
1 - 10). All leads of the western arc terranes plot close to
the average mantle curves of Zartman and Doe (1981) and are
similar to the lead of the ophiolitic rocks found in the Yanbu,
Bir Umq, and Nabitah suture zones (Fig. 5; Pallister et al., in
press). These lead data indicate that rocks of the western arc
terranes formed in an ensimatic oceanic environment away from
the influence of evolved continental crust. On the $^{208}Pb/^{204}Pb$
- $^{206}Pb/^{204}Pb$ graph (Fig. 5), group I data plot below the
average growth curves, and their model lead ages are younger
than their zircon ages (Table 3).

Group II leads are intermediate in composition between groups I
and III and are found in most rocks east of the Nabitah suture.
New data are presented for rocks of the Afif composite terrane
(locs. 22, 24 - 31). Group II leads plot near but above the
group I leads in both diagrams of Figure 5; and most
group II leads plot above the orogene growth curve on the
$^{208}Pb/^{204}Pb$ - $^{206}Pb/^{204}Pb$ diagram. In contrast to group I model
lead ages, group II model lead ages are older than their zircon
ages (Stacey and Stoeser, 1983).

The origin of group II lead can be explained by two different
models: in the first model, lead in the asthenosphere beneath
the entire Shield is assumed to have the same isotopic
composition, and group II lead results from mixing of group I

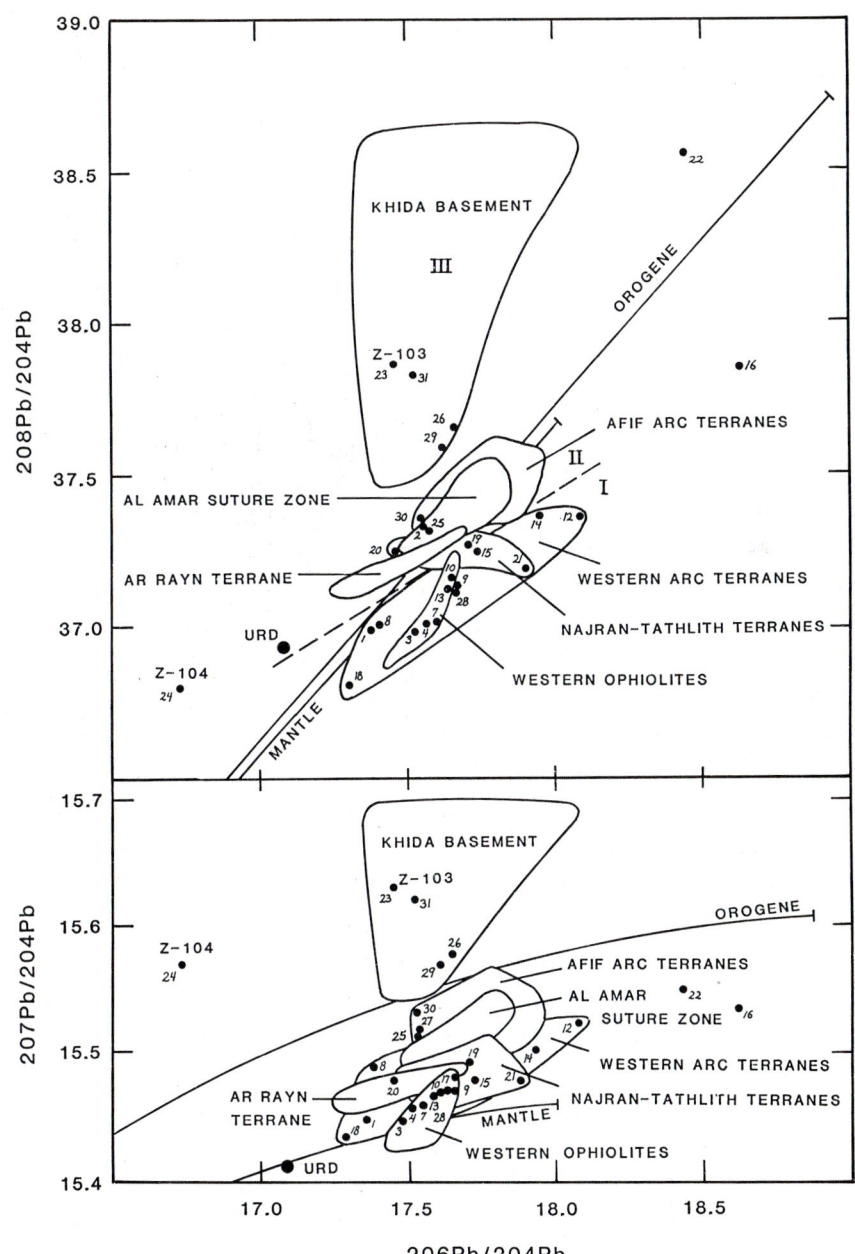

FIGURE 5.

Graph showing common-lead isotope data (Table 3). Terrane field boundaries are compiled from our data and those of Stacey et· al. (1979), Stacey and Stoeser (1983), and Aleinikoff et al. (in press). Average crustal growth curves (orogene) from Stacey and Kramers (1975), and average mantle growth curves from Zartman and Doe (1981).

lead and various proportions of group III continental lead from
the lower crust. In the second model, the isotopic composition
of lead in the asthenosphere beneath the Afif composite terrane
and Ar Rayn terrane is assumed to be different from that
beneath the western arc terranes and the least radiogenic group
II lead, such as that from the Ar Rayn terrane of the Urd
ophiolite (Figs. 4, 5) may be representative of that source.
Regardless of the lead composition of the mantle source, mixing
of mantle and older continental crustal lead is required for
many rocks of the Afif composite terrane. We note, however,
that Nd-Sm isotopic data of Duyverman et al. (1982) for seven
samples from the northern Afif terrane and the Ar Rayn terrane
have ε_{Nd} in the range + 1.6 to + 6.9. These data indicate a
mantle source for the sample materials,but although the authors
maintain otherwise, the data do not permit significant
contamination from an evolved upper crustal source. These
results contrast with the very definitive old continental
signature of ε_{Nd} = -16 at 658 Ma for the Jabal Khida
granodiorite in the south-central Afif terrane (Stacey and
Hedge, 1984).

Group III leads occur within or close to the Khida basement and
clearly reflect the evolved character of that source. New group
III data are presented for the Muhayil anorthosite (loc. 24)
and two granites south of the exposed Khida basement (loc. 26,
29). Similar to group II model lead ages, group III lead model
ages are older than their U-Pb zircon ages, and group III data
plot above the average orogene growth curves (Fig. 5).

Models outlining the evolution of group III leads have been
proposed by Stacey and Stoeser (1983) and Stacey and Hedge
(1984). Group III lead from Khida basement samples such as
granodiorite Z-103 and the Muhayil anorthosite Z-104 (Table 3,
Fig. 5) have high $^{207}Pb/^{204}Pb$ values and comparatively low
$^{206}Pb/^{204}Pb$ that require evolution of the lead in an Archaean
upper crust until the Early Proterozoic. Calculations by Stacey
and Hedge (1984) show that for the 658 Ma Khida sample Z-103,
the $^{238}U/^{204}Pb$ (μ) value would have had an upper crustal value

of at least 13 for evolution between an arbitrarily chosen 2800 Ma and 1628 Ma (the latter being the age of inherited zircons in the sample). At 1628 Ma, lead derived from the Archaean crust was incorporated into a newly forming Khida craton. Then, sometime between 1628 and 658 Ma, a fragment of the Khida basement was detached from this craton, incorporated into the Afif microplate, and transported westward to collide with the western arc terranes. After 1628 Ma, the U-Pb system of the Khida basement fragment remained undisturbed until approximately 658 Ma, at which time remobilization occurred and the Khida granodiorite and Muhayil anorthosite were intruded into the upper crust. Between 1628 Ma and 658 Ma, lead destined for crustal materials, such as the Khida granodiorite, evolved in the lower crust with a depleted μ of approximately 6. In the case of the Muhayil anorthosite, its extremely retarded $^{206}Pb/^{204}Pb$ ratio requires a μ value of only 2 for evolution between 1628 and 658 Ma, which indicates that it was derived from a much deeper crustal level than the Khida granodiorite. These simplistic models demonstrate the history we envisage for lead in group III Khida basement rocks.

The postorogenic granitic rocks of the Afif composite terrane and Ar Rayn terrane presumably represent crustal melts and, as such, derived their lead directly from a source region in the lower crust. The fact that plutonic rocks away from the immediate vicinity of the Khida basement lack group III lead suggests that these regions are not underlain by an older continental basement but rather by oceanic magmatic arc assemblages. If this assumption is correct, then the required mixing of leads in the eastern arc terranes was primarily due to the addition of evolved lead from subducted sedimentary material.

RUBIDIUM-STRONTIUM ISOTOPE RESULTS

Rb-Sr data for some of our samples is presented in Table 4. Except for the younger granites in the Ranyah region, all

Table 4. Rb-Sr whole-rock data for some samples of this study [by kind permission of
C. E. Hedge; analyst, K. Futa. Data normalized to 88Sr/86Sr=0.1194. Analytical
precision for 87Sr/86Sr is ±0.00006 (2σ) between runs. Initial 87Sr/86Sr values
computed using appropriate zircon ages.]

Locality and Sample Number	Rb (ppm)	Sr (ppm)	87Rb/86Sr	87Sr/86Sr	Zircon age (Ma)	Apparent (87Sr/86Sr)$_i$	Sample description

WEST OF NABITAH SUTURE

Southern Al Qarah terrane

1. 111552	0.8	503	0.0522	0.7029	732	0.7024	Talhah tonalite gneiss
2. 112683	0.9	635	0.0452	0.7030	727	0.7025	Suwaydah tonalite gneiss
3. 112805	33.7	838	0.1164	0.7036	675	0.7024	Wadi Arin tonalite
4. 111555	35.4	561	0.1864	0.7042	670	0.7025	Mahda granite
7. 111564	75.2	358	0.6090	0.7083	657	0.7027	Khamis Mushayt granodiorite
8. 111541	42.5	806	0.1526	0.7040	643	0.7026	Al Ar tonalite
10. 111554	101.0	237	1.2386	0.7137	641	0.7024	Jabal Thairwah monzogranite

Northern Al Qarah terrane (Ranyah region)

12. Z-109	106.0	204	1.5110	0.7168	646	0.7026	Shaib Hadhaq monzogranite
15. Z-110	147.0	57	7.4690	0.7696	620	0.7035	Al Jizah monzogranite
16. Z-111	94.6	89	3.0700	0.7308	604	0.7044	Granite

SAMPLES EAST OF NABITAH SUTURE

Najran terrane

19. 128917	38.0	589	0.1906	0.7075	641	0.7029	Wadi Simlal quartz diorite gneiss
20. 128905	49.0	954	0.1486	0.7044	642	0.7031	A'ashiba quartz diorite

Tathlith terrane

22. Z-100	3.9	62	0.1845	0.7054	698	0.7035	Tonalite

South-central Afif composite terrane

23. Z-103*	73.5	509	0.4183	0.71228	658	0.70835	Jabal Khida granodiorite
24. Z-104	0.9	570	0.0047	0.70682	<658	0.70678	Muhayil anorthosite
25. Z-107	86.5	312	0.8032	0.7104	635	0.7031	Granodiorite
26. Z-102	282.0	112	7.3140	0.7669	630	?	Al Bahah granite
27. Z-106	163.0	387	1.2230	0.7140	621	0.7031	Monzogranite
29. Z-108	361.0	10	116.1	1.6867	614	?	Huqban granite
30. Z-101	174.0	198	2.5450	0.7247	592	0.7033	Monzogranite

* Data from Stacey and Hedge (1984).

initial $^{87}Sr/^{86}Sr$ ratios from samples west of the Nabitah suture are less than 0.7027, and initial $^{87}Sr/^{86}Sr$ ratios for samples east of the suture are 0.7029 or greater. At 658 Ma, the Khida basement samples (locs. 23, 24) have elevated initial ratios of 0.7084 and 0.7068, respectively. Furthermore, Stacey and Agar (1985) showed that the $^{87}Sr/^{86}Sr$ for the 1773 Ma Kabid gneiss was 0.714 at 600 Ma. The slightly elevated initial $^{87}Sr/^{86}Sr$ ratio for the arc rocks east of the Nabitah suture relative to those west of the suture can be explained either by the addition of an evolved component to mantle-derived melts or a difference in the mantle source regions. In any case, differences in the strontium data are consistent with the lead data and do not help to distinguish between continental contamination (model 1) and different mantle sources (model 2). In this regard, the 694 ± 8 Ma Urd ophiolite (Fig. 4) from the Ad Dawadimi terrane has an initial $^{87}Sr/^{86}Sr$ ratio of 0.7030 (Stacey et al., 1984) that is slightly elevated relative to those from the western arc terranes, but is not an unreasonable value for mantle derived rock at 700 Ma, and so is also equivocal.

PRE-OROGENIC ARC MAGMATISM

Magmatic arc assemblages are found along both margins of the Nabitah suture (Fig. 6), and each margin will be described separately below.

Nabitah suture-western margin

Three terranes occur along the western margin: Al Qarah, An Nimas, and Hijaz (Fig. 1). For the Al Qarah terrane, we have U-Pb zircon ages for three tonalites of 732 ± 4, 728 ± 5, and 719 ± 9 Ma (loc. 1, 2, 12). In addition, Fleck et al. (1980) have published whole-rock Rb-Sr ages of 724 ± 93 and 723 ± 107 Ma for two quartz diorites and 746 ± 16 and 785 ± 96 Ma for volcanic rocks in the northern part of the

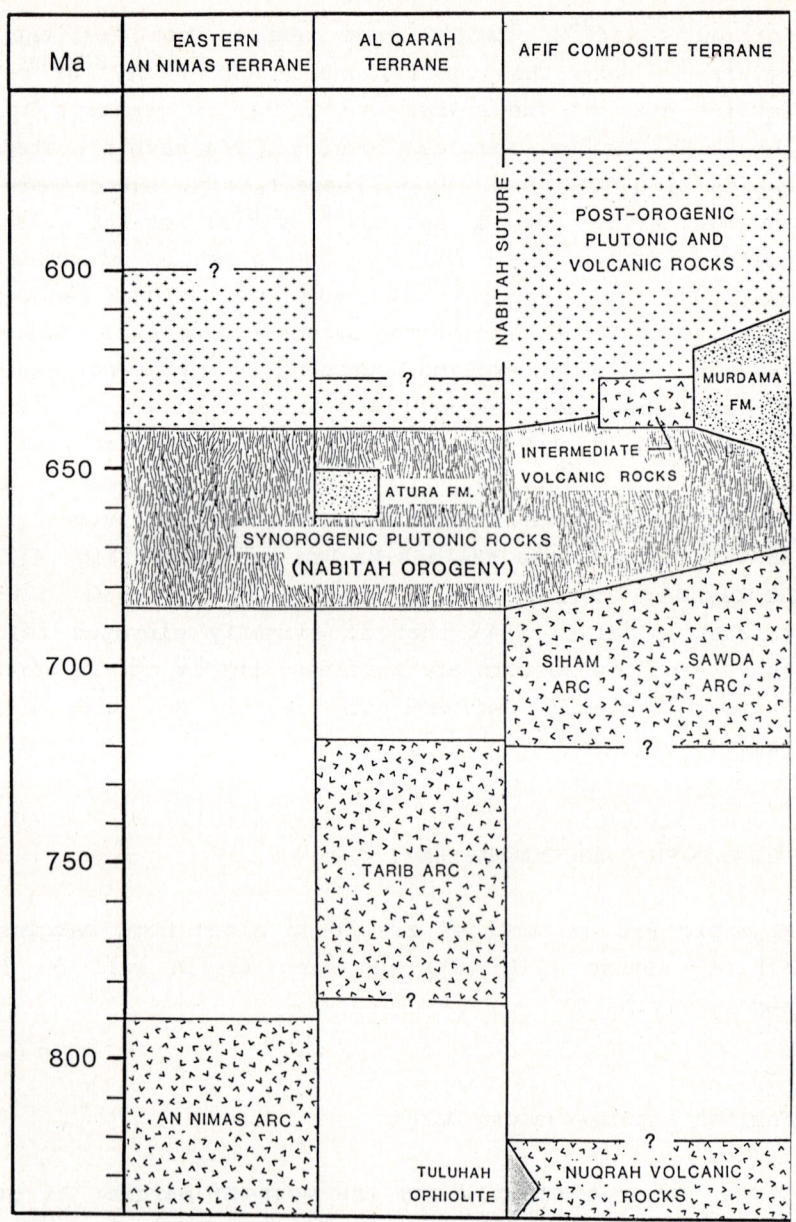

FIGURE 6.

Schematic diagram showing major 840-560 Ma late Proterozoic tectonic events and lithologic units in the southern, central and eastern Arabian Shield. Unit boundaries queried where age limit is inferred. (FM., formation).

AL AMAR SUTURE	AR RAYN TERRANE	TECTONIC PHASE	
		NAJD FAULTING	MILD EXTENSION
			LOCAL EXTENSION OR COMPRESSION
SYNOROGENIC PLUTONIC ROCKS (AL AMAR OROGENY)		COLLISION	COMPRESSION
ABT FORMATION URD OPHIOLITE ?	AL AMAR ARC ?	MAGMATIC ARC FORMATION AND ACCRETION	

terrane. We conclude that the Al Qarah terrane contains a magmatic arc (the Tarib arc of Stoeser and Camp, 1985; Stoeser, 1986) that formed from approximately 785 to 719 Ma (Fig. 6).

Along the central part of the western margin, we have an age of 811 ± 4 Ma for a tonalite from a large north-striking tonalite and trondhjemite batholith (loc. 18). This pluton lies on strike with the An Nimas terrane to the south, which contains 850 - 790 Ma magmatic arc rocks (Cooper et al., 1979; Fleck et al., 1980). Feldspar lead data for the sample are typical of those for ensimatic arc rocks of the southern Asir and, based on these data, we conclude that the ensimatic An Nimas arc extends at least this far north in the Shield. It should be noted, however, that immediately to the west of our sample locality, Calvez and Kemp (1982) have U-Pb zircon and Rb-Sr whole-rock ages of 772 ± 22 and 748 ± 28 Ma for rhyolitic volcanics and 760 ± 10 Ma for a tonalite. They also have a 816 ± 3 Ma zircon age for a tonalite 50 km to the southwest. Thus, along the west-central margin either Tarib-age magmatic rocks intrude a basement of An Nimas age rocks or a younger belt of magmatic rocks lies to the west of our sample locality.

The relationship between the An Nimas and Tarib arcs is unclear. Because there are no ophiolitic rocks in the Junaynah zone and no evidence for any pre-800 Ma basement in the Al Qarah terrane, we conclude that the Wadi Tarib arc formed near the eastern margin of the An Nimas terrane and that only a back-arc basin separated them. This basin may have been closed and deformed when the Asir composite terrane collided with the Tathlith and Najran terranes along the Nabitah suture.

In the northern part of the western Nabitah margin (Hulayfah region), the only data for magmatic arc rocks is the Rb-Sr whole rock age of 720 ± 10 Ma for the Hamra tonalite (Fig. 4) (Calvez et al., 1983).

It appears, therefore, that magmatic arc rocks formed from 811 to 719 Ma along much of the western margin prior to closure on the Nabitah suture. As will be discussed below, the Nabitah suture truncates magmatic belts and no simple relationship exists between the pre-Nabitah western arcs and suture.

Nabitah suture-eastern margin

Arc magmatism along the eastern margin of the Nabitah suture is well studied only in the central part of the belt (Zalm region, Fig. 1). In this area, Agar (1985) and Stacey and Agar (1985) have shown that a belt of basalt and andesite, the Siham arc, formed along the western margin of the Afif microplate. They estimated that the Siham arc formed between 720 + to 690 Ma as a result of eastward-directed subduction from the Nabitah zone prior to suturing.

Because of a lack of adequate geochronologic data for the rest of the eastern margin of the Nabitah suture, it is not known if the Siham arc extends along the entire eastern margin of the suture. South of the Ruwah fault zone, a belt of volcanic and dioritic to tonalitic plutonic rocks lies within the Tathlith terrane. Two tonalites in the northern part of the terrane have ages of 698 ± 3 Ma (loc. 22) and 666 ± 8 Ma (Cooper et al., 1979). The older tonalite has an age similar to the Siham arc rocks which suggests that the Siham arc extends into the Tathlith terrane; the younger tonalite was probably emplaced during the Nabitah orogeny.

To the north in the Nuqrah region (Fig. 1), Delfour (1977) mapped several belts of volcanic rocks. Two "soda-rhyolites" from these rocks have an U-Pb zircon age of 839 ± 23 Ma and a Rb-Sr age of 821 ± 48 Ma (Calvez et al., 1983). The only available age for an ophiolite of the Nabitah suture is an U-Pb zircon age of about 847 - 823 Ma for the Tuluhah ophiolite west of Nuqrah (Fig. 1; Pallister et al., in press). Based on its age, the Tuluhah ophiolite, therefore, is probably related to either the An Nimas magmatic arc (Pallister et al., in press) or the Nuqrah volcanic rocks. Except for the Khida basement, the Nuqrah volcanics are the oldest known rocks within the Afif composite terrane. Because the volcanics of the Nuqrah area appear to be on strike with the Siham volcanics to the south and there are few age data for Nuqrah volcanic rocks, Siham age volcanic rocks may also be present. The one common lead determination for the Nuqrah volcanic rocks belongs to group I (Fig. 4), which along with the age data, suggests that the Nabitah suture may lie to the east of the volcanics rather than to the west as shown in Figure 1.

NABITAH OROGENY

Location of Nabitah suture

The Nabitah suture marks the site of collision of the western arc terranes with the Afif composite terrane (or microplate). Although the Nabitah suture is fairly well defined from the Tuluhah ophiolite at the northern end of the suture to Hamdah at the southern end (Fig. 1), its extension beyond these locations is problematic. At the northern end of the Nabitah belt, the line of ultramafic bodies marking the suture is truncated just north of the Tuluhah ophiolite by the Halaban-Zhargat fault zone and cannot be traced farther north. Schmidt et al. (1979) estimated 125 km of left-lateral displacement along the fault zone, and Cole (1986) calculated a minimum displacement of 105 km, these results suggest that the suture north of the fault zone is concealed beneath Phanerozoic cover or alluvium to the northwest.

South of Hamdah, the location of the Nabitah suture is conjectural. It may continue directly south on what appear to be extensions of the Nabitah fault zone (Johnson et al., 1984; Pallister et al., in press); it may turn southeast along the Mulha fault zone (Warden, 1982), until it either disappears beneath the Phanerozoic cover rocks; or turns south along the Ashara fault zone (Fig. 1; Stoeser and Camp, 1985). The first interpretation is attractive because of its simple geometry;

however, it is not possible to trace any suitable fault zone or tectonic boundary very far to the south, and such a continuation would extend the suture through the pre-suturing Wadi Tarib batholith (Stoeser et al., 1984b). The second interpretation is supported by the fact that at Hamdah the belt of ultramafic complexes turns eastward north of the Mulha fault zone (Fig. 1). The fault zone lies within a 10 to 20 km wide northwest-striking belt of synorogenic intrusive rocks, gneiss domes, deformation and metamorphism as high as the granulite facies that Warden (1982) interpreted as a suture. In support of Warden's suture interpretation, is the fact that the Mulha fault zone does not extend northwestward past the end of the Nabitah fault zone at Hamdah but rather appears to merge with it. The Ashara fault zone interpretation is negated somewhat by the lack of ultramafic complexes indicative of a suture along the zone, however, Conway (1984) reports that volcanic rocks on either side of the zone are different from each other. In addition, as discussed above, the limited feldspar lead data for rocks east of the fault zone (Stacey and Stoeser, 1983) show a transition similar to that seen in the Zalm area, that is, a transition from ensimatic group I lead west of and in the Nabitah suture to group lead II east of the zone (Fig. 4; Stacey and Agar, 1985).

Another fundamental question is if the Afif composite terrane continues south of the Ruwah fault zone and east of the Nabitah suture. Continuation of the terrane is supported by our lead isotopic data as discussed above, the apparent extension of the Shiham arc into the Tathlith terrane, and the report of continentally derived sedimentary rocks northeast of Hamdah (White, 1985) that suggest a continental margin lies to the east. We, therefore, tentatively conclude that the Nabitah suture in the south is coincident with the Nabitah, Mulha, and Ashara fault zones, that the Afif composite terrane extends southwards to include the Tathlith and Najran terranes, and that Khida-type basement may be concealed beneath the Phanerozoic sedimentary cover rocks to the east of these terranes.

Age of Nabitah suture

The only available way to estimate the age of suturing is to bracket it between the cessation of arc magmatism along the margins of the Nabitah suture and the beginning of synorogenic magmatism within the Nabitah orogenic belt.

Based on our data for the Al Qarah terrane, arc magmatism was occurring as late as approximately 720 Ma and synorogenic magmatism had begun by 675 Ma. Therefore, in the south the Nabitah suture closed sometime between 720 and 675 Ma. If the 698 ± 3 Ma tonalite from locality 22 in the northern Tathlith terrane also represents arc magmatism, then closure occurred between about 695 and 670 Ma. Fortunately, the work of Stacey and Agar (1985) on rocks in the central part of the suture allow the time of suturing immediately to the north of the Ruwah fault zone to be fixed within a fairly narrow limit. They show that the Siham arc was active until about 690 Ma and that the earliest synorogenic gneiss was emplaced at 683 ± 9 Ma. Therefore, closure along the Nabitah in the central part of the belt occurred between 690 to 683 Ma.

Nabitah orogenic magmatism

On the basis of field mapping and our age data, we recognize early and late phases of Nabitah synorogenic plutonism in the southern Shield (Stoeser et al., 1984). The early phase, between 675 and 650 Ma, was marked by the intrusion of gneissic leucogranitoid rocks that appear to have been emplaced in a compressional orogenic environment. These early synorogenic plutonic rocks are characterized by gneissic biotite or two-mica trondhjemite, leucogranodiorite, and leucomonzogranite, with granodiorite being the dominant lithology (e.g., locs. 4, 6, 7, Table 2). They are typically weakly peraluminous, and a few are garnet bearing (du Bray, 1983). In addition to the leucogranitoids, some dioritic and tonalitic intrusions were also emplaced during the early phase (locs. 3, 5, Table 2).

Although some of the early phase plutons are simple oval intrusions, most are irregular complexes a hundred or more kilometers long and consisting of multiple intrusions of granodiorite, granite, diorite and gabbro. These complexes typically contain large volumes of remobilized older crust in the form of diffuse to sharply defined, xenolithic masses of amphibolite and dioritic to tonalitic orthogneisses (Schmidt 1981a, b; Stoeser et al., 1984a). In addition to the irregular complexes, the southern Shield also contains classic gneiss domes or antiforms from a few kilometers to tens of kilometers long (Schmidt et al., 1979; Amlas et al., 1984; Stoeser et al., 1984a). Regional metamorphic grade is lower to upper greenschist facies, but the intrusive complexes and gneiss domes have broad metamorphic aureoles typically in the amphibolite and locally in the granulite facies (Schmidt, 1980; Warden, 1982). In addittion, some of the gneiss domes have associated migmatite envelopes up to ten or more kilometers wide (Schmidt et al., 1979; Schmidt, 1980; Stoeser et al., 1984a). These intrusive gneiss complexes are found throughout the Al Qarah, Tathlith, and Najran terranes. In particular, a continuous 450-kilometer-long belt of gneiss complexes extends the length of the Al Qarah terrane (Schmidt et al., 1979; Stoeser et al., 1984b). The early phase of the Nabitah orogeny, therefore, represents a period of massive partial melting and mobilization within the crust.

In the southern Nabitah orogenic belt, the late phase of the Nabitah orogeny (from 650 to 640 Ma) is characterized by a widespread bimodal suite of diorite to mafic tonalite and foliated monzogranite that appear to have been emplaced in a more passive environment in which vertical movement predominated (locs. 8, 9, 13, 19, 20; Stoeser et al., 1984a,b). The late-orogenic plutons probably formed by continued, extensive deep crustal melting and emplacement in the mesozone

after significant compression had ceased but while temperatures
in the crust were still highly elevated.

In the Zalm region of the Nabitah orogenic belt, a
20-kilometer-wide zone of granitoid gneiss lies along the
Nabitah suture. Agar (in press) reports that the gneiss
consists of two suites, an older suite of extremely complex
tonalite-granodiorite gneiss and subordinate diorite and
migmatite and a younger suite of gneissic monzogranite,
diorite, and granodiorite, in which segregation gneisses and
migmatites are rare. Two U-Pb zircon ages are available for the
older suite of gneisses (Stacey and Agar, 1985): 782 ± 4 Ma for
a diorite gneiss from the older complex and 683 ± 9 Ma for a
granodiorite gneiss (Fig. 4). The age and composition of the
older rock suggests that it is remobilized An Nimas arc
intermediate plutonic rock that was incorporated into the
synorogenic gneisses. Stacey and Agar (1985) also present ages
for two gneisses in the western part of the Afif composite
terrane:
696 ± 9 for a granodiorite gneiss and 646 ± 17 Ma for a
foliated monzogranite. The above ages indicate that the timing
of synorogenic plutonism in the central part of the Nabitah
orogenic belt is similar to that in the southern part of the
belt.

No detailed information is available for synorogenic plutonic
rocks of the north-central part of the Nabitah orogenic belt,
and only one age is available for Nabitah-age plutonic rocks in
this region, a Rb-Sr age of 662 ± 31 Ma for the Farqayn
granite, one of the linear granite complexes on the eastern
side of the belt (Calvez et al., 1984). Delfour (1977, 1981)
interpreted an early assemblage of foliated to unfoliated
granitic plutonic and amphibolite-grade metamorphic rocks in
the Nuqrah region (Fig. 1) as older basement. He describes
these rocks as consisting of "migmatite and heterogeneous
quartz diorite and granodiorite" that occur in "belts of
biotite schist and gneiss, together with large areas several
kilometers wide, of more homogeneous granodiorite and granite,

(which) have a very gradational contact with the migmatite".
Delfour also indicates that these complexes typically contain
numerous inclusions of various rock types of which amphibolite
is the most common. Letalenet (1979) also describes a suite of
"synorogenic" heterogeneous leucocratic granodiorite and
granite to the south of Nuqrah that are typically foliated,
schistose, gneissic, or migmatitic. These leucogranitoids
contain biotite and hornblende or muscovite, with or without
garnet.

In the northeastern Shield, an early assemblage of foliated to
unfoliated monzogranite, granodiorite, and monzodiorite
complexes that may be related to the Nabitah orogeny is
widespread throughout the Hail terrane (Fig. 1) (Kellogg,
1983b; Johnson and Williams, 1984; Quick and Doebrich, 1986).
Only two U-Pb zircon ages of ~ 650 Ma and 646 ± 6 Ma are
available for these rocks (Johnson and Williams, 1984; Quick
and Doebrich, 1986).

From these descriptions, it is clear that synorogenic plutonic
rocks extend along the length of the Nabitah belt as originally
proposed by Schmidt et al. (1979). Figure 2 shows the
distribution of granodiorite and monzogranite plutonic rocks
within the Arabian Shield. The Nabitah orogenic belt appears in
the Figure as a broad zone of north-striking linear granitic
complexes and plutons. Stoeser (1986) showed that most granitic
rocks in the eastern half of the Shield are younger than 680 Ma
and that the Shield was formed during and immediately after the
Nabitah orogeny.

Collisional tectonics

Until more detailed structural and stratigraphic studies are
available for the Nabitah orogenic belt, detailed evolution of
the Nabitah orogeny cannot be resolved. Thus, this paper does
not attempt to resolve geometry of closure, collision, and
possible transposition along the Nabitah suture.

Compressional deformation during the Nabitah orogeny, as called for by the present and earlier papers (Schmidt et al., 1979; Stoeser et al., 1984a,b; Agar, 1985; Stacey and Agar, 1985), surely must have involved significant crustal shortening including the development of thrust zones in the upper crust. Recent mapping has begun to identify thrust belts within the eastern Shield. Agar (1985, in press) describes an eastward-vergent thrust belt in the Zalm region along the western margin of the Afif composite terrane, and Delfour (1977) also shows similar thrusting locally along the western margin to the north in the Nuqrah region. To the northeast near the northeastern margin of the Shield, Johnson and Williams (1984) and Cole (1986) mapped major north to northwest vergent thrust fault zones (Fig. 1). In the southern Shield, thrust faults have not been mapped except in the Hamdah area, where Worl and Elsass (1980) have interpreted ultramafic complexes as occurring in highly deformed and eroded thrust sheets. These reports of local thrusting along the eastern margin of the Nabitah suture suggest that detailed mapping along the Nabitah suture may ultimately show that a major thrust belt occurs along much of the orogenic belt.

The early synorogenic gneiss complexes of the Al Qarah terrane (675-650 Ma) are typical of catazonal plutons (Agar, 1986a) as defined by Buddington (1959), whereas the late orogenic plutons (650-640 Ma) are more typical of the mesozone. These rocks are all intruded by postorogenic elliptical discordant granite plutons (640-628 Ma) that lack metamorphic aureoles and are characteristic of the deeper part of the epizone (Stoeser et al., 1984b; Aleinikoff and Stoeser, in press). Even assuming a highly elevated geotherm, catazonal intrusive rocks and migmatites do not form above about 18-20 km depth (Wyllie, 1977; Hyndman, 1981). Massive deep-crustal mobilization and melting, however, may cause catazonal complexes and their metamorphic envelopes to rise into the lower part of the mesozone (Flood and Vernon, 1978).

The intrusion of 675 - 650 Ma catazonal rocks by 640 - 628 Ma epizonal granites implies a major episode of uplift and erosion during which as much as 10 - 20 km of crust was removed during a relatively short period of time. The erosional products of this uplift are preserved in the northern Shield as clastic sediments of the Murdama group (see below), but are rarely found in the southern Shield because of erosion after the emplacement of the postorogenic granites but before deposition of Cambro-Ordovician epicontinental sandstones (Stoeser, 1986). An exception is the Atura formation (Fairer, 1983) in the southernmost part of the Al Qarah terrane (Fig. 6), which is deformed and metamorphosed, dips vertically, and contains stretched pebble conglomerates rich in granitic clasts. The Atura sediments must have been synorogenic, and deposited after Nabitah-age granitic rocks were exposed to erosion but while compressional deformation was still occurring.

Regional compression and uplift is also described by Schmidt et al. (1979), who observed deep-crustal rocks along the western flank of the Al Qarah and Tathlith terranes and shallow upper crustal volcanic and plutonic rocks on the eastern flanks. The gneisses of the Al Qarah belt in particular probably were emplaced in an environment that involved westward mass transport and crustal uplift (Schmidt et al., 1979; Stoeser et al., 1984d). This westward-directed mass transport is interpreted to have resulted in thrusts and nappes that have since largely been removed by erosion. An interpretation of regional gravity data for the southern Shield (Gettings, 1984) supports the differential-crustal-uplift hypothesis of Schmidt et al.

Another tectonic feature of the southern Shield is the manner in which both the Mulha and Ruwah fault zones truncate lithologic belts. The Mulha zone truncates the Malahah belt of volcanic rocks that extends northward from the Yemen border (Fig. 1), and the Ruwah zone truncates the Wadi Tarib magmatic belt. Although these fault zones have a Najd-like trend, they are not simple left-lateral wrench fault zones because the truncated belts do not reappear north of them, moreover, both fault zones appear to turn northward at their northwestern ends and merge with the Nabitah suture. Both zones also appear to have controlled the emplacement of plutons, a characteristic which suggests they penetrate deep into the crust. On the basis of data from a deep seismic refraction line across the Shield (Prodehl, 1985), Kröner (1985) suggests that several shallow dipping, low-velocity zones represent interstacking of thick crustal slices. One possible interpretation of the truncation of the Malahah and Tarib belts is that the crust to the east of the Nabitah suture has overridden the truncated belts and that the Mulha and Ruwah zones early in their history may have been analogous to the tear faults found in thrust belts. Although at present there is little evidence to support interpretations of horizontal thick-skinned tectonics within the eastern part of the Shield, we follow Kröner in thinking that crustal over- or underthrusting related to collision may have occurred within the Shield and should be considered in future tectonic studies.

AL AMAR OROGENY

The Al Amar suture zone in the eastern Shield is bounded on the west by the Afif composite terrane and on the east by the Ar Rayn terrane (Fig. 1). Stoeser and Camp (1985) and Cole (1986) propose that a belt of 689 - 667 Ma dioritic and plutonic rocks along the eastern margin of the Afif composite terrane represent a magmatic arc. These rocks are only slightly younger than and may be related to the Urd ophiolite of 694 ± 8 Ma age just to the east (Figs. 4, 6; Stacey et al., 1984). Magmatic-arc-type intermediate volcanic rocks are present within the Ar Rayn terrane to the east of the suture (Coulomb et al., 1981; Le Bel and Laval, 1986), but their age is only known to be older than 670 Ma (Calvez et al., 1984; Stacey et al., 1984). Arc magmatism within the Ar Rayn terrane stopped about 670 Ma and synorogenic leucocratic tonalite to monzogranite intrusions were emplaced within the Al Amar suture zone and the Ar Rayn terrane between 670 and 640 Ma (Coulomb et al., 1981; Calvez et al., 1984; Stacey et al., 1984; Le Bel and Laval, 1986). This intrusive activity was followed by the widespread emplacement of late orogenic diorite and tonalite in the Ar Rayn terrane 640 and 630 Ma. The Al Amar orogeny thus occurred from 670 to 630 Ma and was mostly contemporaneous with the Nabitah orogeny. Orogenesis in the eastern half of the Shield was localized along the Nabitah belt between 680 and 670 Ma, but spread throughout that region between 670 and 640 Ma. Because concordia lower intercepts for Khida basement rocks typically are in the range 690 to 620 Ma, we assume that the Khida basement was largely remobilized during this orogenesis (Stacey and Hedge, 1984; Stacey and Agar, 1985; Stacey, unpublished data).

MURDAMA SEDIMENTATION

The Murdama group of the central and northern Shield consists of coarse to fine-grained clastic sedimentary rocks, intercalated locally with minor limestone and volcanic rocks (Delfour, 1977; Letalenet, 1979; Greene, 1983; Cole, 1986; Wallace and Rowley, in press). Murdama rocks occur in one major basin, the Murdama basin, and in a smaller basin, the southern Murdama basin (Fig. 1), as well as in a number of other minor basins. Both the major and southern basins are on the eastern flank of the Nabitah orogenic belt (Fig. 1).

Wallace and Rowley (in press) have studied the southern part of the major basin and believe sedimentation occurred between 660 and 620 Ma. They indicate that the lower part of the Murdama group consists dominantly of coarse volcaniclastic sediments and significant intercalated intermediate volcanic rocks and that the upper part of the Murdama consists chiefly of fine clastic sediments and contains few volcanic rocks. They show that the direction of sediment transport was to the east and that the environment of deposition was probably subaerial in the west and marine in the east. They were unable to determine the total thickness of the Murdama group but based on

available literature estimate it between 2 and 6 km. After
deposition, the Murdama was deformed and thrust eastward along
the eastern margin of the Afif composite terrane.

Cole (1986) indicates that in the northeastern part of the
Murdama basin, the Murdama was deposited in a marine
environment between 670 and 650 Ma. Subsequent to 642 Ma, the
Murdama group was deformed during several phases of regional
compression and thrust to the north and northwest over 632 Ma
intermediate composition volcanic rocks (Johnson and Williams,
1984; Cole, 1986).

Schmidt (1981b) and Kellogg (1983a), propose that early
sediments of the southern Murdama basin were derived from
highlands to the west that had been uplifted during the
culminant orogeny of Schmidt et al. (1979) and Schmidt and
Brown (1982; our Nabitah orogeny). The total thickness of
sediments is not known but is at least 2 to 3 km (Schmidt,
1981). These sediments were involved in at least one major
episode of compressional deformation subsequent to the Nabitah
orogeny but prior to about 610 to 600 Ma (Schmidt, 1981b;
Kellogg, 1983a). We note that our 635 to 630 Ma granites
(loc. 25, 26) in the south-central Afif composite terrane are
deformed, whereas, our 621 to 614 Ma granites (loc. 27, 29)
from the same region are not. Deformation of the Murdama in the
southern Murdama basin, therefore, may have been prior to
620 Ma.

The above data indicate that the early part of Murdama was
synorogenic and resulted from erosion of the Nabitah highlands
along the Nabitah orogenic belt. Many workers also note that
the Murdama has been deformed into steep folds in many places
(e.g. Letalenet, 1979; Schmidt, 1981b), and involved in
thrusting along the eastern and northern margins of the main
Murdama basin (Johnson and Williams, 1984; Cole, 1986; Wallace
and Rowley, in press). Although some of this deformation
probably occurred during the later part of the Nabitah orogeny,

compressional tectonism continued throughout parts or all of the central and northern Shield between 640 and 620 Ma.

NAJD FAULT SYSTEM

Although the fundamental nature of the Najd fault system has been known since the earliest days of geological research on the Arabian Shield, its genesis and age of inception is still poorly understood. It is primarily a major set of northwest-striking left-lateral transcurrent faults within the northern part of the Shield (Brown, 1972; Moore, 1979). North of the Halaban-Zarghat fault zone, however, Najd-style faults are insignificant. South of the Ruwah fault zone, significant but less prominent Najd faults are found as far south as latitude 18°N. Early workers interpreted the Najd fault system as having formed very late in the Shield (typically 620 - 510 Ma, e.g. Brown, 1972), but recent work indicates it is older.

Cole (1986) interpreted two main episodes of left lateral displacement along the central part of the Halaban-Zarghat fault zone, an early phase at about 670 Ma that has approximately 40 km of displacement and a late phase between about 650 and 640 Ma after deposition of the Murdama group with an additional displacement of 65 km. In the southern Afif composite terrane, Agar (1986b) has concluded that Najd faulting began about 640 Ma with approximately 20 Ma of right-lateral displacement related to extensional grabens, after which faulting was left lateral. Davies (1982, 1984) interpreted a group of granitic plutons in the northwestern part of the Shield as syntectonic with Najd faults of that region. Hedge (1984) dated several of these plutons and obtained U-Pb zircon ages of 676 ± 4, 672 ± 30, and 660 ± 4 Ma. As previously noted, syntectonic plutonic rocks are also associated with the Mulha fault zone in the southern Shield. All of the above data suggest the Najd fault system was active by the end of the Nabitah orogeny (i.e., 640 Ma).

The Ruwah fault zone of Thieme (in press) is a spectacular 15 to 20-kilometer-wide shear zone. Early movement on the zone was left lateral; and was followed by a later episode of vertical tectonism within the zone involving metamorphism to the amphibolite grade, intense deformation, ductile emplacement of fault controlled gneiss domes, and intrusion of syntectonic granitic plutons (Schmidt, 1981b; Kellogg, 1983b). Our sample Z-101 (loc. 30) of a cataclastic hornblende monzogranite within the zone, and its age of 592 ± 4 Ma should date this late episode of Najd tectonism.

Postorogenic granites were emplaced throughout much of the Shield between 640 and 570 Ma concurrent with Najd faulting (see reviews by Jackson, 1986; Ramsay, 1986; Stoeser, 1986). Most postorogenic plutonism south of the Ruwah fault zone had stopped by 625 Ma, whereas in the northern part of the Shield significant plutonism and volcanism continued to about 570 Ma. This postorogenic magmatism was mostly silicic but also involved cogenetic gabbro intrusions, particularly in the southern Shield, and a major episode of intermediate volcanism and plutonism throughout much of the Afif composite terrane between approximately 645 and 620 ma (Fig. 6; Roobol et al., 1983; Johnson and Williams, 1984; Stacey et al., 1984; Agar, 1986b; Cole, 1986; Theime, in press). Between 600 and 570 Ma, most magmatism was restricted to highly evolved peraluminous to peralkaline granitic intrusions which probably formed in an environment of mild extension. In particular, within the eastern part of the Shield, peralkaline granites are largely restricted to the Nabitah orogenic belt (Stoeser, 1986). Displacement on the Najd faults appears to have continued during this time, but movement was minor. After 570 Ma, tectonic activity within the Shield waned, and peneplanation of the Shield was followed by the deposition of Cambro-Ordovician epicontinental sands from both the north and south (Brown, 1972).

CONCLUSIONS

In this paper, we recognize a continuous period of orogenesis and tectonism in the central and eastern Arabian Shield between 680 and 600 Ma, that is subdivided into two phases: (1) an early phase, between 685 and 630 Ma, during which compressional orogenesis involving crustal shortening and synorogenic plutonism occurred as a result of collisions along the Nabitah and Al Amar sutures, and (2) a late phase, between 630 Ma and approximately 600 Ma, during which continued compressional stress from the east resulted in a major left-lateral transcurrent fault system, the Najd fault system.

Caby (1984) and Bentor (1985) have asserted that no continental collision has occurred within the Arabian Shield because of a lack of evidence for major thrust faulting and nappes to indicate crustal shortening, and by the overall low grade of metamorphism exhibited within the Shield. In addition, Almond (1984) and Bentor (1985) have also cited a lack of evidence for uplift related to orogenic mountain building. Earlier in this paper, we presented evidence that the early phase of orogenesis resulted in the Nabitah orogenic belt, a 100 to 200-kilometer-wide north-striking belt of: (1) early synorogenic gneiss domes and complexes of leucocratic metaluminous to peraluminous granodiorite, trondhjemite, and monzogranite and a late synorogenic suite of monzogranite, diorite, and tonalite; (2) regional lower to upper greenschist grade metamorphism, including broad amphibolite to granulite grade aureoles and widespread migmatite; and (3) associated major basins of thick detrital sedimentary rocks resulting from orogenic uplift and erosion. In addition, we presented evidence that up to 10 - 20 km of crust had been removed by uplift and erosion prior to 640 Ma. We conclude that these features of the Nabitah orogenic belt are typical of orogenes produced by continental collision. High grade metamorphic rocks are only common in the southern part of the Nabitah orogenic belt, which may reflect either a greater depth of erosion in the south (Stoeser, 1986), or a greater degree of orogenesis, or both. The problem of crustal shortening is more difficult. Earlier in the paper we attempted to show that there is a growing body of evidence for widespread thrusting throughout the Shield east of the Nabitah suture. The presence of nappes have not been demonstrated, although they have been proposed by Schmidt et al. (1979). To date no structurally oriented studies have been published on the Nabitah orogenic belt and regional interpretations have had to largely rely on reconnaissance quadrangle mapping.

The Nabitah orogeny (685 - 640 Ma) resulted from the collison between the Afif composite terrane and a composite terrane of accreted ensimatic magmatic arcs in the west. It was partly contemporaneous with the Al Amar orogeny (670 - 630 Ma) of the eastern Shield that resulted from collision between the Ar Rayn and Afif composite terranes. Collectively, these collisions caused crustal remobilization throughout the eastern half of the Arabian Shield.

During the late or Najd tectonic phase (630 - 600), local stress regimes within the eastern central and northern Shield varied and involved both compressional and extensional deformation. This resulted in folding of the Murdama sediments, the formation of pull-apart basins along the Najd faults (Moore, 1979), and extensional grabens in at least the central Shield (Agar, 1986b). During the late phase, widespread postorogenic granites were emplaced throughout the Shield, presumably related to crustal thickening as a result of the earlier collisions (Stoeser, 1986). Minor tectonic activity continued from 600 to 570 Ma, and included minor movement on the Najd fault system and A-type anorogenic peraluminous to peralkaline silicic magmatism.

Some geologists have concluded that the Najd fault system resulted from a collision between the Arabian Shield and a continental plate to the east now concealed beneath Phanerozoic sedimentary cover (Greenwood et al., 1976, 1982; Schmidt et al., 1979; Fleck et al., 1980; Davies, 1984).On this basis they applied the rigid indentor model for the Himalayan collisional orogene of Tapponnier and Molnar (1976) to explain the formation of the Najd fault system. Stern (1985) has objected that the Najd fault system is not related to continental collison based on the orientation of the Najd faults, the asynchroneity of Najd faulting relative to Shield orogenesis, and the lack of evidence for major uplift during the time of Najd faulting. As we have discussed earlier in our paper we do not consider the Najd faulting as a part of the collisional orogeny, but rather as a tectonic development at the end of this orogeny (i.e. 640 Ma). Stern concluded that orogenesis ended at 640 Ma within the Shield and that the Najd system was not initiated until 620 Ma. As we have shown above this conclusion is invalid, and that more current work than cited by Stern has shown that the Najd system was active by 640 Ma. Given that regional stress was being taken up by lateral displacement along the Najd fault system after 640 Ma, we would not expect significant vertical uplift at this time.

Although we do not view the Najd fault system as a part of the 690 - 630 ma collisonal orogenesis, we note that Stoeser and Camp (1985) and Burke and Sengör (in press) have suggested that the Najd fault system is the product of extrusion tectonics similar to the model proposed by Tapponier et al. (1982) to explain the eastwards displacement of eastern Asia as a result of the Indus-Tibet collision. As applied to the Arabian Shield, the model assumes that by 640 Ma the Shield had accreted to and was being buttressed by the African craton, such that compressional stress due to continued plate convergence was relieved by extrusion of the northern part of the Shield northwestward along the faults of the Najd system into a nonbuttressed region.

Stern (1985) has also objected to the continental collisional hypothesis for the Arabian Shield on the basis of the lack of older continental crust east of the Al Amar (or Idsas) suture. We agree that it is still not proven that a major continental plate was involved in the Arabian collisional regime. Such a plate is not exposed within the Arabian Shield, and if it exists is concealed beneath the Phanerozoic cover rocks of central and eastern Saudi Arabia. The Ar Rayn terrane has been interpreted as the leading edge of such a continental plate (Schmidt et al., 1979; Fleck et al., 1980; Davies, 1984), but isotopic studies by Calvez et al. (1984), Stacey and Stoeser (1983), and Stacey et al. (1984) fail to sustain this proposal. Instead, the Ar Rayn terrane appears to be a late Proterozoic ensimatic magmatic arc without an older basement. Because significant orogenesis occurred when the Ar Rayn plate collided with the Afif plate along the Al Amar suture, the Ar Rayn terrane may have been a marginal arc on the leading edge of a major plate to the east. In addition, development of the Najd fault system requires some form of significant plate interaction to the east after 640 Ma. We would also note that the continental collisional hypothesis does not require an older evolved continental plate to the east, only a large plate of upper crust. A plate composed primarily of accreted island arcs and of late Proterozoic age, for example, would do just as well.

There is little direct evidence for a continental plate east of the Al Amar suture zone: there are no basement exposures in the central Arabian Peninsula, no bore holes which reach crystalline basement, and no available deep seismic surveys. The only exposures of Precambrian basement in the eastern part of the Arabian Peninsula are in the Muscat region of Oman. In that region, a 834 Ma granite near Jabal Ja'alan intrudes a metasedimentary basement. Stern (1985) has claimed that the age of the Ja'alan granite demonstrates the lack of older continental crust in the Muscat region. It is not the age of the granite which is the critical issue, however, but the nature of the crust underlying the granite. Lead isotopic data for feldspar from the granite and a vein galena from the metasedimentary rocks are similar to those for samples of the Khida basement (Stacey et al., 1980; J.S. Stacey and J.S. Pallister, unpublished data), indicating that there is an older evolved basement beneath eastern Oman. In addition, Stacey and Stoeser (1983) report lead data for a Tertiary galena from northeastern Yemen, that has lead isotopic ratios similar to those in samples from Oman and the Khida basement. We also note our earlier conclusion that the Tathlith and Najran terranes represent a continental margin similar to the western margin of the Afif composite terrane. We, therefore, tentatively infer that at least the southern part of the Arabian Peninsula is underlain by an evolved continental basement of Archaean to early Proterozoic age that may at least in part be similar in character to the Khida basement. The Khida basement in turn may represent a fragment derived from that source region.

We conclude by noting that although the general tectonic framework of the Arabian Shield is now fairly well known, very little detailed stratigraphic or structural information is presently available for the terranes, terrane boundaries, and suture zones of the Shield. Until such studies are conducted, regional tectonic interpretations for the Shield will remain highly speculative.

ACKNOWLEDGEMENTS

The work on which this report is based was performed in accordance with an agreement between the Saudi Arabian Ministry of Petroleum and Mineral Resources and the U.S. Geological Survey. R.C. Greene, M.R. Brock, and J.S. Pallister aided in sample selection in the Ranyah and south-central Afif regions. Analytical work was efficiently carried out by L.B. Fischer, and the tedious work of mineral separations was shared by J. Waldhoff, R. Ogg and D. Plume. The manuscript was greatly improved by the reviews of D.L. Schmidt, F.S. Simons, J.L. Wooden, T. Reischmann, W. Todt, and an anonymous reviewer.

REFERENCES

Agar, R.A., 1985. Stratigraphy and paleogeography of the Siham
    group: direct evidence for a Late Proterozoic continental
    microplate and active continental margin in the Saudi
    Arabian Shield. J. Geol. Soc. London, 142, 1205-1220.
Agar, R.A., 1986a. Structural geology of felsic plutonic rocks
    in the Arabian Shield: styles, modes and levels of
    emplacement. J. African Earth Sci., 4, 105-121.
Agar, R.A., 1986b. The Bani Ghayy Group: sedimentation and
    volcanism in pull-apart grabens of the Najd Strike-slip
    orogen, Saudi Arabian Shield. Precambrian Res. 31,
    259-274.
Agar, R.A., in press: Geology of the Zalm quadrangle, sheet
    22F, Kingdom of Saudi Arabia. Saudi Arabian Deputy
    Ministry for Mineral Resources Geoscience Map GM-89, scale
    1 : 250 000.
Almond, D.C., 1984. The concept of "Pan-African Episode" and
    "Mozambique Belt" in relation to the Geology of East and
    North-east Africa. Faculty Earth Sci., King Abdulaziz
    Univ. Jiddah, Bull. 6, 71-88.
Al-Rehaile, M., and A.J. Warden, 1980. Comparison of the Bi'r
    Umq and Hamda ultramafic complexes, Saudi Arabia. In:
    Evolution and Mineralization of the Arabian-Nubian Shield,
    Al-Shanti, A.M.S., (Editor), King Abdulaziz Univ. Inst. of
    Applied Geol. Bull. 3, v. 4, Pergamon Press, Oxford-New
    York, 143-156.
Al-Shanti, A.M.S., and A.H.G. Mitchell, 1976. Late Precambrian
    subduction and collision in the Al Amar-Idsas region,
    Arabian Shield, Kingdom of Saudi Arabia. Tectonophysics,
    30, T41-T47.
Al-Shanti, A.M.S., and I.G. Gass, 1983. The Uper Proterozoic
    ophiolite melange zones of the easternmost Arabian Shield.
    J. Geol. Soc. London, 140, 867-876.
Aleinikoff, J.N., and D.B. Stoeser, in press. Zircon morphology
    and U-Pb geochronology of seven Pan-African metaluminous
    and peralkaline granite complexes of the Arabian Shield.
    Saudi Arabian Deputy Ministry of Mineral Resources Open-
    File Report.
Amlas, M., A.W. Basahel, and S.R. Divi, 1984. Polyphase
    deformation in a dome-and-mushroom structure near Khamis
    Mushyat, Southern Arabian Shield. Faculty Earth Sci., King
    Abdulaziz Univ., Jiddah, Bull. 6, 409-420.
Bakor, A.R., I.G. Gass and C. Neary, 1976. Jabal al Wask,
    northwest Saudi Arabia: An Eocambrian back-arc ophiolite.
    Earth Planet. Sci. Lett. 30, 1-9.
Baubron, J.C., J. Delfour and Y. Vialette, 1976.
    Geochronological measurements (Rb/Sr; K/Ar) on rocks of
    Saudi Arabia. Bureau Rech. Geol. Min. Open-File Report
    76-JED-22, 152 p.
Bentor, Y.K., 1985. The crustal evolution of the Arabo-Nubian
    massif, with special reference to the Sinai Peninsula.
    Precambrian Res. 28, 1-74.
Bokhari, F.Y. and J.C. Kramers, 1981. Island arc character and
    later Precambrian age of volcanics at Wadi Shwas, Saudi
    Arabia: geochemical and Sr and Nd isotopic evidence. Earth
    Planet. Sci. lett. 54, 409-422.

Brown, G.F., 1972. Tectonic map of the Arabian Peninsula. Saudi Arabian Directorate General of Mineral Resources Map Arabian Peninsula AP-2.

Brown,. G.F. and R.G. Coleman, 1972. The tectonic framework of the Arabian Peninsula. Int. Geol. Congr. XXIV, Sec. 3, 300-305.

Buddington, A.F., 1959. Granite Emplacement with special reference to North America. Geol. Soc. Amer. Bull. 70, 671-747.

Burke, K. and C. Sengör, in press. Tectonic escape in the evolution of the continental crust. In: Deep Structure of the Earth's Crust, Geodynamics Series, v. 14, L. Brown and M. Barazangi (Editors), Amer. Geophys. Union.

Caby, R., 1984. Pan-African crustal evolution of the Touareq Shield (Central Sahara) and the Arabian Shield: a comparison. Faculty Earth Sci., King Abdulaziz Univ., Jiddah, Bull. 6, 89-100.

Calvez, J-Y., C. Alsac, J. Delfour, J. Kemp and C. Pellaton, 1983. Geological evolution of western, central and eastern parts of the northern Precambrian Shield, Kingdom of Saudi Arabia. Saudi Arabian Deputy Ministry for Mineral Resources Open-File Report BRGM-OF-03-17, 57 p.

Calvez, J-Y., C. Alsac, J. Delfour, J. Kemp and C. Pellaton, 1984. Geological evolution of western, central and eastern parts of the northern Precambrian Shield, Kingdom of Saudi Arabia. Faculty Earth Sci., King Abdulaziz Univ., Jiddah, Bull, 6, 24-49.

Calvez, J-Y. and J. Kemp, 1982. Geochronological investigations in the Mahd Adh Dhahab quadrangle, Central Arabian Shield. Saudi Arabian Deputy Ministry for Mineral Resources Technical Record BRGM-TR-02-5, 41 p.

Camp, V.E., 1984. Island arcs and their role in the evolution of the western Arabian Shield. Geol. Soc. Amer. Bull. 95, 913-921.

Cole, J.C., 1986. Geology of the Aban Al Ahmar quadrangle, Sheet 25F, Kingdom of Saudi Arabia. Saudi Arabian Deputy Ministry for Mineral Resources, Open-File Report USGS-OF-04-9, 86 p., scale 1 : 250 000.

Coney, P.J., D.L. Jones and J.W.H. Monger, 1980. Cordilleran suspect terranes. Nature, 288, 329-333.

Conway, C.M., 1984. Geology and regional setting of the Al Masane ancient mine area, southeastern Arabian Shield, Kingdom of Saudi Arabia. Saudi Arabian Deputy Ministry for Mineral Resources Open-File Report USGS-OF-04-53, 97 p.

Cooper, J.A., J.S. Stacey, D.B. Stoeser and R.J. Fleck, 1979. An evaluation of the zircon method of isotopic dating in the southern Arabian craton. Contrib. Mineral. Petrol. 68, 429-439.

Coulomb, J.J., J. Felenc and J. Testard, 1981. Volcanisme et mineralisations a Zn-Cu de la ceinture d'Al Amar (Royaume d'Arabie Saoudite). Bull. Bureau Rech. Géol. Min., Paris (Deuxième Serie), 2, 41-71.

Davies, F.B., 1982. Pan-African granite intrusion in response to tectonic volume changes in a ductile shear zone from northern Saudi Arabia. J. Geol. 90, 467-483.

Davies, F.B., 1984. Strain analysis of wrench faults and
    collision tectonics of the Arabian-Nubian Shield. J. Geol.
    92, 37-54.
Delfour, J., 1977. Geology of the Nuqrah Quadrangle, 25E,
    Kingdom of Saudi Arabia. Saudi Arabian Directorate General
    of Mineral Resources Geologic Map GM-28, scale
    1 : 250 000.
Delfour, J., 1981. Geologic, tectonic and metallogenic
    evolution of the northern part of the Precambrian Arabian
    Shield (Kingdom of Saudi Arabia). Bull. Bureau Rech. Géol.
    Min. (deuxième serie) nos. 1-2, Section II, 1980-1981,
    1-19.
Du Bray, E.A., 1983. Petrology of muscovite-bearing granitiods
    in the eastern and southeastern Arabian Shield. Deputy
    Ministry for Mineral resources Open-file Report,
    USGS-OF-03-10, 36 p.
Duyverman, H.H.,N.B.W. Harris and C.J. Hawkesworth, 1982.
    Crustal accretion in the Pan African: Nd and Sr isotope
    evidence from the Arabian Shield. Earth Planet. Sci. Lett.
    59, 315-326.
Duyverman, H.J., 1984. Late Precambrian granitic and volcanic
    rocks and their relation to the cratonization of the
    Arabian Shield. Faculty Earth Sci., King Abdulaziz Univ.,
    Jiddah, Bull. 6, 50-70.
Fairer, G.M., 1983. Geology of the Wadi Baysh quadrangle, sheet
    17F, Kingdom of Saudi Arabia. Saudi Arabian Deputy
    Ministry for Mineral Resources Open-File Report
    USGS-OF-03-57, 49 p.
Fischer, L.B., 1986. Microwave dissolution of geologic
    materials: Application to isotope dilution analysis.
    Analytical Chemistry, Jan., 261-263.
Fleck, R.J., R.G. Coleman, H.R. Cornwall, W.R. Greenwood,
    D.G. Hadley, D.L.Schmidt, W.C. Prinz and J.C. Ratte, 1976.
    Geochronology of the Arabian Shield, western Saudi Arabia:
    K-Ar results. Geol. Soc. Amer. Bull. 87, 9-21.
Fleck, R.J., W.R. Greenwood, D.G. Hadley, R.E. Anderson and
    D.L. Schmidt, 1980. Rubidium-strontium geochronology and
    plate-tectonic evolution of the southern part of the
    Arabian Shield. U.S. Geol. Surv. Professional paper 1131,
    38 p.
Flood, R.H. and R.H. Vernon, 1978. The Cooma granodiorite,
    Australia: an example of in situ crustal anatexis?.
    Geology, 6, 81-84.
Frisch, W. and A. Al-Shanti, 1977. Ophiolite belts and the
    collision of island arcs in the Arabian Shield.
    Tectonophysics, 43, 293-306.
Gass, I.G., 1981. Pan-African (Upper Proterozoic) plate
    tectonics of the Arabian-Nubian Shield. In: Precambrian
    Plate Tectonics, A. Kröner (Editor), Elsevier, Amsterdam,
    387-405.
Gettings, M.E., 1984. The isostatic gravity anomaly field of
    southwestern Saudi Arabia and its interpretation. Saudi
    Arabian Deputy Ministry for Mineral Resources Technical
    Record USGS-TR-04-22, 107 p.
Goldich, S.S. and L.B. Fischer, 1986. Air-abrasion experiments
    in U-Pb dating of zircon. Chemical Geology, 58, 195-215.

Greene, R.C., 1980. Reconnaissance geology of the Ranyah Quadrangle, sheet 21/42D, Kingdom of Saudi Arabia. Directorate General of Mineral resources, Ministry of Petroleum and Mineral Resources, Jiddah, Saudi Arabia. Scale 1 : 100 000.

Greene, R.C., 1983. Stratigraphy of the Murdama Formation between Afif, Halaban, As Sawadah, Kingdom of Saudi Arabia. Saudi Arabian Deputy Ministry for Mineral Resources Open-File Report DM-OF-03-2, 35 p.

Greenwood, W.R., D.G. Hadley, R.E. Anderson, R.J. Fleck and D.L. Schmidt, 1976. Later Proterozoic cratonization in southwestern Saudi Arabia. Philos. Transact. Royal Soc. London, Series A, 280, 517-527.

Greenwood, W.R., D.B. Stoeser, R.J. Fleck and J.S. Stacey, 1982. Late Proterozoic island-arc complexes and tectonic belts in the southern part of the Arabian Shield, Kingdom of Saudi Arabia. Saudi Arabian Deputy Ministry for Mineral Resources Open-File Report USGS-OF-02-8, 46 p.

Hedge, C.E., 1984. Precambrian geochronology of part of northwestern Saudi Arabia. Saudi Arabian Deputy Ministry for Mineral Resources Open-File Report USGS-OF-04-31, 12 p.

Hyndman, S.W., 1981. Controls on source and depth of emplacement of granitic magma. Geology, 9, 244-249.

Jackaman, B., 1972. Genetic and environment factors controlling the formation of massive sulfide deposits of Wadi Bidah and Wadi Wassat. Saudi Arabian Directorate of Mineral Resources Technical Record TR-2-1, 244 p.

Jackson, N.J., J.N. Walsh and E. Pegram, 1984. Geology, geochemistry and petrogenesis of late Precambrian granitoids in the central Hijaz region of the Arabian Shield. Contrib. Mineral. Petrol., 87, 205-219.

Jackson, N.J., 1986. Petrogenesis and evolution of Arabian plutonic rocks. J. African Earth Sci. 4, 47-59.

Johnson, P.R., E. Scheibner and E.A. Smith, 1984. Mineral Occurrence base map of the Saudi Arabian Shield. Riofinex Saudi Arabian Mission, scale 1 : 1 000 000.

Johnson, P.R. and G.J. Vranas, 1984. The origin and development of late Proterozoic rocks of the Arabian Shield: an analysis of terranes and mineral environments. Saudi Arabian Deputy Ministry for Mineral Resources Open-File Report RF-OF-04-32.

Johnson, P.R. and P.L. Williams, 1984. Geology of the Jabal Habashi quadrangle, Sheet 26F, Kingdom of Saudi Arabia: Saudi Arabian Deputy Ministry for Mineral Resources Open File Report USGS-OF-10, 87 p.

Kellogg, K.S., 1983a. Geology of the Precambrian rocks of the Wadi Tathlith Quadrangle, Sheet 20G, Kingdom of Saudi Arabia. Saudi Arabian Deputy Ministry for Mineral Resources Open-File Report USGS-OF-04-1, 31 p., scale 1 : 250 000.

Kellogg, K.S., 1983b. Reconnaissance geology of the Qafar Quadrangle, Sheet 27/41D, Kingdom of Saudi Arabia: Saudi Arabian Deputy Ministry for Mineral Resources Open-File Report USGS-OF-04-2, 35 p.

Krogh, T.E., 1973. A low contamination method for hydrothermal decomposition of zircon and extraction of U and Pb for isotopic age determinations. Geochim. Cosmochim. Acta, 37, 485-491.

Kröner, A., 1983. Evolution of tectonic boundaries in the Late Proterozoic Arabian-Nubian Shield of Northeast Africa and Arabia. International Symposium on Precambrian Crustal Evolution, Chinese Academy of Geological Sciences, Beijing, 24-25.

Kröner, A., 1985. Ophiolites and the evolution of tectonic boundaries in the late Proterozoic Arabian-Nubian shield of northeastern Africa and Arabia. Precambrian Res., 27, 277-300.

Le Bel, L. and M. Laval, 1986. Felsic plutonism in the Al Amar-Idsas area, Kingdom of Saudi Arabia. J. African Earth Sci., 4, 87-98.

Le Metour, J., V. Johan and M. Tegyey, 1983. Geology of the ultramafic-mafic complexes in the Bi'r Tuluhah and Jabal Malhijah areas. Saudi Arabian Deputy Ministry for Mineral Resources Open-File Report BRGM-OF-03-40, 47 p.

Letalenet, J., 1979. Geologic map of the Afif Quadrangle, sheet 23F, Kingdom of Saudi Arabia. Saudi Arabian Directorate General of Mineral Resources Geologic Map GM-47A, scale 1 : 250 000.

Ludwig, K.R., 1982. Calculation of uncertainties of U-Pb data. Earth Planet. Sci. Lett., 46, 212-230.

Ludwig, K.R. and L.T. Silver, 1977. Lead isotope inhomogeneity in Precambrian igneous K-feldspars. Geochim. Cosmochim. Acta, 41, 1457-1471.

Marzouki, F.M.H., N.J. Jackson, C.R. Ramsay and D.P. Darbyshire, 1982. Composition, age, and origin of two Proterozoic diorite-tonalite complexes in the Arabian Shield. Precambrian Res., 19, 31-50.

Moore, J.M., 1979. Tectonics of the Najd transcurrent fault system, Saudi Arabia. J. Geol. Soc. London, 136, 441-454.

Pallister, J.S., J.S. Stacey, L.B. Fischer and W.R. Premo, in press. Precambrian opiolites of Arabia: U-Pb geochronology, Pb isotopic chracteristics, and implications for microplate accretion. Saudi Arabian Deputy Ministry for Mineral Resources Open-File Report.

Prodehl, C., 1985. Interpretation of a seismic-refraction survey across the Arabian Shield in western Saudi Arabia. Tectonophysics, 111, 247-282.

Quick, J.E. and J.L. Doebrich, 1986. Geology of the Wadi Ash Shu`bah Quadrangle, sheet 26E, Kingdom of Saudi Arabia. Saudi Arabian Deputy Ministry for Mineral Resources Open-File Report USGS-OF-04-11, scale 1 : 100 000.

Ramsay, C.R., 1986. Specialized felsic plutonic rocks of the Arabian Shield and their precursors. J. African Earth Sci., 4, 153-168.

Reischmann, Th., A. Kröner and A. Basahel, 1984. Petrography, geochemistry and tectonic setting of metavolcanic sequences from the Al Lith area, southwestern Arabian Shield. Faculty Earth Sci., King Abdulaziz Univ., Jiddah, Bull. 6, 366-378.

Roobol, M.J., C.R. Ramsay, N.J. Jackson and D.P.F. Darbyshire, 1983. Late Proterozoic lavas of the central Arabian Shield - evolution of northeastern Africa. Earth Planet. Sci. Lett., 39, 109-117.

Schmidt, D.L., 1980. Geology of the Wadi al Miyah Quadrangle, sheet 20/42B, Kingdom of Saudi Arabia. U.S. Geological Survey Saudi Arabian Mission Technical Record 12, 87 p., scale 1 : 100 000.

Schmidt, D.L., 1981a. Geology of the Jabal al Qarah quadrangle, sheet 20/43C, Kingdom of Saudi Arabia. U.S. Geological Survey Miscellaneous Document 31, 52 p., scale 1 : 100 000.

Schmidt, D.L., 1981b. Geology of the Jabal Yafikh Quadrangle, sheet 20/43B, Kingdom of Saudi Arabia. U.S. Geological Survey Saudi Arabian Mission Miscellaneous Document 39, 99 p., scale 1 : 100 000.

Schmidt, D.L. and G.F. Brown, 1982. Major-element chemical evolution of the late Proterozoic shield of Saudi Arabia. Saudi Arabian Deputy Ministry for Mineral Resources Open-File Report USGS-OF-02-88, 34 p.

Schmidt, D.L., D.G. Hadley, W.R. Greenwood, L. Gonzalez, R.G. Coleman and G.F. Brown, 1973. Stratigraphy and tectonism of the southern part of the Precambrian Shield of Saudi Arabia. Saudi Arabian Directorate General of Mineral Resources Bulletin 8, 13 p.

Schmidt, D.L., D.G. Hadley and D.B. Stoeser, 1979. Late Proterozoic crustal history of the Arabian Shield, southern Najd province, Kingdom of Saudi Arabia. King Abdulaziz Univ., Inst. Applied Geol. Bull, 3, v. 2, Pergamon Press, Oxford-New York, 41-58.

Sillitoe, R.H., 1979. Metallogenic consequences of late Precambrian suturing in Arabia, Egypt, Sudan, and Iran. King Abdulaziz Univ., Inst. Applied Geol. Bull. 3, v. 2, Pergamon Press, Oxford-New York, 110-120.

Stacey, J.S. and R.A. Agar, 1985. U-Pb isotopic evidence for the accretion of a continental microplate in the Zalim region of the Saudi Arabian Shield. J. Geol. Soc. London, 142, 1189-1203.

Stacey, J.S., M.H. Delevaux, J.W. Gramlich, B.R. Doe and R.J. Roberts, 1980. A lead isotopic study of mineralization in the Arabian Shield. Contrib. Mineral. Petrol., 74, 175-188.

Stacey, J.S. and C.E. Hedge, 1984. Geochronologic and isotopic evidence for early Proterozoic continental crust in the eastern Arabian Shield. Geology, 12, 310-313.

Stacey, J.S. and J.D. Kramers, 1975. Approximation of terrestrial lead evolution by a two stage model. Earth Planet. Sci. Lett., 26, 207-221.

Stacey, J.S. and D.B. Stoeser, 1983. Distribution of oceanic and continental leads in the Arabian-Nubian Shield. Contrib. Mineral. Petrol., 84, 91-105.

Stacey, J.S., D.B. Stoeser, W.R. Greenwood and L.B. Fischer, 1984. U-Pb zircon geochronology and geologic evolution of the Halaban-Al Amar region of the eastern Arabian Shield, Kingdom of Saudi Arabia. J. Geol. Soc. London, 141, 1043-1055.

Stern, R.J., 1985. The Najd fault system, Saudi Arabia and Egypt: A Late Precambrian rift-related transform system? Tectonics, 4, 497-511.

Stoeser, D.B., 1986. Distribution and tectonic setting of plutonic rocks of the Arabian Shield. J. African Earth Sci., 4, 21-46.

Stoeser, D.B. and V.E. Camp, 1985. Pan-African microplate accretion of the Arabian Shield. Geol. Soc. Amer. Bull., 96, 817-826.

Stoeser, D.B., R.J. Fleck and J.S. Stacey, 1984a. Geochronology and origin of an early tonalite gneiss of the Wadi Tarib batholith and the formation of syntectonic gneiss complexes in the southeastern Arabian Shield. Faculty Earth Sci., King Abdulaziz Univ., Jiddah, Bull. 6, 351-364.

Stoeser, D.B., J.S. Stacey, W.R. Greenwood and L.B. Fischer, 1984b. U/Pb zircon geochronology of the southern portion of the Nabitah mobile belt and Pan-African continental collision in the Saudi Arabian Shield. Deputy Ministry for Mineral Resourcces Technical Record USGS-TR-04-05, 88 p.

Streckeisen, A., 1976. To each plutonic rock its proper name. Earth-Sci. Rev., 12, 1-33.

Tapponnier, P. and P. Molnar, 1976. Slip-line field theory and large-scale continental tectonics. Nature, 264, 319-324.

Tapponnier, P., G. Peltzer, A.Y. Le Dain, R. Armijo, 1982. Propagating extrusion tectonics in Asia: new insights from simple experiments with plasticine. Geology, 10, 611-616.

Thieme, J.G., in press. Geology of the Jabal Khida quadrangle, sheet 21G, Kingdom of Saudi Arabia. Deputy Ministry for Mineral Resources Geoscience Map, scale 1 : 250 000.

Vail, J.R., 1985. Pan-African (late Precambrian) tectonic terrains and the reconstruction of the Arabian-Nubian Shield. Geology, 13, 839-842.

Wallace, C.A. and P.D. Rowley, in press. Reconnaissance study of the Murdama group, central and northern Murdama basin, Kingdom of Saudi Arabia: Saudi Arabian Deputy Ministry for Mineral Resources Open-File Report.

Warden, A.J., 1982. Reconnaissance geology of the Markas quadrangle, sheet 18/43B, Kingdom of Saudi Arabia. Saudi Arabian Deputy Ministry for Mineral Resources Open-File Report USGS-OF-02-41, 58 p., 1 : 100 000-scale.

White, D.L., 1985. The significance of continental derivation of the Proterozoic Mahanid Formation, southeastern Arabian Shield. J. Geol. Soc. London, 142, 1235-1238.

Worl, R.G. and F.E. Elsass, 1980. Evaluation of mineralization in serpentinite and enclosing rocks in the Hamdah area, Kingdom of Saudi Arabia. U.S. Geological Survey Saudi Arabia Mission Rep. 276, 33 p.

Wyllie, P.J., 1977. Crustal Anatexis: An experimental review. Tectonophysics, 43, 41-71.

Zartman, R.E. and B.R. Doe, 1981. Plumbotectonics - the model. Tectonophysics, 75, 135-162.

Chapter 10

Ophiolites, Sutures, and Micro-Plates of the Arabian-Nubian Shield: A Critical Comment

W. R.Church
Departement of Geology, University of Western Ontario, London Ontario, Canada N6A 5B7

Keywords: Proterozoic, Pan-African, NE Africa, Arabia, ophiolite, suture, ophiolite formation, micro-plates, strike-slip tectonics

Abstract: The practice of identifying intra-arc oceanic crust and microplate boundaries in terms of the distribution of ophiolite complexes in the Arabian-Nubian Shield may produce too simple a picture of the late Proterozoic paleogeography of northeast Africa. Some ultramafic-mafic complexes of the Arabian-Nubian Shield considered to represent oceanic crust are more likely to be intrusive complexes within arc sequences, whereas others, although correctly identified as having formed at a spreading centre, do not appear to have the characteristics of 'normal' MOR material, and may rather have formed in a fore-arc environment during the early stages of arc development. The tensional stress (spreading) regime required for ophiolite formation in a fore-arc environment can perhaps be generated by oblique subduction, which may also induce strike-slip transportation of slivers of the fore-arc ophiolite so produced to locations far from the site of their formation. Such strike-slip displacements may have been more important in the assembly of the various arc terranes of the Arabian-Nubian Shield than has generally been considered, and the extreme width of the Shield owes more perhaps to microplate assembly involving lateral migration than to arc amalgamation by frontal collision.

INTRODUCTION

It is now commonplace to represent the Arabian-Nubian Shield as
a cratonized assemblage of oceanic and continental margin arcs
located between the Nile craton to the west and some other
presently ill-defined craton to the east of the Ar Rayn
terrane, the most easterly volcanic terrane within the exposed
part of the Arabian Shield (Roobol et al., 1983). Most models
of the Shield assume that the accretionary process involved
face-on collision of a succession of arcs, and that
consequently the collision zone would be marked by a trail of
the remnants of the subducted ocean crust. Definition of the
five arc terranes (microplates) currently recognised in the
Arabian-Nubian Shield (Stoeser and Camp, 1985; Vail, 1985)
(Fig. 1b) is based on this assumption, and depends particularly
on the supposed presence of ophiolite-decorated suture zones in
the Sudan (Embleton et al. 1982), and the long distance
correlation of the ophiolites of Bir Umq and Jabal Thurwah
(Delfour, 1982), and of Jabal Al Wask and Sol Hamed (Duyverman,
1984; Nassief et al., 1984). The microplate scheme envisioned
by Vail (1985) is similar to that of Stoeser and Camp (1985)
but includes the definition of two additional arc terranes; one
in the Sinai region, the other the result of the division of
the Asir terrane of Stoeser and Camp into two independent
microplates. Vail has also raised the status of Camp and
Stoeser's microplate distribution map to that of a palinspastic
map, implying that the present position of the microplates is
essentially that which they occupied at the time of their
amalgamation. Although in substantial agreement over the
location of suture zones, Vail, and Camp and Stoeser
nevertheless disagree concerning the polarity of subduction;
Vail, along with Kröner (1985), preferring a subduction
direction to the west, and Camp and Stoeser a subduction
direction to the east.

Irrespective of the difficulty of determining subduction
polarity, defining plate boundaries in terms of ophiolite
trails, while a simple elegant, and apparently rational

procedure, may however not be legitimate. Firstly, as is illustrated in Figure 1, defining plate boundaries by joining ophiolite occurrences appears to be an arbitrary procedure; secondly, it is assumed that ophiolites represent intra-arc oceanic crust, which may not be true; and thirdly, available geological data suggests that the evolution of the Saudi-Nubian Shield was more complicated than current microplate models suggest. In the subsequent parts of this paper, following a brief review of the 'ophiolite problem', the status of some of the currently favoured suture zone sites will be examined from the latter two points of view.

THE OPHIOLITE PROBLEM

It is clear that many ophiolites formed at spreading centers analogous to those of mid ocean ridges. Nevertheless, the origin of many ophiolites as examples of MOR-type oceanic crust has long been questioned, and it has been suggested that they variously represent the roots of primitive island arcs, back-arc marginal basins, and, more recently (Pearce et al. 1984), spreading centres formed above subduction zones at the inception of arc development - and therefore located within the fore-arc parts of the subsequently evolved arc. Basaltic rocks formed at mid-ocean ridges are usually plagioclase-phyric, exhibit $TiO2$ (wt.%) values numerically equivalent to the FeOt/MgO ratio of the rocks, are LREE and LIL depleted, have Ti/Zr ratios of about 100, and only rarely (Bouvet Fracture zone of the Southwest Indian Ridge, Le Roex et al. 1984) exhibit negative Nb-Ta chondrite-factorized anomalies. Cumulates in ophiolites with these characteristics (e.g. Macquarie Island, Vale, 1972; Bay of Islands, Church and Riccio, 1977; the 'Alps', Serri, 1981) characteristically include troctolites, and the crystallisation sequence at low pressures is olivine-plagioclase-clinopyroxene-orthopyroxene. It should be noted however that anomalous low TiO_2 basaltic rocks have been found in the vicinity of transform fault zones (0.5 wt.%; Bryan, 1979, ARP74 stations 31 and 33), as well as

at one site in the Somali Basin (Frey et al. 1980; site 236).
Basalts taken from back-arc basins such as the Scotia Sea
include MOR, LIL-enriched ocean island, and arc types, as well
as mixed MOR/arc rocks exhibiting LREE enrichment and Nb
depletion relative to La and Ba-Rb-K (Saunders and Tarney,
1984). In contrast, fore-arc basement rocks (e.g. Marianas,
Crawford, 1981) include arc tholeiites and rocks of the
boninite series. Cumulate rocks formed in this environment,
such as - according to Pearce et al. (1984) - those of the
Troodos ophiolite (Desmet, 1977; Robinson et al., 1983; Murton,
1986), crystallise in the sequence olivine (chromite; high Cr)-
clinopyroxene-orthopyroxene-plagioclase, whereas boninitic
volcanic rocks with extremely low TiO_2 values (0.2 wt.%) and
concave upwards REE patterns crystallized in the sequence
olivine-orthopyroxene-clinopyroxene-plagioclase. The association of such rocks with a ductile fault zone in the Troodos
ophiolite (Arakapas zone; Murton, 1986) and with a spreading
centre in the case of the Betts Cove ophiolite of Newfoundland
(Coish and Church, 1979; Coish et al., 1982; Church, 1987)
might suggest that ophiolites with these characteristics owe
their preservation in part to their origin as strike-slip fault
slivers detached from the frontal part of arcs as a result of
oblique subduction. If such ophiolitic slivers are transported
by strike slip movement prior to obduction and arc assembly,
they may be less useful in the delineation of arc boundaries
than has tended to be assumed, although of course they do
indicate the one-time existence of active subduction.

In the Arabian-Nubian Shield petrographic descriptions and
limited chemical data are available for the Al Wask (Bakor et
al. 1976), Jebel Ess (Shanti, 1984), Jebel Thurwah (Nasseef et
al. 1984), and Al Amar Idsas (Nawab, 1979; Church, 1980)
ophiolites in Saudi Arabia, the Sol Hamed (Hussein, 1981;
Fitches et al. 1983) and Wadi Onib (Hussein et al. 1984)
ophiolites in the Sudan, and the Fawkhir (Stern, 1979; Dixon,
1979), Wadi Ghadir (El Bayoumi, 1980) and Sabahiya (Basta,
1983) ophiolites in Egypt. Although plagio-phyric lavas occur
in the high-Ti (>3 wt.%) 'within-plate' Wadi Ghadir ophiolite,

in no case have troctolitic cumulates been recorded, and for this reason alone the interpretation of any of the above ophiolites as major ocean or back-arc oceanic crust is presently uncertain.

THE SUTURE PROBLEM

Historical Perspective

The earliest mention of ophiolitic rocks in the Arabian-Nubian Shield and their interpretation as fractionated products of deep-sea mafic magmatism can be attributed to Rittman (1958), but the first references to the Arabian-Nubian ophiolites in the context of plate tectonic theory are contained in papers by Garson and Shalaby (1974; 1976), Bakor, Gass and Neary (1976), Neary, Gass, and Cavanagh (1976) (Fig. 1a), and Al Shanti and Mitchell (1976). Garson and Shalaby considered the ophiolites to mark oceanic sutures separating a series of continental margin arcs that developed episodically above a long-lived - since the Archean - westward dipping subduction zone. Bakor et al. and Neary et al., however, following the suggestion of Greenwood et al. (1975) that the Arabian Shield could be considered a cratonized island-arc developed above an easterly dipping subduction zone, proposed that the ophiolites represent oceanic remnants of as many as seven northwest-trending back-arc marginal basins, five of which could be recognized within the Nubian Shield of the Eastern Desert of Egypt. This idea was also taken up by Frisch and Al-Shanti (1977), who described the development of the Arabian Shield in terms of a complex of arcs and back-arc basins which were sequentially closed along slip planes dipping generally to the east, and further developed by Gass (1977) in his proposal that the whole of continental North Africa east of the West African Craton was formed of cratonized oceanic island arcs of Late Proterozoic age.

In contrast Kazmin et al. (1978) explained the development of the Arabian-Nubian Shield in terms of the opening and closing

of an intracontinental Proto-Red Sea basin with dimensions
approximately defined by the present distribution of
greenstones in the Arabian-Nubian Shield, whereas Shackleton
(1979), in linking the Egyptian and Saudi Arabian suture zones
of Bakor et al. to ophiolite occurrences in southern Sudan,
Ethiopia, and Kenya (Fig. 1a), suggested that oceanic crust was
periodically extended by the southward propagation of spreading
ridges of oceanic crust in the north into continental crust in
the south; there was therefore a transition from crust formed
by arc accretion in the north to crust characterised by
Himalayan-type collision in the south. Some support for these
views was provided by Dixon's discovery of Archean-age zircons
in quartzite clasts in a conglomerate located within the
ophiolite zone of the southern Eastern Desert of Egypt
(Dixon, 1979).

In the Eastern Desert of Egypt mention of ultramafic rocks as
obducted slabs of mantle and oceanic material was made by
Abdel-Khalek (1979), and the back-arc hypothesis was reinforced
by the discovery of typical ophiolite sequences at Wadi Ghadir
(El-Sharkawy and El-Bayoumi, 1979) and Bir Fawkhir. The
representation of the Eastern Desert ophiolites as marking a
series of intra-arc suture zones was however questioned by
Church (1979; 1980; 1983), who argued that the ophiolitic
material of the Eastern Desert occurred in association with
exogeosynclinal deposits, and its primary distribution was
therefore lithostratigraphically controlled; that is, the
ophiolitic material represents olistostromal debris derived
from an 'internal'oceanic source undergoing east towards west
obduction in a manner similar to that invoked in the case of
the emplacement of the early Ordovician flysch and associated
ophiolite sheets of the western margin of the Appalachian
system. On this basis Church concluded that the ophiolitic
zones of the Eastern Desert did not mark the location of in
situ marginal basin oceanic crust, but rather, that the present
distribution of the ophiolitic 'geosynclinal sequence' was
controlled by secondary deformation structures. In terms of
modern plate systems analogy was drawn with the Pacific region

between Australia and New Zealand. Stern (1979) also proposed that the Arabian-Nubian orogen originated as an intra-continental rift, and Engel et al. (1980) drew attention to similarities between the Arabian-Nubian Shield and the development of Archean systems.

The nature of the ophiolite-bearing melange of the Eastern Desert was described by El Sharkawi and El Bayoumi (1979) and by Shackleton et al. (1980), and petrographic details of the ophiolite at Fawkhir were given by Nasseef et al. (1980). El Bayoumi (1980) interpreteted the melange as having originated in a trench environment, and the tectonic history of the Eastern Desert in terms of a rift ocean basin which was closed by consumption of oceanic crust along a westerly dipping subduction zone. In the Arabian shield, arc development as a result of westerly subuction was also favoured by Nawab (1979), Schmidt et al. (1979), and Hadley and Schmidt (1980), whereas Gass (1979) considered the Shield to have developed above one or more easterly-inclined subduction zones. An exogeosynclinal arc-obduction model for the development of the ophiolite-bearing terrane of the Eastern Desert was espoused by Ries et al. (1983), who, however, suggested that ophiolitic melange material was deposited on both continental and oceanic crust close to the interface between an arc and a continental margin about to enter into collision with one another. The melange received debris from the arc, by westward sliding of oceanic material into a trench from thrusts in the fore-arc prism, as well as from the continent, and was tectonically imbricated with slabs of serpentinite during a thrusting event - coeval with the extrusion of the Dokhan volcanics - that brought the melange terrane over shelf sediments of the subducting continental margin.

Following abandonment of the view that the ultramafic rocks of the Eastern Desert delineated in situ zones of back-arc ocean collision, Embleton et al. (1984) defined a new set of northeast trending suture zones in the Sudan, whereas in Saudi Arabia Delfour (1982) depicted the Jabal Thurwah ophiolite as

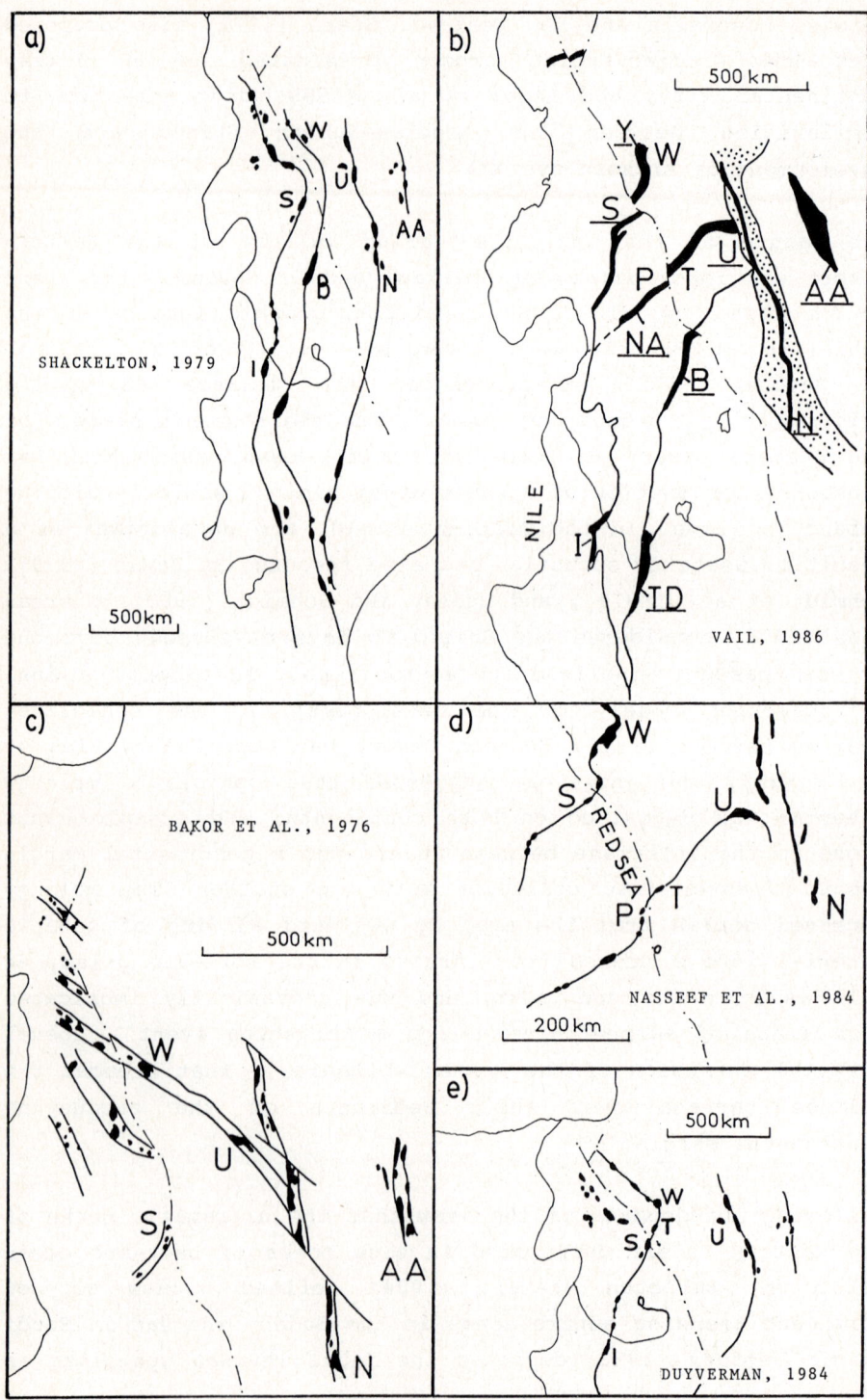

a) SHACKELTON, 1979

500km

W
S U
AA
B N
I

b) VAIL, 1986

500 km

Y W
S
P T U
NA AA
B
N
NILE
I
TD

c) BAKOR ET AL., 1976

500 km

W
U
S
AA
N

d) NASSEEF ET AL., 1984

W
S REDSEA U
P T N
200 km

e) DUYVERMAN, 1984

500km

W
S T U
I

FIGURE 1.

Ophiolite sutures of the Arabian Nubian Shield

a) Shackleton, 1979;

b) Vail, 1985;

c) Bakor et al., 1976;

d) Nasseef et al., 1984;

e) Duyverman, 1984.

Ophiolite complexes/sutures:

T - Jebel Thurwah; W - Jabal Al Wask; P - Port Sudan;

U - Bir Umq; S - Sol Hamed; I - Ingessana;

B - Baraka; N - Nabitah; AA - Al Amar.

N - Nakasib; Y - Yanbu; TD - Tulu Dimitri;

The dotted area on Figure 1b is the Nabitah orogenic belt. The underlined letters on Figure 1b are the names of the sutures delineated by Vail (1985).

lying in the same northeast trending zone as the Bir Umq ophiolite. The northeast trending sets of sutures on either side of the Red Sea were then correlated by Duyverman (1984) (Fig. 1e) and Nassief et al. (1984) (Fig. 1d); that is, the original northeast trend suggested for the ophiolite-decorated zones was abandoned in favour of a northwest trend.

Following extensive mapping in Sinai, Shimron (1984) interpreted the geology of the Wadi Kid area in terms of northwards (westwards) subduction of oceanic crust beneath Proterozoic continental crust and the resulting development of accretionary prisms marginal to an Andean type margin - although the existence of ophiolites in the Wadi Kid and surrounding areas of the northern Eastern Desert was denied by Stern et al. (1985). El Bayoumi and Greiling (1984) and El Ramly et al. (1984) also considered the Nubian Shield to have formed above a west dipping subduction zone following the development of an island arc and a marginal basin arranged such that the latter separated the arc from the Nile craton. Following closure of various oceani systems to the east, rocks of the marginal basin and the arc were thrust over the continental margin. In this model the ophiolitic material of the melange unit(s) forms the basal units of a series of imbricate slices and are therefore considered to have been tectonically incorporated into the melange rather than formed as olistostromal units.

Stacey and Hedge (1984) confirmed the presence of continental crust beneath the eastern part of the Arabian-Nubian Shield - proving that 'accretion' and 'intracratonic' are not mutually exclusive concepts - and, adopting the suture distribution of Nasseef et al. (1984), Stoeser and Camp (1985) have attempted to describe the development of the Arabian Shield in terms of the amalgamation of three ensimatic island arc terranes and two microplates with continental affinities. Vail (1985) has now extended this concept into the Red Sea Hills region of the Sudan. Stoeser and Camp also revert to the view of Greenwood et al. (1976) in considering closure of the main ocean basin to

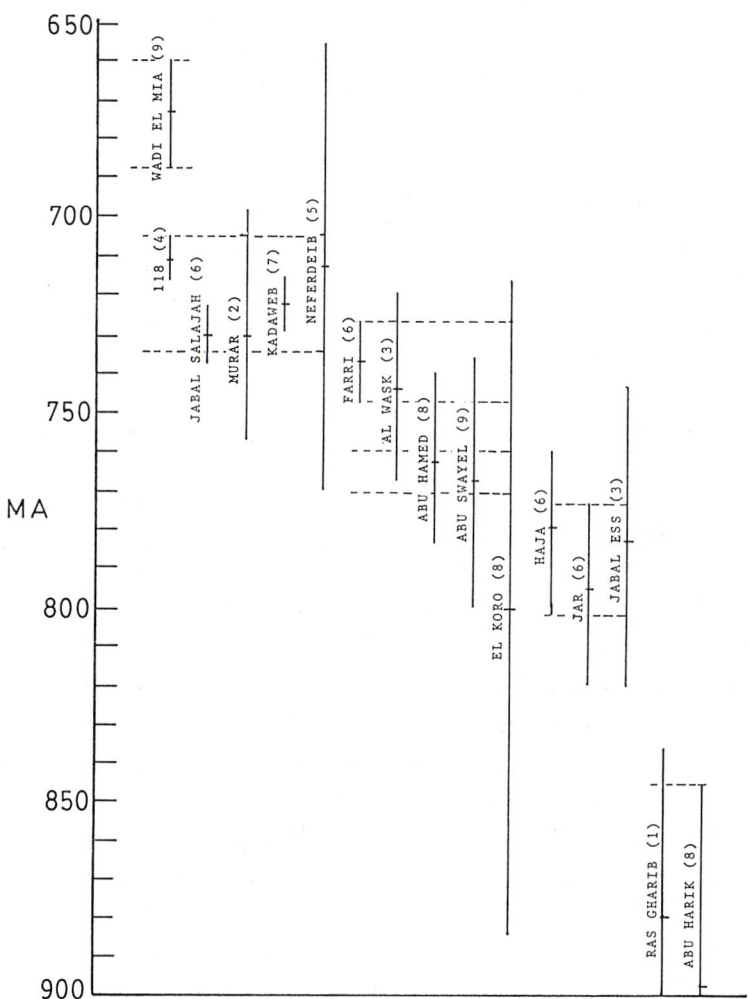

FIGURE 2.

Isotopic ages of volcanic and plutonic rocks in the age range 650 - 900+ Ma, from the Eastern Desert of Egypt, Red Sea Hills of Sudan, and the Al Ays region of Saudi Arabia. Data from: 1) Abdel Rhaman, 1986; 2) Calvez, 1984; 3) Claesson et al., 1984; 4) Dixon, 1981; 5) Fitches et al., 1983; 6) Kemp et al., 1982; 7) Klemenic, 1985; 8) Ries et al., 1985; 9) Stern and Hedge, 1985;

Vertical bars are the uncertainty ranges.
Horizontal dashed lines divides the data set into six age ranges: 660-690; 705-735; 730-750; 760-770; 770-805; >850.

the northwest to have involved subduction towards the
southeast. Kröner (1985) and Vail (1985) on the other hand
invoke a pattern of subduction closure opposite to that adopted
by Stoeser and Camp.

The Eastern Desert

The 'ultramafic rocks as sutures' point of view has previously
been debated (Church, 1979, 1981; Nassief and Gass, 1977) with
respect to the serpentinites of the Eastern Desert of Egypt,
and it is now generally conceded that in the latter instance,
other than Sinai (Vail, 1985, the ultramafic rocks are indeed
allochthonous. Work since that time in the Marsa Alam - Gebel
Zabara - Hafafit and Meatiq regions of the Eastern Desert
(El Bayoumi, 1980, 1984; Basta, 1983; Sturchio et al. 1984;
Habib et al. 1985) has also shown that the ophiolite/arc
debris-bearing terranes (including low-Ti high MgO 'boninitic'
units) constitute a set of east-facing thrust sheets, down
through wich there appears to be an incremental increase in
strain, style of deformation, and degree of metamorphism
(Church, 1983). The geology of the higher sheets is further
complicated by the structural imbrication of apparently younger
volcanic and volcano-sedimentary units. In the southern part of
the Egyptian Desert, continental margin sediments,
autochthonous or parallochthonous relative to Nubian basement
rocks, are possibly represented by marbles beneath the schists
and overlying ultramafic sheets and associated olistostromal
mudstones of the Abu Swayel region (e.g. at Um Krush in the Abu
Swayel region; AS, Fig. 3). In this respect it is conceivable
that the whole of the Eastern Desert is a composite
allochthonous sheet, the leading edge of which is located along
the line of the Nile, with the Nuba Mountains (El Ageed and
El Rabaa, 1981) and Ingessana (Kabesh, 1961) ophiolites of the
Sudan to the south occupying an external positon analogous to
that of the Bay of Islands ophiolite of the Appalachian system.

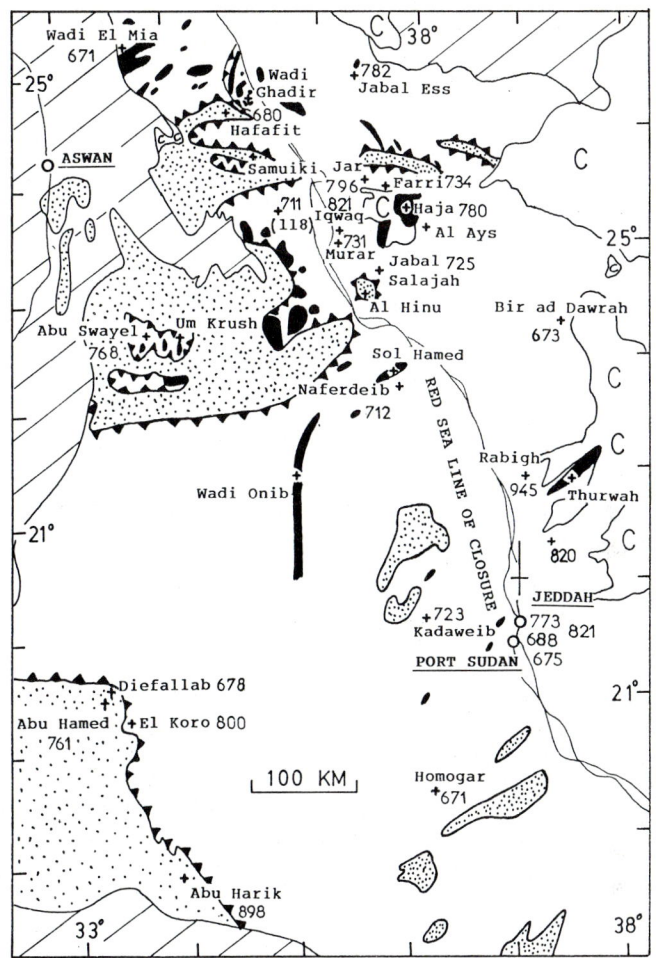

FIGURE 3.

Location and geologic sketch map of the southern Eastern Desert of Egypt and northwest Saudi Arabia.

Ornament: black - ophiolites; dotted - metasediments/metavolcanics beneath the supposed ophiolitic-melange thrust sheet; oblique lines - Nubian Sandstone; C - Cenozoic volcanic.

Numbers: isotopic ages in Ma, data from Al Shanti et al., 1984b; Calvez et al., 1984; Claesson et al., 1984; Dixon, 1981; Duyverman, 1984; Fitches et al., 1983; Kemp et al., 1982; Klemenic, 1985; Kröner et al., 1984; Ries et al., 1985; Stern and Hedge, 1985; Sturchio et al., 1974.
(118) - tonalite dated by Dixon (1981).

The isotope studies of Harris et al. (1984) and Ries et al.
(1985) indicate that metasediments (Rahaba Group) of the Bayuda
Desert and southern Eastern Desert of Egypt (Wadi Haimur in the
Abu Swayel region), while intruded by mantle derived igneous
rocks and metamorphosed at c. 900 Ma., include the erosion
products of older early-middle Proterozoic continental crust.
Furthermore, in the northern part of the Eastern Desert of
Egypt Abdal Rhaman (1986) has determined an age of 881 ± 44 for
a tonalite suite at Ras Gharib. The high R_i value   of these
rocks suggests the existence of older crust at depth beneath
this part of the Eastern Desert. On the other hand, the very
low R_i values of some rocks (e.g. Abu Swayel, Stern, 1979;
Bayda and Jizl, Kemp et al. 1982; Um Aud diorite, unpublished
data) and the Nd/Nd data of Bokhari and Kramers (1981),
Claesson et al. (1984) and Harris et al. (1984) indicate that
the mantle beneath at least parts of the Arabian-Nubian Shield
is anomalously LIL depleted, and in this respect the problem of
the presence or absence of thinned old continental crust
beneath the Eastern Desert of Egypt and the Red Sea Hills of
the Sudan is therefore not entirely resolved. It is also
puzzling that the Abu Hamed Quartzite unit of the Bayuda
Desert, although dominated by seemingly shallow water
carbonates and quartzites and therefore similar to the Wadi
Haimur sediments of the Abu Swayel region, does not have a
strong older crustal isotopic signature comparable to the Wadi
Haimur and Rahaba rocks. Nevertheless, the available isotopic
data are compatible with the view that the Eastern Desert is in
part composed of continental margin slope and rise deposits,
including thick marbles at Abu Swayel, and pelitic and
quartzitic rocks, intercalated with ophiolitic melange units,
at Hafafit, Wadi Miya (chiastolite and calc-siliceous rocks),
and Meatiq; intrusive arc (?) rocks emplaced at about 900 Ma;
and a thrust cover of ophiolitic rocks unconformably overlain
by arc volcanics (Abu Swayel 768 ± 30, Neferdeib 712 ± 60,
El Koro 800 ± 80, Kadaweib 723 ± 6; Figs. 2 and 3) with an age
uncertainty range of 700-800 Ma (Fig. 2), all intruded by
tonalite at 711 ± 6 Ma (Dixon, 1979). Other than in the
northern Eastern Desert, the isotopic data indicate that 'old'

basement rocks if present are of considerably reduced thickness beneath the Eastern Desert. According to Sturchio et al. (1984) thrusting continued until at least 614 Ma. It is uncertain therefore whether the Abu Swayel-Naferdeib volcanics were thrust with the ophiolitic rocks or were formed after ophiolite emplacement. The ophiolites were however emplaced prior to 680 Ma, the age of diorite (Sturchio et al. 1984) intrusive into deformed ophiolitic material of the Hafafit area, and probably prior, to 711 Ma (Dixon, 1981).

A further problem concerning the source of material in the melange sheet relates to the origin of pebbles of older Precambrian gneiss and granitoid material, and of quartzites and arkoses containing Archean age detrital zircons (Dixon, 1979). Since the melange is almost certainly derived from the east the presence of this material implies either that foundered continental crust, now considerably thinned and basified, and its cover of slope and rise sediments extended well to the west of the present Red Sea, or that a continental microplate was at one time interspersed with the arcs of the Arabian Shield, or that at one time the Afif zone was much closer to the Eastern Desert of Egypt. Clearly, further field and isotopic studies are required to resolve these problems.

The Yanbu Suture

The Yanbu suture is defined by Camp (1984) in terms of the Karig model of accretionary prism development, with the Farri Group representing the accretionary prism and the Al Ays Group the overlying fore-arc basin. Camp also suggests that the Yanbu suture zone was the site of origin of the ophiolitic melange of the Eastern Desert of Egypt, thereby implying that formation of the melange post-dates the initial phase of development of the Al Ays fore-arc basin, and is therefore likely older than 725 ± 12 Ma, the age of the Jabal Salajah tonalite. However, while the geology of the Al Ays area is indeed complex, it is not certain that available descriptions and age relationships are compatible with the accretionary prism model.

<u>The Farri Group: an accretionary prism;</u>

Whereas Camp (1984) refers to the Farri Group as being formed of highly deformed ophiolite-bearing accretionary prism deposits, Kemp et al. (1982) describe the group as being composed of thick, welded keratophyric tuff (locally peralkalic) with an age of 742 ± 6 Ma (Calvez, 1983, oral communication quoted by Claesson et al., 1984), mixed sedimentary and volcanic rocks, andesites and epiclastic breccias, turbiditic sediments, and pillowed basalt. Furthermore, the presence of two-pyroxene gabbros, volcanics with low TiO_2 relative to P_2O_5, and absence of a sheeted diabase unit in the Jabal Al Wask ophiolite (Bakor et al., 1976) suggests that the Al Wask complex may represent part of an arc rather than ocean crust separating adjacent arcs.

The Al Wask ophiolite occurs in association with the upper basalt unit of the Farri Group and has a radiometric age (Sm/Nd, 743 ± 24 Ma, Claesson et al., 1984), identical to that of the underlying felsic volcanic rocks of the Farri Group. The age of the ophiolite and Farri Group would therefore appear to be well constrained at c. 730-750. However, zircons from the Haja plagiogranite supposedly associated with the ophiolite have an age of 780 ± 20 Ma (Calvez, 1983, oral communication quoted by Claesson et al., 1984). Therefore, either the Haja granophyre and Al Wask ophiolite have an 'overlap' age of 760-770 Ma (Fig. 2) and are older than the Farri Group, or the granophyre represents remobilized material of the 796 ± 23 Ma old Jar tonalite suite melted by the intrusion of Al Wask mafic liquids. In either case, rocks mapped by Kemp et al. (1982) as Farri but cut by the 796 ± 23 Ma old Jar tonalite (Kemp et al. 1982) likely belong to a volcanic unit older than the Farri Group. These rocks, along with the gneissic Al Hinu Formation and the gneissic rocks of the domes to the north of Al Wask could be equivalents of the lower thrust units exposed in the

Hafafit (El Ramly et al., 1984) and Um Samuiki (Shukri and Mansour, 1980) domes of the Eastern Desert (Fig. 3).

Since the Farri-Al Wask rocks are unconformably overlain by Al Ays sediments (which are cut by the 725 Ma old Salajah tonalite), they could well represent the older (750-725 Ma) part of a southeasterly migrating arc (Al Ays Group) presently located between pre-800 Ma rocks to the northwest (older Farri) and 805 Ma (Hedge pers. comm. to Stoeser and Camp, 1984) and 945 ± 45 Ma (Al Shanti et al. 1984) intrusive-volcanic complexes (Birak and Rabigh rocks) to the southeast.

The root zone of the Eastern Desert and Jabal Ess ophiolites

The well preserved Jabal Ess ophiolite described by Shanti and Roobol (1979) and Shanti (1984) is characterised by the enigmatic presence of both low-Ti, 'normal' Ti (TiO_2 = $FeOt/MgO$), and 'within-plate' basaltic liquids. It therefore represents an example of an ophiolite complex in which interlayered basaltic and dike sequences exhibit both oceanic and arc characteristics. In terms of its low degree of metamorphism and deformation and its association with melange, turbidites, and arc volcanics, the Jebel Ess ophiolite is comparable to the adjacent (after closure of the Red Sea) uppermost ophiolite - melange unit of the Wadi Ghadir - Marsa Alam - Hafafit nappe pile of the Eastern Desert of Egypt. In as much as it is the most easterly ophiolite in this zone, it may therefore represent the highest structural unit of the nappe. In the southern part of the Eastern Desert relatively undeformed volcanic rocks (Abu Swayel volcanics, 768 ± 31; Stern, 1979) overlie intensely deformed allochthonous sheets of ultramafic rock and associated distal mudstones and minor olistostromal units that have been thrust over garnet-amphibolite grade layered metasedimentary and metavolcanic schists and gneisses. The age of the Abu Swayel volcanics relative to the Jebel Ess ophiolite and Jar tonalite is not certain due to the large ± values associated with the isotopic ages of these rocks (Fig. 2), but an age of 782 or older for the Jabal Ess ophiolite (Claesson et al. 1984) is compatible with an emplacement age for the Eastern Desert ophiolites of 760 to 780 Ma, slightly later than the intrusion of the Jar tonalite suite in the Al Wask area, and the Birak und Dhukhr tonalite suites to the southeast. Arc plutonic and volcanic rocks equivalent in age to the Jar tonalite could therefore be the source of the abundant felsic plutonic and volcanic debris in the Eastern Desert melange; which would imply that the surface outcrop of the boundary separating the melange from its arc source lies east of the Jabal Ess ophiolite and west of the Iqwaq granodiorite (821 ± 40 Ma). However, if the gneiss domes of the Al Wask region are analogous to those of Hafafit and Um Samuiki of the Eastern Desert, it is conceivable that the Jar tonalite suite, if not also the Farri and Al Ays groups, has been thrust to the west of the location of the root zone of the suture. Locating the suture may therefore be an intractable problem.

The extension of the Yanbu suture into the Sudan

Nasseef et al. (1984; Fig. 1d) and later Vail (1985; Fig. 1b) extended the Yanbu suture into the northern Red Sea Hills of Sudan to link up with the supposed Sol Hamed - Wadi Onib suture. However, as discussed above, the Al Wask complex may not represent oceanic crust. Furthermore, given the clearly allochthonous nature of the ophiolites in southern Egypt (e.g. Um Krush in the Abu Swayel area, Fig. 3) it is more likely that the Sol Hamid ophiolite belt constitutes the leading edge of a major northwesterly directed thrust than the actual zone of closure of an oceanic basin. Embleton et al. (1984) have proposed that a second more southerly suture be recognised to include the Nakasib belt of the Sudan and the Jabal Thurwah ophiolite of Saudi Arabia. It should be noted however that serpentinites occur southwest of Muhammed Qol in the area between the Sol Hamid and Nakasir belts - and others may remain to be discovered. It is therefore not inconceivable that ultramafic material underlies more area between Sol Hamid and the region to the south than just the supposed suture zones. Furthermore, the presence of wehrlite and lherzolite cumulates in the Sol Hamed ophiolite (Fitches et al., 1983) as well as perhaps also the Wadi Onib ophiolite, and the absence of troctolitic cumulates in any of these ophiolites, suggests that rather than representing intra-arc oceanic crust they may well be strike-slip fault slices of primitive suprasubduction zone fore-arc crust.

The Bir Umq - Jabal Thurwah Suture

The Jebel Thurwah ophiolite (Nasseef et al., 1984) is petrologically (presence of orthopyroxene-bearing cumulates) and chemically (low Ti) unlike 'normal' oceanic crust. Rather, as suggested by Nasseef et al. (1984), it is more similar to oceanic crust developed within the fore-arc segment of a primitive arc (Troodos, Cyprus; Betts Cove, Appalachians). The ophiolite can be interpreted as an allochthonous sheet emplaced onto the Samran Formation from either the northwest or southeast from beyond the present outcrop of the Samran, or from beneath the Samran. Nor is it proven that it is in continuity with either the Bir Umq ophiolite or with the widely spaced and poorly developed ultramafic rocks in the vicinity of Port Sudan in the Sudan. Even in terms of Camp's attempt (Camp, 1984) to model the At Taif metamorphics, Baish volcanics, and An Nimas plutonics in terms of a frontal arc, back arc basin, and remnant arc system, it is apparent that the trend of the marginal basin and remnant arc is north-south, the purported northeast trend of the suture being evident only in terms of a supposedly primary elongation of the much younger Fatimah volcanic basin, the northeast trend of faults cutting sediments and volcanics of the supposed arc - trench gap, and the supposed recognition of an upper slope discontinuity. The status of the Jebel Thurwah ophiolite as marking a suture zone is therefore highly uncertain. It is of interest to note that the boundary that does separate zones containing rocks with an age greater than 900 Ma from terranes of less than 900 Ma within the Asir terrane is not marked in Saudi Arabia by a suture.

The Nabitah suture - Nabitah mobile belt

Interpretation of the ultramafic-mafic complexes of the Nabitah
zone as ophiolites is not universally accepted (Caby, 1984;
Agar, 1985). Although a plate boundary may well therefore lie
somewhere within the Nabitah zone, it is not necessarily
demarcated by an 'ophiolite'-decorated fault zone. According to
Stacey and Agar (1985) the suture is occupied by the
syntectonic Subay igneous suite, in which case strike-slip
juxtaposition of the adjacent arc terranes remains a
possibility. It might also be noted in this respect that Caby
(1984) concluded that continental accretion of the Arabian
Shield was essentially vertical and suggested that the absence
of evidence for large scale horizontal movements precluded
involvement of the Shield in any major collisional event.
Furthermore, Stoeser et al. (1984) state that the synorogenic
emplacement of 650 to 690 Ma old granodiorite plutons in areas
marginal to the supposed Nabitah suture zone reflects major
westward-directed compressional orogeny related to a
continental collisional event possibly located somewhere to the
east of the exposed southeastern Arabian Shield.

The Al Amar suture

A reasonable case can be made for situating the Al Amar
ophiolitic rocks (Al Shanti and Mitchell, 1976; Nawab, 1979;
Jackson et al. 1980) in proximity to a suture. However,
irrespective of the number of geologic problems concerning
correlation and age (Al Shanti et al., 1984; Calvez et al.
1985) that require clarification, there is considerable
uncertainty about the oceanic nature of the ophiolites (Le
Metour et al. 1982), which, as in the case of the Thurwah
ophiolite, do not appear to represent 'normal' oceanic crust
(Church, 1980a). Since the Al Amar ophiolites are likely
allochthonous, perhaps emplaced in association with the
deposition of flysch representing the products of the erosion
of the overthrusting plate (Church, 1980b), their present
location does not determine the suture zone nor the nature of
movement on the suture. The metallogenic zonation - copper to
the east, silver-lead-zinc to the west, might imply subduction

from east to west leading to the development of a Sea of Japan type back-arc basin as suggested by Nawab (1979), but since copper and gold zones also occur to the west of the polymetallic zone, subduction could equally well have been towards the east.

CONCLUSIONS

Although some ultramafic-mafic complexes of the Arabian-Nubian shield likely formed in a spreading centre environment, whether fore-arc, back arc or continental margin rift (Tasman-sea-type), their present locations do not delineate with certainty the original sites of the ocean basins. The practice of linking widely separated ultramafic-mafic complexes to define suture zones cannot therefore be accepted without question. Furthermore, given that the width of the Arabian-Nubian shield from the Nile to the Nabitah boundary zone is of the same order of distance as that across the strike-slip amalgamated collage of the Canadian Cordillera, the possibility must be allowed not only that the microplates of the Nubian Arabian shield have been laterally rafted into position, but that the plate boundaries may have been considerably modified as a result of collision. Any assessment of crustal growth rates in the Arabian-Nubian shield should take this point into account.

Notwithstanding the considerable uncertainty concerning the arc amalgamation process in the Shield, geochronological data obtained by Stacey and Hedge (1985) and Stacey and Agar (1985) now indicate that the Arabian Shield - as might be guessed from the distribution of zinc-lead mineral showings - is bordered to the east by older continental crust. The isotopic studies of Harris et al. (1984), Ries et al. (1985), and Abdel Rahman (1986) suggest that the western margin of the Nubian Shield is formed of an about 900 Ma continental margin volcanic arc over which the Egyptian-Sudanese ophiolite terrain has been thrust from the east. The results of recent studies in southern East Africa (Maboko et al., 1985) also suggest that the granulite belts of western Tanzania were formed 715 Ma ago and may have been thrust (Sacchi et al., 1984), along with 'ophiolite'-type rocks, to the southwest over a basement of 1100 Ma continental margin arc rocks (R_i .7027) and its cover of 900-1000 Ma continental-derived sediments (R_i .7091-.713). If the Hijaz ocean was continuous along the length of east Africa during the late Proterozoic, it would appear therefore that either considerable overthrusting has caused the loss from view of the southern arc equivalents of the Arabian-Nubian Shield or the southerly arc elements have migrated laterally northwards and presently form part of the Arabian-Nubian amalgamated terrane. In this context, a more comprehensive examination of the potential role of strike-slip movements in the assembly of the Arabian-Nubian arcs may prove to be more profitable than is perhaps currently thought.

REFERENCES

Abdel-Khalek, M.L., 1979. Tectonic evolution of the basement rocks in the southern and central Eastern Desert. Bull. Inst. Appl. Geol. Jeddah, v. 3, (1), p. 53-62.

Abdel-Rahman, A-F., 1986. Plutonism and tectonic evolution of the Ras Gharib segment of the northern Nubian Shield. Ph.D. thesis. McGill University.

Agar, R.A. 1985. Stratigraphy and palaeogeography of the Siham group: direct evidence for a late Proterozoic continental microplate and active continental, margin in the Saudi Arabian Shield. J. Geol. Soc. London, 142, p. 1205-1220.

Al Shanti, A.M.S., A.A. Abdel-Monem and F.H. Marzouki, 1984a. Geochemistry, petrology, and Rb-Sr dating of trondjemite and granophyre associated with the Jabal Tays ophiolite, Idsas area, Saudi Arabia. Precambrian Res., 24, p. 321-334.

Al Shanti, A.M.S., A.A. Abdel-Monem and A.A. Radain, 1984b. Rb-Sr dating and petrochemistry of Um Gerig granitic rocks (Rabigh Area), western Saudi Arabia. Faculty of Earth Sciences King Abdulaziz University Bulletin 6, p. 233-248.

Al Shanti, A.M.S. and A.H.G. Mitchell, 1976. Late Precambrian subduction and collision in the Al Amar-Idsas region, Arabian Shield, Kingdom of Saudi Arabia. Tectonophysics, v. 30, p. T41-T47.

Bakor, A.R., I.G. Gass and C.R. Neary, 1976. Jabal al Wask, northwest Saudi Arabia: an Eocambrian back-arc ophiolite. Earth Planet. Sci. Lett., v. 30, p. 1-9.

Basta, F.F., 1983. Geology and geochemistry of the ophiolitic melange and other rock units in the area around and west of Gebel Ghadir, Eastern Desert, Egypt. Ph.D. thesis, Cairo University, 137p.

Bokhari, F.Y. and J.D. Kramers, 1981. Island arc character and late Precambrian age of volcanics at Wadi Shwas, Hijaz, Saudi Arabia: geochemical and Sr and Nd isotopic evidence. Earth Planet. Sci. Lett., v. 54, p. 409-422.

Bryan, W.B., 1978. Regional variation and petrogenesis of basalt glasses from the FAMOUS area, Mid-Atlantic ridge. J. Petrol., v. 20, p. 293-325.

Caby, R., 1984. Pan-African evolution of the Tuareg Shield (Central Sahara) and the Arabian Shield: a comparison. Faculty of Earth Sciences King Abdulaziz University Bulletin 6, p. 23-25.

Calvez, J.Y., C. Alsac, J. Delfour, J. Kemp and C. Pellaton, 1984. Geological evolution of the western, central and eastern parts of the Northern Precambrian Shield, Kingdom of Saudi Arabia. Faculty of Earth Sciences, King Abdulaziz University, Bulletin 6, p. 23-48.

Calvez, J.Y., J. Delfour and J.L. Feyhesse, 1985. 2000 million-yr old inherited zircons in plutonic rocks from the Al Amar region: New evidence for an early Proterozoic basement in the Eastern Arabian Shield; Saudi Arabian Deputy Ministry for Min. Res. Rep. BRGM-OF-05-11, 27p.

Camp, V.E., 1984. Island arcs and their role in the evolution of the western Arabian Shield. Geol. Soc. America Bull., v. 95, p. 913-921.

Church, W.R., 1979. Granitic and metamorphic rocks of the Taif area, western Saudi Arabia; discussion. Geol. Soc. America Bull., v. 90, p. 893-896.

Church, W.R., 1980a. Late Proterozoic Ophiolites. Orogenic Mafic and Ultramafic Association, Colloques Internationaux du Centre National de la Recherche Scientifique No. 272, p. 105-118.

Church, W.R., 1980b. Geology of the Jabal Idsas-Jabal Tays-Jabal Zriba areas in the Eastern Arabian Shield. Newsletter - "Pan-African Crustal Evolution in the Arabian-Nubian Shield", v. 3, p. 53-57.

Church, W.R., 1983. Late Precambrian evolution of Afro-Arabian crust from ocean arc to craton; discussion. Geol. Soc. America Bull., v. 94, p. 679-681.

Church, W.R., 1987. Discussion of I.O. Oshin and J.H. Crocket, 1985. The geochemistry and petrogenesis of ophiolitic volcanic rocks from Lac de l'Est, Thetford Mines Complex, Quebec, Canada. Canadian J. Earth Sci., v. 23, p. 202-213 (in press).

Church, W.R. and L. Riccio, 1977. Fractionation trends in the Bay of Islands ophiolite of Newfoundland: polycyclic cumulate sequences in ophiolites and their classification. Can. J. Earth Sci., v. 14, p. 1156-1165.

Claesson, S., J.S. Pallister and M. Tatsumoto, 1984. Samarium-neodymium data on two late Proterozoic ophiolites of Saudi Arabia and implications for crustal and mantle evolution. Contrib. Mineral. Petrol., v. 85, p. 244-252.

Coish, R.A. and W.R. Church, 1979. Igneous geochemistry of mafic rocks in the Betts cove ophiolite, Newfoundland. Contrib. Mineral. Petrol., v. 70, p. 29-39.

Coish, R.A., R. Hickey and F.A. Frey, 1982. Rare element geochemistry of the Betts Cove ophiolite, Newfoundland: complexities in ophiolite formation. Geochim. Cosmochim. Acta, v. 46, p. 2117-2134.

Crawford, A.J., L. Beccaluva and G. Serri, 1981. Tectono-magmatic evolution of the West Philippine region and the origin of boninites. Earth Planetary Sci. Letters, v. 54, p. 346-356.

Delfour, J., 1982. Geologic, tectonic and metallogenic evolutions of the northern part of the Precambrian Arabian Shield (Kingdom of Saudi Arabia). Bull. BRGM (deuxieme serie) 2, p. 1-19.

Desmet, A., 1977. Contribution a l'etude de la croute oceanique mesozoique de Mediterranee orientale: Les Pillow-lavas du Troodos (Cypre). D.SST thesis, l'Universite de Nancy, 221p.

Dixon, T.H., 1979. The evolution of continental crust in the Late Precambrian Egyptian Shield. Ph.D. thesis, University of California, San Diego, 230p.

Dixon, T.H., 1981. Age and chemical characteristics of some pre-Pan-African rocks in the Egyptian Shield. Precambrian Res., v. 14, p. 119-133.

Duyverman, H.J., 1984. Late Precambrian granitic and volcanic rocks and their relation to the cratonisation of the Arabian Shield. Faculty of Earth Sciences King Abdulaziz University Bulletin 6., p. 50-69.

El Ageed, A.I. and S.M. El Rabaa, 1981. The geology and structural evolution of the northeastern Nuba Mountains, southern Kordofan Province, Sudan. Bull. Geol. Miner. Res. Dept. Sudan, p. 1-50.

El Bayoumi, R.M., 1980. Ophiolites and associated rocks of Wadi Ghadir, east of Gebel Zabara, Eastern Desert, Egypt. Ph.D. thesis, Cairo University, 227p.

El Bayoumi, R.M., 1984. Ophiolites and melange complex of Wadi Ghadir area, Eastern Desert of Egypt. Faculty of Earth Sciences King Abdulaziz University Bulletin 6, p. 329-342.

El Bayoumi, R.M.A. and R. Greiling, 1984. Tectonic evolution of a Pan-African plate margin in southeastern Egypt - a suture zone overprinted by low angle thrusting. In: J. Klerkx and J. Michot (eds). Geologie africaine - African Geology, p. 47-56.

El Ramly, M.F., R. Greiling, A. Kröner and A.A.A. Rashwan, 1984. On the tectonic evolution of the Hafafit area and environs, Eastern Desert of Egypt. Faculty of Earth Sciences King Abdulaziz University Bulletin 6, p. 113-126.

El Sharkawi, M.A. and R.M. El Bayoumi, 1979. The ophiolites of Wadi Ghadir area, Eastern Desert, Egypt. Annals Geol. Surv. Egypt, v. 9, p. 125-135.

Embleton, J.C.B., D.J. Hughes, P.M. Klemenic, S. Poole and J.R. Vail, 1984. A new approach to the stratigraphy and tectonic evolution of the Red Sea Hills, Sudan. Faculty of Earth Sciences King Abdulaziz University Bulletin 6, p. 101-112.

Engel, A.E.J., T.H. Dixon and R.J. Stern, 1980. Late Precambrian evolution of Afro-Arabian crust from ocean arc to craton. Geol. Soc. America Bull., v. 91, p. 699-706.

Fitches, W.R., R.H. Graham, I.M. Hussein, A.C. Ries, R.M. Shackleton and R.C. Price, 1983. The late Proterozoic ophiolite of Sol Hamed, NE Sudan. Precambrian Res., v. 19, p. 385-411.

Frey, F.A., J.S. Dickey Jr., G. Thompson, W.B. Bryan and H.L. Davies, 1980. Evidence for heterogeneous primary MORB and mantle sources, NW Indian Ocean. Contr. Mineral. Petrol., v. 74, p. 387-402.

Frisch, W. and A.M.S. Al-Shanti, 1977. Ophiolite belts and the collision of island arcs in the Arabian Shield. Tectonophysics, v. 43, p. 293-306.

Garson, M.S. and I.M Shalaby, 1976. Precambrian-Lower Palaeozoic plate tectonics and metallogenesis in the Red Sea region. Special Paper Geological Association of Canada, v. 14, p. 573-96.

Gass, I.G., 1977. The evolution of the Pan-African crystalline basement in NE Africa and Arabia. J. Geol. Soc. London, v. 134, p. 129-38.

Gass, I.G., 1979. Evolutionary model for the Pan-African crystalline basement. Bull. Inst. Appl. Geol. Jeddah, v. 3, (1), p. 11-20.

Greenwood, W.R., D.G. Hadley, R.E. Anderson, R.J. Fleck and D.L. Schmidt, 1975. Late Proterozoic cratonization in southwestern Saudi Arabia. U.S. Geol. Survey Saudi Arabian Project Rept. 196, 23p.

Greenwood, W.R., D.G. Hadley, R.E. Anderson, R.J. Fleck and D.L. Schmidt, 1976. Late Proterozoic cratonization in southwestern Saudi Arabia. Phil. Transact. R. Soc., Ser. A, v. 280, p. 517-527.

Habib, M.E., A.A. Ahmed and O.M. El Nady, 1985. Two orogenies in the Meatiq area of the Central Eastern Desert, Egypt. Precambrian Res., v. 30, p. 83-111.

Hadley, D.G. and D.L. Schmidt, 1980. Proterozoic sedimentary rocks and basins of the Arabian Shield and their evolution. Bull. Inst. Appl. Geol. Jeddah, v. 3, (4), p. 26-50.

Harper, G.D., 1984. The Josephine ophiolite, northwestern California. Geol. Soc. America. Bull., v. 95, p. 1009-1026.

Harris, N.B.W., C.J. Hawkesworth and A.C. Ries, 1984. Crustal evolution in North-East and East Africa from model Nd ages. Nature, v. 309, p. 773-776.

Hussein, I.M., 1981. An outline of the geology and structure of the Sol Hamed ophiolite of the Halaib area, northern Red Sea Hills, Sudan. Newsletter - "Pan-African Crustal Evolution in the Arabian-Nubian Shield", v. 4, p. 19-25.

Hussein, I.M., A. Kröner and St. Dürr, 1984. Wadi Onib - a dismembered Pan African ophiolite in the Red Sea Hills of the Sudan. Faculty of Earth Sciences King Abdulaziz University Bulletin 6, p. 319-328.

Jackson, N., A. Kröner, W.R. Church and A. Hashad, 1980. Notes on some stratigraphic relationships in the J.I. area. Newsletter. Pan-African Crustal Evolution in the Arabian-Nubian Shield, No. 3, p. 16-17.

Kabesh, M.L., 1961. The geology and economic minerals and rocks of the Ingessana Hills. Bull. Geol. Surv. Sudan, v. 11, p. 1-61.

Kazmin, V., A. Shifferaw and T. Balcha, 1978. The Ethiopian Basement: stratigraphy and possible manner of evolution. Geol. Rundschau, v. 67 (2), p. 531-546.

Kemp, J., C. Pellaton and J.-Y. Calvez, 1982. Cycles in the chelogenic evolution of the Precambrian shield in part of north western Saudi Arabia. Prof. Paper Saudi Arabian Dir. Gen. Mineral. Res., v. 1, p. 27-41.

Klemenic, P.M., 1985. New geochronological data on volcanic rocks from Northeast Sudan and their implication for crustal evolution. Precambrian Res., v. 30, p. 263-276.

Kröner, A., 1985. Ophiolites and the evolution of tectonic boundaries in the Late Proterozoic Arabian-Nubian Shield of northeast Africa and Arabia. Precambrian Res., v. 27, p. 277-300.

Kröner, A., M. Halpern and A. Basahel, 1984. Age and significance of metavolcanic sequences and granitoid gneisses from the Al-Lith area, southwestern Arabian Shield. Faculty of Earth Sciences King Abdulaziz University Bulletin 6, p. 380-388.

Le Metour, J., V. Johan and M. Tegyey, 1982. Relationships between ultrabasic to basic complexes and volcanic sedimentary series in the Precambrian of the Arabian Shield. Saudi Arabian Deputy Ministry for Miner. Res. Open-File Report BRGM-OF-02-21, 42p.

Maboko, M.A.H., N.A.I.M. Boelrijk, H.N.A. Priem and E.A.Th. Verdurmen, 1985. Zircon U-Pb and biotite Rb-Sr dating of the Wami River granulites, eastern granulites, Tanzania: evidence for approximately 715 Ma old granulite-facies metamorphism and final Pan-African cooling approximately 475 Ma ago. Precambrian Res., v. 30, p. 361-378.

Murton, B.J., 1986. Anomalous oceanic lithosphere formed in a leaky transform fault: evidence from the western Limassol Forest complex, Cyprus. J. Geol. Soc. London, v. 143, p. 845-854.

Nasseef, A.O. and I.G. Gass, 1977. Granitic and metamorphic rocks of the Taif area, Western Saudi Arabia. Geol. Soc. America Bull., v. 88, p. 1721-1730.

Nassief, M.O., R. MacDonald and I.G. Gass, 1984. The Jebel Thurwah Upper Proterozoic Ophiolite complex, western Saudi Arabia. J. Geol. Soc. London, v. 141, p. 537-546.

Nawab, Z.A., 1979. Geology of the Al-Amar-Idsas region of the Arabian Shield. Bulletin Institute of Applied Geology, Jeddah, v. 3, (4), p. 29-40.

Neary, C.R., I.G. Gass and B.J. Cavanagh, 1976. Granitic association of northeastern Sudan. Geol. Soc. America Bull., v. 87, p. 1501-12.

Pearce, J.A., S.J. Lippard and S. Roberts, 1984. Characteristics and tectonic significance of supra-subduction zone (ssz) ophiolites. In: B.P. Kokelaar, and M.F. Howells (eds). Special Publication of the Geological Society, v. 16, p. 77-94.

Ries, A.C., R.M. Shackleton, R.H. Graham and W.R. Fitches, 1983. Pan-African structures, ophiolites and melange in the Eastern Desert of Egypt: a traverse at 26 N. J. Geol. Soc. London, v. 140, p. 75-95.

Ries, A.C., R.M. Shackleton and A.S. Dawoud, 1985. Geochronology, geochemistry and tectonics of the NE Bayuda Desert, N. Sudan: implications for the Western Margin of the Late Proterozoic fold belt of NE Africa. Precambrian Res., v. 30, p. 43-62.

Rittmann, A., 1958. Geosynclinal volcanism, ophiolite and Barramiya rocks. Egyptian Journal of Geology, v. II, p. 61-65.

Robinson, P.T., W.G. Melson, T. O'Hearn and H.-U. Schmincke, 1983. Volcanic glass compositions of the Troodos ophiolite, Cyprus. Geology, v. 11, p. 400-404.

Le Roex, A.P., H.J.B. Dick, A.J. Erlank, A.M. Reid, F.A. Frey and S.R. Hart, 1984. Geochemistry, mineralogy, and petrogenesis of lavas erupted along the Southwest Indian ridge between the Bouvet Triple Junction and 11 degrees east. Journal Petrology, v. 24, p. 267-318.

Roobol, M.J., C.R. Ramsay, N.J. Jackson and D.P.F. Darbyshire, 1983. Late Proterozoic lavas of the Central Arabian Shield - evolution of an ancient volcanic system. J. Geol. Soc. London, v. 140, p. 185-202.

Sacchi, R., J. Marques, M. Costa and C. Casati, 1984. Kibaran events in the southernmost Mozambique belt. Precambrian Res., v. 25, p. 141-159.

Saunders, A.D. and J. Tarney, 1984. Geochemical characteristics of basaltic volcanism within back-arc basins. In: B.P. Kokelaar and M.F. Howells (eds). Special Publication of the Geological Society, v. 16, p. 59-76.

Schmidt, D.L., D.G. Hadley and D.B. Stoeser, 1979. Late Proterozoic crustal history of the Arabian Shield, southern Najd Province, Kingdom of Saudi Arabia. Bull. Inst. Appl. Geol. Jeddah, v. 3, (2), p. 41-58.

Serri, G., 1981. The petrochemistry of ophiolite gabbroic complexes: a key for the classification of ophiolites into low-Ti and high-Ti types. Earth Planetary Science Letters, v. 52, p. 203-212.

Shackleton, R.M., 1979. Precambrian tectonics of North-East Africa. Bull. Inst. Appl. Geol. Jeddah, v. 3, (2), p. 1-7.

Shackleton, R.M., A.C. Ries, R.H. Graham and W.R. Fitches, 1980. Late Precambrian ophiolitic melange in the eastern desert of Egypt. Nature, v. 285, p. 472-474.

Shanti, M., 1984. The Jabel Ess ophiolite. Faculty of Earth Sciences King Abdulaziz University Bulletin 6, p. 289-318.

Shanti, M. and M.J. Roobol, 1979. A late Proterozoic ophiolite complex at Jabal Ess in northern Saudi Arabia. Nature, v. 279, p. 488-91.

Shimron, A.E., 1984. Evolution of the Kid Group, southeast Sinai Peninsula: thrusts, melanges, and implications for accretionary tectonics during the late Proterozoic of the Arabian-Nubian Shield. Geology, v. 12, p. 242-247.

Shukri, N.M and M.S. Mansour, 1980. Lithostratigraphy of Um Samuiki district, Eastern Desert, Egypt. Bull. Inst. Appl. Geol. Jeddah, v. 3, (4), p. 83-94.

Stacey, J.B. and A. Agar, 1985. U-Pb isotopic evidence for the accretion of a continental microplate in the Zalm region of the Saudi Arabian Shield. J. Geol. Soc. London, v. 142, p. 1189-1204.

Stacey, J.S. and C.E. Hedge, 1984. Geochronologic and isotopic evidence for early Proterozoic crust in the eastern Arabian Shield. Geology, v. 12, p. 310-313.

Stern, R.J., 1979. Late Precambrian ensimatic volcanism in the Central Eastern Desert of Egypt. Ph.D. thesis, University of California, San Diego, 210p.

Stern, R.J., D. Gottfried and C.E. Hedge, 1985. Discussion of A.E. Shimron, 1984. Evolution of the Kid Group, southeast Sinai Peninsula: thrusts, melanges, and implications for accretionary tectonics during the late Proterozoic of the Arabian-Nubian Shield. Geology, v. 13, p. 155.

Stern, R.J. and C.E. Hedge, 1985. Geochronologic and isotopic constraints on late Precambrian crustal evolution in the Eastern Desert of Egypt. American J. Sci., v. 285, p. 97-127.

Stoeser, D.B. and V.E. Camp, 1985. Pan-African microplate accretion of the Arabian Shield. Geol. Soc. America Bull., v. 96, p. 817-826.

Stoeser, D.B., R.J. Fleck and J.S. Stacey, 1984. Geochronology and origin of an early Tonalite Gneiss of the Wadi Tarib Batholith and the formation of syntectonic gneiss complexes in the Sotheastern Arabian Shield, Kingdom of Saudi Arabia. Faculty of Earth Sciences King Abdulaziz University Bulletin 6, p. 351-364.

Sturchio, N., M. Sultan, P. Sylvester, R. Batiza, C. Hedge, E.M. El Shazly and A. Abdel-Meguid, 1984. Geology, age, and origin of the Meatiq dome: implications for the Precambrian stratigraphy and tectonic evolution of the Eastern Desert of Egypt. Faculty of Earth Sciences King Abdulaziz University Bulletin 6, p. 127-144.

Vail, J.R., 1985. Pan-African (late Precambrian) tectonic terrains and the reconstruction of the Arabian-Nubian shield. Geology, v. 13, p. 839-842.

Varne, R. and M.J. Rubenach, 1972. Geology of Macquarie island and its relationship to oceanic crust. In: D.E. Hayes (ed). Antarctic Oceanology II: The Australian-New Zealand sector. Antarctic Research Series, v. 19, p. 251-266.

Metallogenesis

Metallogenesis, as discussed in Chapter 11 by Pohl, is one of the most important aspects of Pan-African geology, both for economic and tectonic reasons.

The interdependence of metallogenesis and (plate) tectonic evolution is convincingly demonstrated by Pohl. For example, particular elements can be effectively looked for only in specific tectonic environments. On the other hand, particular 'metal provinces' (e.g. tin) are indicative of a possible pre-existing continental domain and may help to distinguish between juvenile Pan-African and reworked, Mozambiquian terrains.

Economically, it is inevitable to apply the tectonic, and structural, know-how to mineral exploration in order to satisfy growing demands for resources.

Further aspects of metallogenesis are also treated by El-Gaby et al. in Chapter 2.

Chapter 11

Precambrian Metallogeny of NE-Africa

W. Pohl

Institute of Geology, Technical University, P. O. Box 3329, D-3300 Braunschweig, FRG

Keywords: Proterozoic, Pan-African, NE Africa, Metallogeny, Mozambique belt, Arabian-Nubian Shield, Mineralisation, syngenetic, Mineralisation, ophiolite related; Mineralisation, magmatogenic; Mineralisation, metamorphogenic

Abstract: The Precambrian of NE-Africa comprises essentially two major units: the Arabian-Nubian volcano-sedimentary greenschist-ophiolite assemblage, and the Mozambique belt sensu lato. Radiometric dating reveals increasingly, that both units evolved synchronously within a vast area of convergent tectonics from about 950 Ma to roughly 600 Ma. Both contain ophiolite belts, separating different terranes, and magmatic arcs. Different, however, is the Mozambique belt in respect to its important component of older crust and epicontinental (volcano-)sedimentary cover rocks, its high grade of regional metamorphism and strong, polyphase folding. All this is pre-dating late Pan-African westerly overthrusting of greenschist-assemblage rocks observed in the Eastern Desert, Egypt.

Likewise, the metallogenic analysis of both units reveals comparable and differing features, the former including ophiolite and magmatic arc related mineralisation, while the latter are characterized by the typically Mozambiquian metamorphogenic deposits, and the post-orogenic lithophile

element mineralisation associated with A-type magmatism as well
as gold-quartz veins which occur mainly within the greenschist
terrains. The ensimatic, central part of the Arabian Shield is
characterized by Nb, Zr, Y, REE, U and Th occurrences, while Sn
(exploited earlier), W and Be are more frequent in a marginal
setting with continental influence. This appears to confirm the
rôle of tin as an indicator of old sialic crust.

INTRODUCTION

The formation of metalliferous deposits depends basically on
the availability of a suitable source and a variety of
geological processes which liberate, transport and concentrate
the respective element(s). Obviously, at the scale of single
deposits it is most important to investigate the conditions of
transport and local concentration, in order to guide
intelligently exploration and exploitation. Metallogenic
research addresses the more fundamental - and evasive -
problems concerning the source and migration of metals in space
and time. Usually it is often tacitly assumed that a previous
geochemical enrichment is a precondition for "productive"
source regions. Such regions were called metal domains
(Routhier, 1980) or metal provinces (Petrascheck, 1965), while
areas of comparable geological evolution and mineral deposits
constitute metallogenic provinces.

The accumulating evidence that continents are formed by
accretion of plates, microplates and terranes of variable
constitution and pre-accretion history invalidates the concept
of ancient and persistent metal domains transsecting major
tectonic boundaries or whole continents (as drawn for example
by Schuiling, 1967). Consequently, "ordinary" rocks must be
suitable sources for many metals, but of course only within
their geochemical characteristics. Of special significance in
this context are for example lithophile elements as their
enrichment will in most cases indicate a source consisting of
continental crust; alternatively, a very low degree of partial
melting of mantle rocks or oceanic crust may be a feasible
process based on geochemical models.

The geological evolution of NE-Africa produced a large number
of mineral deposits, although few of them are important in
economic terms (Pohl, 1984). Their variability, however, offers
a unique possibility to test metallogenic theory and to compare
current geological models with metallogenic evidence.

PROTEROZOIC GEOLOGY OF NE-AFRICA

To the East of the Archean-Lower Proterozoic Tanzania, Uganda,
and East Sahara cratons (Almond, 1984) occur essentially two
major geological units, one comprising multiply deformed,
partly epicontinental meta-sediments, meta-volcanics, and
gneisses of high metarmorphic grade, and the other consisting
of meta-volcanics and intrusive rocks of mainly ensimatic
character and generally of very low to low metamorphic grade
(the volcano-sedimentary-ophiolite greenschist assemblage of
Vail, 1983). The high grade rocks appear in discontinous
outcrops to the West of the low-grade assemblage which forms
most of the Arabian-Nubian neocraton (ANNC; Stoeser and Camp,
1980), but also in tectonic windows inside its western (for
example: Hafafit area, southern Egypt; Elbayoumi and Greiling,
1984) and southern parts (Eritrea: Mohr, 1979; Jemen and
Somalia: Warden, 1981; Warden and Daniels, 1984; etc.). Farther
south, in Eastern Uganda, Kenya and Tanzania comparable rocks
are widely exposed within the northern part of the Mozambique
belt. There is a general consensus now that the boundary
between the two major units is tectonic, probably produced by
low angle thrusting (nappes), and that it represents a major
structural and metamorphic break. This situation is variably
interpreted as an Andean type continental margin of the East
Sahara Craton nearly contemporaneous with the volcanic arcs of
the ANNC (Elbayoumi and Greiling, 1984), or as a Lower to
Middle Proterozoic basement which was only passively involved
in the Pan-African evolution (Almond, 1984; Stoeser and Camp,
1985). This argument will be further discussed below.

In view of the near-continuity of high-grade metamorphics from
Kenya to southern Egypt (Vail, 1983) it is tempting to see the
Egyptian outcrops as the northern continuation of the
Mozambique belt along strike (Cahen, 1961; Hepworth, 1979;
Pohl, 1979). For comparison, a concise summary of Mozambique
belt geology in Kenya is here presented.

THE MOZAMBIQUE BELT AND ITS MINERALISATION

Geology

Although the Mozambique belt was first defined in 1948 by Holmes (1951), it is incompletely understood until now in spite of appreciable mapping and scientific endeavor carried out since.

Main descriptive features (see also Table 1 and Fig. 1) of the Mozambique belt include the ubiquitous Pan-African K/Ar-arges, (~600 Ma), the easterly dipping thrusts along its western orogenic front against the Tansania craton, the high grade and polyphase metamorphism generally of upper amphibolite facies, the presence of granulites, charnockites and anorthosites within the belt representing a lower crustal environment, and characteristic polyphase folding. Incipient migmatisation is ubiquitous, but associated deep level granites are rare. Post-tectonic high level Pan-African granites appear only in the South (Mozambique) and in the North.

In SE-Kenya, a variegated miogeosynclinal shelf succession with meta-carbonates, -volcanics and -clastic rocks (Kurase Group) may be differentiated lithologically from monotonous gneisses of intermediate composition possibly representing meta-volcanics and meta-greywackes including intrusives (Kasigau Group; Pohl et al., 1980). Orthoamphibolites are frequent in both groups, while felsic meta-volcanics are rare in the first and not mapped until now in the Kasigau Group. Intermediate magmatic rocks are more widespread than thought earlier. The sediments may have been deposited in the late Middle and early Upper Proterozoic II. Elsewhere, the occurrence of large tracts of pre-Mozambiquian basement comprising mainly Usagaran-Ubendian rocks metamorphosed at about 1900 Ma in Tanzania (Gabert and Wendt, 1974) and Kibaran granulites and gneisses in Mozambique (R. Sacchi et al., 1984) has to be noted.

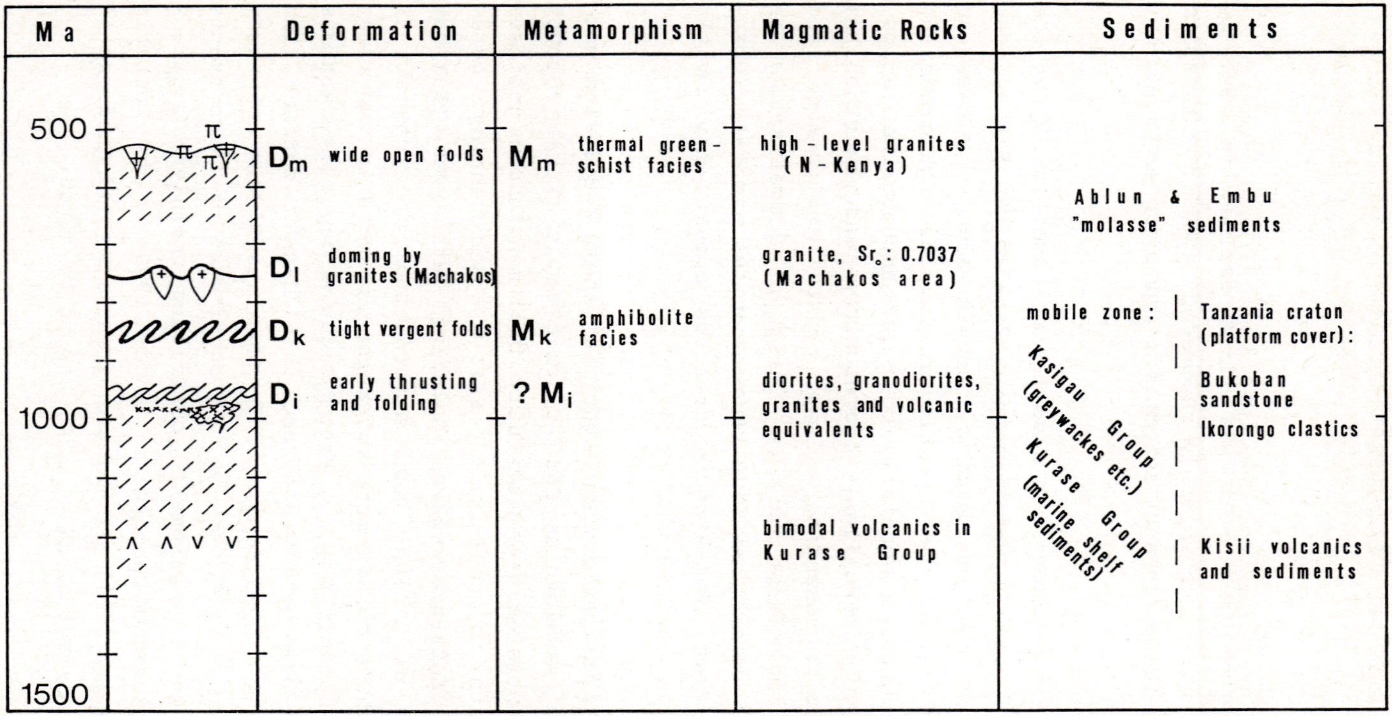

FIGURE 1.

Schematic time column of Mozambique belt development in

E-Africa (age data mainly from Cahen et al., 1984)

The main deformation and metamorphism of the Mozambique belt in East-Africa appear to have occurred at about 840 Ma, although this date is not well defined (Cahen et al., 1984). It was apparently followed by tectono-thermal events around 774± 14 Ma (accompanied by granite intrusion) and 600 Ma (ibidem).

The Mozambique belt contains numerous usually small ultramafic bodies, frequently associated with mafic rocks and generally concordant to strike and dip of the country rocks. In SE-Kenya, these occurrences form long pearl-strings which may be interpreted as marking folded thrusts or sutures (Pohl and Niedermayr, 1979). These rocks are deformed and metamorphosed to the same degree as their country rocks. Some of them may be tectonically reduced ophiolites (Shackleton, 1977; Pohl, 1979; Pohl et al., 1980; Vearncombe, 1983; Frisch and Pohl, in print) while others, which are associated with granulites, are possibly derived from subcontinental mantle (Prochaska and Pohl, 1983).

Lithologies, admittedly rare Rb/Sr-ages (Cahen et al., 1984), deformational pattern, metamorphic history, and mineral deposits (see below) are remarkably similar in Kenya (Pohl et al., 1980), in the Sudanese Bayuda desert (Meinhold, 1979; Ries et al., 1985), and in windows of high grade rocks exposed within the low-grade domains (e.g. Elbayoumi and Greiling, 1984). Hesitations to call these latter areas "Mozambiquian" are mainly due to conflicting interpretations of age and nature of the Mozambique belt (Almond, 1984). This should not be allowed, however, to prevent recognition of the virtual identity (Cahen, 1971; see also Stoeser and Camp, 1985).

Mineralisation

In view of the heterogeneous nature of the Mozambique belt, mineral deposits will be of widely varying age and origin. Economically important mineralisation is rare, but this may be partly due to a relatively low level of modern exploration. Prospective geological environments are numerous as well as minor mineralisations, and this should encourage further mineral exploration programmes.

Minerals of possible economic significance in the NE-African Mozambique belt include (see also Pohl, 1984):

Table 1: Main descriptive features of the Mozambique belt in
          E-Africa

LITHOLOGY

Metamorphic rocks representing basement (?) and Mozambiquian
sediments; the latter include
a) variegated shelf-sediments with bimodal volcanics (Kurase
   Group)
b) belts of meta-greywackes (?) with basic-intermediate
   magmatic rocks (Kasigau Group)

METAMORPHISM

Polyphase metamorphism of amphibolite and - locally - granulite
facies; followed by high T / low P (sillimanite) and thermal
greenschist facies phase

STRUCTURES

D_i poorly preserved early ductile thrusting and folding; D_k
nearly homoaxial overprinting of folds with E-dipping axial
planes; D_l doming by granites (Machakos area); D_m wide open
folds with northerly plunge

ULTRAMAFICS

a) ophiolites near the western orogenic front (Vearncombe 1983)
b) numerous small metamorphosed tectonically reduced ophiolites
   (?) along thrusts
c) meta-dunites in granulite areas may be derived from subcontinental mantle

GRANITOIDS

Synorogenic diorites, granodiorites, granites are followed by
granite (gneiss) domes around 774 ± 14 Ma; post-orogenic
granites and pegmatites cluster around 600 Ma

AGES

Sediments may be late Middle Proterozoic II; main metamorphism
and deformation about 840 Ma

MINERALIZATION

Cr, Pt, Ni and magnesite in ultramafics; syngenetic Fe-Mn-Ba-
basemetal occurrences; Cu-occurrences in intermediate rocks;
metamorphogenic graphite, kyanite, gemstones and asbestos;
pegmatites with Nb/Ta, REE, Be, Li, muscovite and gemstones.

1. Syngenetic stratiform ores
   - Fe-quartzites in Kenya and Ethiopia;
   - magnetite associated with basic metavolcanics of the
   - Kasigau group (Kenya);
   - banded iron formations (Bur region, Somalia: Warden,
     1981);
   - Mn-quartzites with elevated base metal contents (Kenya,
     Red Sea Hills, Sudan); and
   - pyritiferous quartzites ± Cu.

2. Ophiolite related ores and minerals
   - Cr, Ni, Pt and Cu occurrences are known from a number of
     ultramafic/mafic bodies (for example West Pokot: Pulfrey
     and Walsh, 1969)
   - magnesite veinlets and stockwork bodies in dunites
     (Kinyiki Hill, Kenya; etc.);

3. Ores related to intermediate magmatic rocks
   - Cu-occurrences associated with dioritic-rhyolitic gneiss
     in subduction related arcs
     (for example Voi area: Frisch and Pohl, 1985)
   - magnetite/ilmenite in charnockitic gneiss, anorthosite and
     associated pegmatites (Pare Mts., Tanzania).

4. Metamorphogenic minerals
   These comprise a number of economically important
   commodities: flake graphite and kyanite in meta-sediments,
   amphibole asbestos in ultramafics; green gem-quality
   vanadium grossularite in calcsilicate graphite schists, and
   tanzanite (blue zoisite) in impure marbles (Pohl et al.,
   1980).

5. Pegmatite ores and minerals
   Mozambique belt pegmatites comprise clearly two different
   groups, including conformable, folded bands and layers of
   metamorphic mobilisates, and undeformed (but often
   cataclased) crosscutting veins. It is quite probable, that
   even within these sub-groups several periods of formation

will be discovered by future research, although at present most known ages of cross-cutting pegmatites cluster around 600 Ma (Cahen et al., 1984). Minerals and ores formerly exploited in numerous small mines include (Du Bois and Walsh, 1970):

- beryl, columbo-tantalite, Mn-tantalite, microlite, cassiterite, lepidolite, spodumene, bismuth, bismuthinite and U/REE minerals
- muscovite, vermiculite, quartz, feldspar; and, presently only of economic significance:
- coloured gemstones including tourmaline, amazonite, emerald, ruby and sapphire (Tanzania, Kenya: Pohl and Niedermayer, 1979; red corundum at Hafafit, Egypt).

THE ARABIAN-NUBIAN NEOCRATON AND RELATED UNITS

Geology

The Arabian shield consists of an assemblage of at least five microcontinents of predominantly ensimatic origin separated by belts of ophiolites (Stoeser and Camp, 1985). Some of these structural units have been identified in the Nubian Shield (Vail, 1983 and 1984).

Several of the terranes in the East and South appear to have continental cores, possibly of Middle or Lower Proterozoic age (Stacey and Hedge, 1984), in addition to the widespread volcano-sedimentary and plutonic rocks of ensimatic or Andean arc character. This continental basement should better not be called Mozambiquian, as in Stoeser and Camp (1985), because it is probably pre-Mozambiquian in view of the emerging Late Proterozoic age of the Mozambique belt s.s. (Cahen et al., 1984; etc.).

Rifting of this older basement and the opening of Arabian-Nubian oceanic basins occurred before 950 Ma, when the first subduction related rocks were formed (see Fig. 2). Basic and

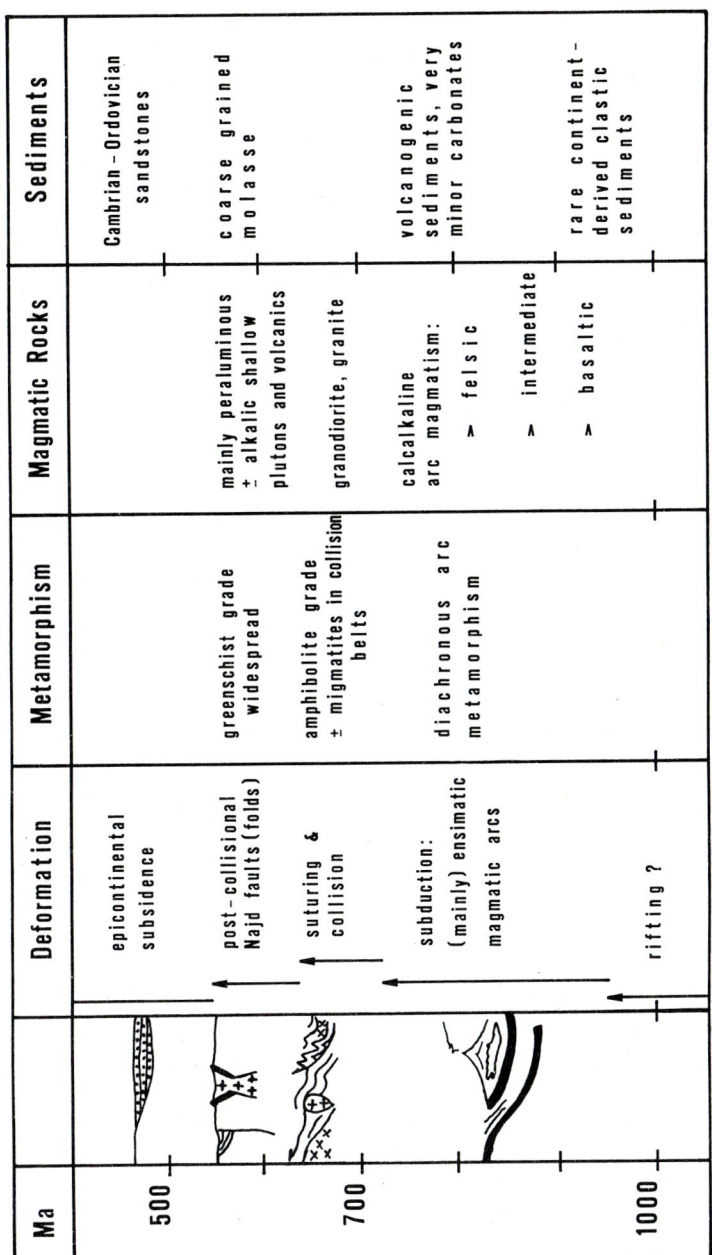

FIGURE 2.

Panafrican evolution of the Arabian-Nubian Shield (mainly after Stoeser and Camp, 1985)

intermediate lithologies are predominant. Arc activity ceased
by accretion and suturing of the microcontinents and
continental collision in the East and West during the period
between 715-640 (610 in Egypt?) Ma producing orogenic belts
marginal to the ANNC. Post-collisional features comprise
folding, transcurrent faulting (Najd system), greenschist
metamorphism, molasse sedimentation, and intracratonic
magmatism including siliceous peraluminous and alkalic volcanic
and plutonic rocks dated from 640-550 Ma. Little later,
epicontinental Cambrian and Ordovician sandstones covered the
region (Stoeser and Camp, 1985).

The coincidence of many ages of Mozambiquian and Arabian-Nubian
arc rocks suggest at least partially coeval evolution. In the
Nubian Shield, epicontinental sedimentary basins, passive and
active continental margins (Meinhold, 1979) must have existed
prior to about 610 Ma, when continental collision ceased and
cratonisation was achieved. The amphibolite grade metamorphism
and the early deformation in the Hafafit area, Egypt, for
example, is clearly older than the nappes with ophiolites and
arc components overthrust from the East (Elbayoumi and
Greiling, 1984). In the Bayuda desert, the upper amphibolite
grade metamorphism of metasediments has been dated at
761 ± 22 Ma (Ries et al., 1985), which coincides with a phase
of granite intrusion, metamorphism and deformation in Tanzania
and Kenya (Maboko et al., 1985; Shibata and Suwa, 1979).
Obviously, the evolution of the Mozambique belt and the ANNC
was independent prior to their collision before about 600 Ma.
However, both units reflect a generally convergent tectonic
regime from 950 to 600 Ma, which produced before about 700 Ma
subduction derived, mainly ensimatic rocks in the ANNC and ±
coeval collisional features in the Mozambique belt.

Mineralisation

At present the Arabian-Nubian Shield is not an important source
of mineral commodities. Most notable of course, is the
historical significance of gold (Hume, 1937), with new

activities promising a revival of former mining. Some platinum
(Ethiopia) and chromite (Ingessana, Sudan) are also produced.
A large number of prospects and mineral occurrences are known,
however, and many will certainly be developed when metal
markets will have recovered from the present depression.

More important types of mineral deposits in the Arabian-Nubian
Shield include

1. Sygenetic stratiform ores and minerals
   - magnetite/hematite with BIF characteristics, probably of
     volcanogenic-hydrothermal origin (Egypt, Saudi Arabia);
     some of these ferruginous banded cherts contain gold
     (Johnson and Vranas, 1984);
   - magnesite in sedimentary carbonates (Saudi Arabia: Bokhari
     et al., 1981);
   - Mn-Zn-Cu-barite lenses in continent derived meta-sediments
     (Southern Arabian Shield: Johnson and Vranas, 1984);

2. Ophiolite related ores and minerals
   - Cr and Pt in ultramafic members (Egypt, Sudan, Ethiopia);
   - magnesite veinlets and stockwork bodies in dunite and
     serpentinite;
   - talc, both as huge masses of low-grade talc-carbonate
     rock, and forming high-grade deposits locally in shear
     zones (Egypt).

3. Deposits related to arc magmatism
   Ensimatic and Andean type arcs, although only partially
   mapped in time and space, are the most prominent
   metallogenetic environment of the Arabian-Nubian Shield.
   Some of the deposits associated with them are
   - replacement magnetite ($\pm$ Cu) ores and veins at contacts of
     gabbros, diorites and granodiorites;
   - stockwork and proximal massive sulfide deposits associated
     with acidic subvolcanic domes and breccias containing
     Cu, Zn, Pb, Au and Ag (for ex. Jabal Sayid, Saudi Arabia);

- more distal, lenticular Zn-Pb-Cu sulfide beds of a more
  hydrothermal-sedimentary character and with bands of
  exhalites, calc-dolomite and graphitic tuffs (e.g. Nuqrah,
  Saudi Arabia);
- quartz veins and stockwork bodies with Au-Ag
  mineralisation within or surrounding acidic subvolcanic
  intrusions (e.g. Mahd Ad Dhahab, Saudi Arabia).

4. Mineralisation associated with post-orogenic magmatism
   In view of the complex evolution of the shield, post-
   orogenic conditions may have been reached in one area, while
   elsewhere orogenic deformation was still active. Roughly,
   however, accretion and suturing were accomplished at about
   640 Ma. Magmatic rocks formed from 640 to about 550 Ma
   include granodiorites, monzogranites, alkali-feldspar
   granites including alkali granite, and syenites, often with
   the equivalent volcanic rocks; especially in the Southern
   Shield, layered gabbroic massifs are frequent (Stoeser,
   1986). Intrusion levels were shallow, and because of
   restricted erosion over much of the area many still form
   subvolcanic ring complexes (often bimodal - Vail, 1985b) or
   other caldera related structures (Roobol and White, 1986).
   During the same period, the Najd fault system was
   intermittently active, greenschist metamorphism took place,
   and coarse molasse sediments were deposited.

   Important types of mineralisation are:
   - Ta/Nb, Sn, Be and W associated with small cupolas of
     highly evolved and/or altered (albitisation,
     greisenisation) granites; the deposits include
     disseminations within the cupolas, marginal pegmatitic
     phases, and external quartz veins and stockworks (e.g. Abu
     Dabbab, Egypt);
   - occurrences of Nb, Zr, Y, REE, U and Th are related to
     alkali granites and their pegmatitic-hydrothermal suite
     (for ex. Jabal Said, Saudi Arabia: Hackett, 1986);
   - ilmenite/magnetite in layered mafic complexes;

- scheelite with quartz and calc-silicates in hornblendite and amphibolite of the Baish group occurs at Wadi as Salile (Saudi Arabia: Johnson and Vranas, 1984). The deposit is in immediate vicinity of a post-tectonic muscovite-biotite granite, and resembles in many respects the Mittersill-Felbertal deposit of the Eastern Alps (Höll, 1975). There, as at Salile, a genetic relation to acidic magmatism has been questioned, but emerges as the more probable model.

- Au-quartz (-carbonate) veins with pyrite (± Ag, Cu, As, Pb, Zn etc.) are widespread in the shield; they are hosted by intrusive, volcanic and ophiolitic rocks, including post-tectonic granites, and ordinarily a direct relation to cooling intrusives is not visible (Al Shanti et al., 1978). Single veins are usually thin (<1m) and several 100m long, often, however, they form systems with a length of several kilometers and strong tectonic control. This is usually ductile or brittle shearing, which may have boudinaged the veins in some cases. The origin must be sought in large hydrothermal systems either induced by metamorphism or by cooling unexposed intrusions (Almond et al., 1984). Of special interest are Au-quartz veins transsecting ophiolites with well developed listwaenitisation of ultramafic country rocks (Barramiya, Egypt) indicating high CO_2-contents within the fluids (Buisson and Leblanc, 1985). This appears to confirm an origin according to the greenschist metamorphogenic model (Boyle, 1979).

The regional distribution pattern of mineral deposits in the Arabian-Nubian Shield was described at different scales by a number of authors (among others: Garson and Shalaby, 1976; Al Shanti et al., 1978; Sillitoe, 1979; Pohl, 1979 and 1984; Delfour, 1980; Jackson, 1986). Especially important appears to be the observation, that some lithophile element concentrations (Sn, W, Be etc.) occur in a marginal zone only comprising parts of the Nubian, Southern and Eastern Arabian Shields surrounding the central shield, which itself is deficient in these metals

but characterised by Cu (Zn, Ag, Au, etc.) and Th, U, REE and
Nb. A similar pattern could be deduced concerning the existence
of older continental crust or of continent-derived sediments
(Stoeser and Camp, 1985). Obviously, this coincidence is not
unexpected. Based on the recognition of the terrane
configuration in the Arabian Shield (Stoeser and Camp, 1985),
some of these were outlined across the Red Sea and metallogenic
domains described (Vail, 1985a).

DISCUSSION AND CONCLUSIONS

A comparative analysis of main mineralisation types in the
Mozambique belt and the volcano-sedimentary greenschist-
ophiolite assemblage of the Arabian-Nubian Shield produces
a number of similarities as well as differences (Fig. 3).
Mineralisations restricted to one of both large units include:
- the Mozambiquian metamorphogenic group;
- Ti/magnetite in anorthosites and charnockites of the
  Mozambique belt;
- part of the pegmatites (gemstones like ruby and sapphire, but
  also mica deposits are only exploited in the Mozambique
  belt), and
- the mineralisation associated with A-type post-orogenic
  magmatism in the Arabian-Nubian Shield, although some overlap
  into Mozambiquian terrain occurs in Sudan, Ethiopia and
  Somalia.

Broadly comparable, however, are the following metallogenic
environments:
- stratiform and partly BIF-type iron ores;
- stratiform Mn-barite-basemetal occurrences within continent-
  derived, shelf type sedimentary suites, suggestive of rift or
  passive margin situations;
- the ophiolite association, not only concerning the
  - necessarily - limited list of ores and minerals, but also
  in view of their comparable low grades and quantities:

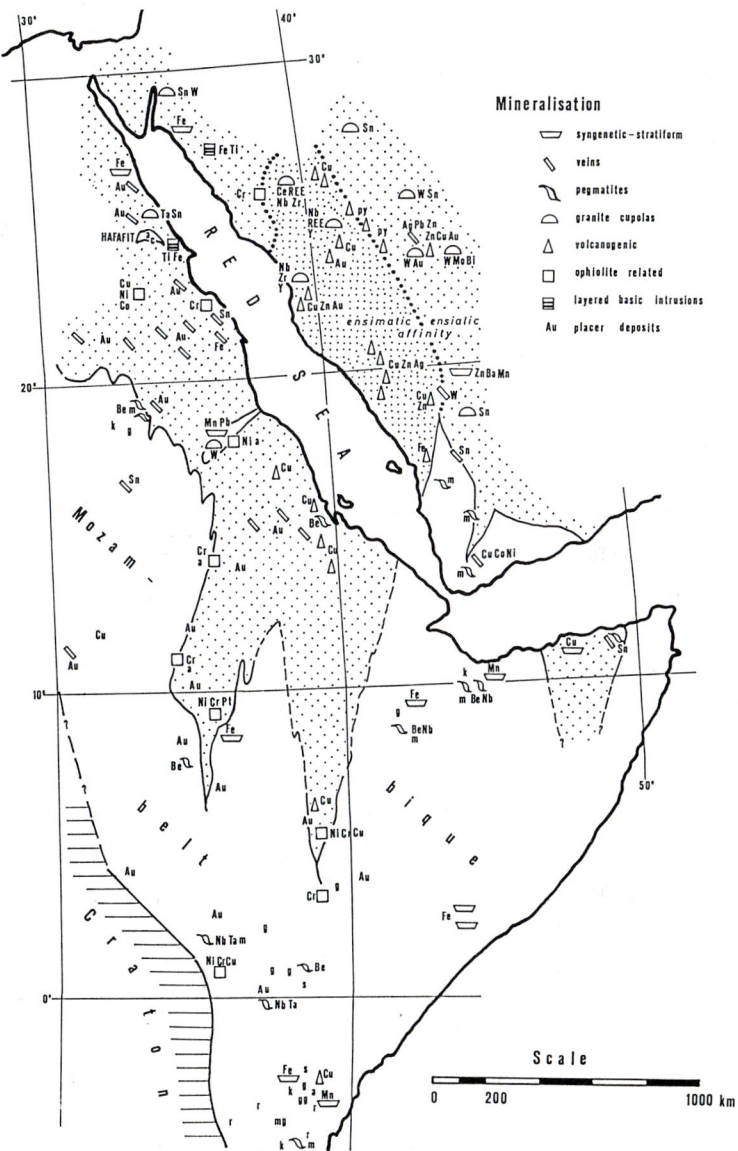

FIGURE 3.

Metallogenic sketch map of NE-Africa
(data from Pohl, 1984; Jackson, 1986; Johnson and Vranas, 1984)
Abbreviations: a - asbestos, c - corundum, g - graphite, gg - gem grossularite, k - kyanite, m - mica, py - pyrite, r - ruby, s - sapphire; element symbols for metallic ores.

Stippled: volcano-sedimentary   greenschist-ophiolite assemblage

- subduction related intermediate/acidic magmatism with related
  ores is better known in the Arabian Shield at present, but
  there are indications that this environment is much more
  frequent and prospective in the Mozambique belt than thought
  earlier (Frisch and Pohl, 1985); and
- part of the pegmatites containing Be, Sn, Nb/Ta, U, Th, REE,
  etc.

These observations obviously qualify earlier models proposing
widely uncomparable metallogenic features of both units (Pohl,
1979 and 1984; Vail 1985a). They strengthen, however, the
argument for close geodynamic relations between the Mozambique
belt and the Arabian-Nubian greenschist assemblage.

Based on the sketchy knowledge of the vast Mozambique area, the
belt is at present best understood as a mosaic of various
terranes and microplates, of which many contain older basement,
with active and passive margins, and a network of sutures
marked by ophiolites (Pohl, 1985; Frisch and Pohl, 1985;
Shackleton, 1986). Polyphase amphibolite and granulite
metamorphism occurred from Tanzania (Maboko et al., 1985)
through Kenya (Shibata and Suwa, 1979) into Sudan (Ries et al.,
1985) between 900-700 Ma, well within the time of Arabian-
Nubian arc activity. This generally convergent tectonic regime
would have ended earlier in the Mozambique belt, thus allowing
erosion before overthrusting of low grade volcano-sedimentary
units and ophiolites from the (present) East (Elbayoumi and
Greiling, 1984). Any near-surface mineral deposits within the
Mozambique belt would have been removed by this erosion, and
this may be part of the explanation for its paucity of major
metal mining. Comparison with other Mid- to Late Proterozoic
fold belts (Australia, Canada, Finland), however, refutes this
argument, and encourages future prospecting.

The spatial distribution of the A-type granitoid suite and
associated mineralisation within and surrounding the Arabian-
Nubian greenschist assemblage, but unknown in the Kenya-
Tanzanian Mozambique belt may be understood by late Proterozoic

collisoional crustal thickening of the first area and consequent radiogenic heating (Sillitoe, 1979) or perhaps more probable, as a consequence of the formation and further evolution of megaliths (Ringwood, 1986) in the upper mantle of the area. One group of lithophile element concentrations (Sn, Be, W, etc.) forms a ring around an area in the central Arabian Shield, where Nb, Zr, Y, REE, U and Th are more often enriched. This appears to confirm the long-standing metallogenic model of a close association of Sn-deposits with evolved continental crust, and is in agreement with petrogenetic indications, that A-type granites have deep sources in the lower crust. Upper crustal S-type granites (Chappel and White, 1974) have only been indentified in the eastern part of the Arabian Shield (Stoeser, 1986). Obviously the Arabian-Nubian Shield acquired a weak continental metallogenic signature in spite of its mainly ensimatic origin through an evolution spanning some 350 Ma. The size of its lithophile element deposits is generally small, and it is tempting to relate this to the short and monocyclic history of much of the Arabian-Nubian continental crust.

The understanding of regional metallogenic units (provinces, districts, areas) has advanced considerably within the Arabian-Nubian Shield (Vail, 1985a). Much less detailed is the knowledge concerning the Mozambique belt (Pohl, 1984). Accordingly, mineral exploration as well as scientific efforts should be directed towards the latter, especially the magmatic arc environments prospective for copper, basemetals and gold. Within metamorphic volcano-sedimentary suites suggestive of crustal extension, stratiform manganese-basemetal-barite deposits of the Broken Hill-type (Australia) constitute an interesting target. Curiously, Cyprus-type sulfide deposits appear to be insignificant in ophiolites of both Mozambique belt and the Arabian-Nubian greenschist assemblage, as well as carbonate-hosted lead, zinc, etc. within the vast areas of Mozambique shelf suites. The latter may also warrant exploration for phosphate and Veitsch-type magnesite (Pohl and Siegel, 1986).

REFERENCES

Almond, D.C., 1984. The concepts of "Pan-African Episode" and "Mozambique Belt" in relation to the geology of East and North-East Africa.-Bull.Fac.Earth Sci. K.A.U. 6 (1983), 71-87.

Almond, D.C., F. Ahmed and M.Z. Shaddad, 1984. Setting of gold mineralization in the Northern Red Sea Hills of Sudan. Economic Geol. 79, 389-392.

Al-Shanti, A.M.S., W. Frisch, W. Pohl and M.M. Abdel Tawab, 1978. Precambrian Ore Deposits in the Nubian and the Arabian Shields and their Correlation across the Red Sea.-Österr. Akad. Wiss. Schriftenr. Erdwiss. Komm. 3, 37-44, Wien.

Bokhari, M.M., P.L. Binda and J.H. Schellekens, 1981. Evidence of former evaporites in late Precambrian carbonate rocks of Jabal Rokham, Saudi Arabia.-Bull.Fac.Earth Sci. K.A.U. 4 (1981), 149-158.

Boyle, R.W., 1979. The geochemistry of gold. - Geol. Survey Can. Bull. 280, 584 pp.

Buisson, G. and M. Leblanc, 1985. Gold in carbonatized ultramafic rocks from ophiolite complexes. Economic Geol. 80, 2028-2020.

Cahen, L., 1961. Review of geochronological knowledge in Middle and Northern Africa. In: Geochronology of Rock Systems. Ann. New York Acad. Sci. 91, 535-566.

Cahen, L., N.J. Snelling, J. Delhal and J.R. Vail, 1984. The geochronology and evolution of Africa. Oxford Science Publ., Clarendon Press, Oxford.

Chappell, B.W. and A.J.R. White, 1974. Two contrasting granite types. Pacific Geol. 8, 173-174.

Delfour, J., 1980. Geologic, tectonic and metallogenic evolution of the northern part of the Precambrian Arabian Shield (Kingdom of Saudi Arabia). - Bull. BRGM II/1-2, 1-20, Orleans

Du Bois, C.G.B. and J. Walsh, 1970. Minerals of Kenya.-Geol. Survey Kenya Bull. 11, 82 pp, 24 maps, Nairobi.

Elbayoumi, R.M.A. and R. Greiling, 1984. Tectonic evolution of a Panafrican plate margin in sotheastern Egypt -suture zone overprinted by low-angle thrusting. In: African Geology, J. Klerkx and Michot, eds., 47-56, Tervuren.

El-Shatouri, H.M. and M.L. Al-Eryani, 1979. Notes on the mineral distribution maps of the Arab Republic of Yemen.-Bull.Fac.Earth Sci. K.A.U. 3/1, 121-129.

Frisch, W. and W. Pohl, 1985. Petrochemistry of some mafic and ultramafic rocks from the Mozambique belt, SE-Kenya. Mitt.Öst.Geol.Ges. 78 (1985), 1986. Festschrift W.E. Petraschek, 97-114, 1986.

Gabert, G. and I. Wendt, 1974. Datierung von granitischen Gesteinen im Dodoman und Usagaran System und in der Ndembera Serie (Tanzania).-Geol.Jb. 11, 3-55, Hannover.

Garson, M.S. and I.M. Shalaby, 1976. Precambrian-Lower Paleozoic plate tectonics and metallogenesis in the Red Sea region.-Geol.Ass.Canada Sp.P. 14, 573-596.

Habib, M.A., A.A. Ahmed and O.M. Elnady, 1985. Two orogenies in the Meatiq area of the central Eastern Desert, Egypt. Precambrian Research 30, 83-111.

Hackett, D., 1986. Mineralized aplite-pegmatite at Jabal Sa'id, Hijaz region, Kingdom of Saudi Arabia. J.Afr.Earth Sci. 4, 257-267.

Hepworth, J.V., 1979. Does the Mozambique Orogenic Belt continue into Saudi Arabia? Bull.Fac.Earth Sci. 3/1, 39-52.

Holmes, A., 1951. The sequence of Pre-Cambrian orogenic belts in south and central Africa.-Int.Geol.Congr. XIIIV, 14, 254-269.

Höll, R., 1975. Die Scheelitlagerstätte Felbertal und der Vergleich mit anderen Scheelitvorkommen in den Ostalpen. Abh. Bayer. Akad. Wiss. math.naturw. Kl. N.F. 157, A-B, 114 pp, München.

Hume, W.F., 1937. The minerals of Economic Value. Vol. II part II of Geology of Egypt, Cairo.

Jackson, N.J., 1986. Mineralization associated with felsic plutonic rocks in the Arabian Shield. J.Afr. Earth Sci. 4, 213-227.

Johnson, P.R. and G.J. Vranas, 1984. The geotectonic environments of Late Proterozoic mineralisation in the southern Arabian Shield. Precambrian Research 25, 329-348.

Maboko, M.A.H., N.A.I.M. Boelrijk, H.N.A. Priem and E.A.Th. Verdurmen, 1985. Zircon U-Pb and biotite Rb-Sr dating of the Wami River granulites, Eastern granulites, Tanzania: Evidence for approximately 715 Ma old granulite facies metamorphism and final Panafrican cooling approximately 475 Ma ago. Precambrian Research 30, 361-378.

Meinhold, K.-D., 1979. The Precambrian basement complex of the Bayuda desert, Northern Sudan.-Rev.Géol.Dyn. Géogr.Phys. 21, 395-401, Paris.

Mohr, P., 1979. Lithology and structure of the Precambrian rocks of Eritrea. Bull.Fac.Earth Sci. K.A.U. 3/2, 7-16.

Petrascheck, W.E., 1965. Typical features of metallogenic provinces. Economic Geol. 60, 1620-1634.

Pohl, W., 1979. Metallogenic/minerogenic analysis - contribution to the differentiation between Mozambiquian basement and Panafrican superstructure in the Red Sea region. - Ann.Geol.Survey Egypt 9, 32-44, Cairo.

Pohl, W., 1981a. Is the Kashebib Group older basement to the Greenschist Assemblage; - Some thoughts resulting from the Port Sudan meeting. IGCP-164 Newsletter, 4, p. 44-48, Jeddah.

Pohl, W., 1984. Large-scale metallogenetic features of the Precambrian in North-East Africa and Arabia. Bull.Fac.Earth Sci. K.A.U. 6, (1983), 591-601.

Pohl, W., 1985. Geologie Zentral- und Ostafrikas, ein neuer Anfang; Mitt. TU Carolo-Wilhelmina Braunschweig, 20, 1, 33-37.

Pohl, W., A. Horkel, W. Neubauer, G. Niedermayr, R.E. Okelo, J.K. Wachira and W. Werneck, 1980. Notes on the geology and mineral resources of the Mtito Andei - Taita Hills area (Southern Kenya). Mitt. Öst. geol. Ges. 73, 135-152, Wien.

Pohl, W. and G. Niedermayr, 1979. Geology of the Mwatate Quadrangle and the Vanadium Grossularite Deposits of the Area (with a geological map 1:50.000 by W. Pohl).-Kenya Geol.Survey Rpt. 101, 55 pp, 13 figs., Nairobi.

Pohl, W. and W. Siegl, 1986. Sediment-hosted magnesite deposits. In: Handbook of strata-bound and stratiform ore deposits, K.H. Wolf ed., Vol. 14, pp 223-310, Elsevier V., Amsterdam.

Prochaska, W. and W. Pohl, 1983. Petrochemistry of some mafic and ultramafic rocks from the Mozambique Belt, Northern Tanzania. J. Afr. Earth Sci. 1, 3/4, 183-191.

Ries, A.C., R.M. Shackleton and A.S. Dawoud, 1985. Geochronology, geochemistry and tectonics of the NE Bayuda desert, N Sudan: implications for the western margin of the late Proterozoic fold belt of NE Africa. Precambrian Research 30, 43-62.

Ringwood, T., 1986. Dynamics of subducted lithosphere and implications for basalt petrogenesis. Terra Cognita 6, 67-77.

Roobol, M.J. and D.L. White, 1986. Cauldron-subsidence structures and calderas above Arabian felsic plutons; a preliminary survey. J.Afr.Earth Sci. 4, 123-134.

Routhier, P., 1980. Où sont les metaux pour l'avenir; Mém.BRGM 105, 1-408, 97 figs., 14 tables, Orléans.

Sacchi, R., J. Marques, M. Costa and C. Casati, 1984. Kibaran events in the southernmost Mozambique Belt. Precambrian Research 25, 141-159.

Schuiling, R.D., 1967. Tin belts on the continents around the Atlantic Ocean. Economic Geol. 62, 540-550.

Shackleton, R.M., 1977. Possible late-Precambrian ophiolites in Africa and Brazil. Ann. Rep. Res. Inst. Afr. Geol. Univ. Leeds 20, 3-7.

Shackleton, R.M., 1986. Precambrian collision tectonics in Africa. In: M.P. Coward & A.C. Ries (eds), 1986, Collision Tectonics, Geol. Soc. Spec. Public. 19, 329-349.

Shibata, K. and K. Suwa, 1979. A geochronological study on granitoid gneiss from the Mbooni Hills, Machakos area, Kenya. 4th Prelim. Rept. Afr. Studies, Nagoya Univ., 163-167.

Sillitoe, R.H., 1979. Metallogenic consequences of Late Precambrian suturing in Arabia, Egypt, Sudan, and Iran. Bull.Fac.Earth Sci. KAU 3/1, 110-120.

Stacey, J.S. and C.E. Hedge, 1984. Geochronologic and isotopic evidence for early Proterozoic continental crust in the eastern Arabian Shield. Geology 12, 310-313.

Stoeser, D.B., 1986. Distribution and tectonic setting of plutonic rocks of the Arabian Shield. J.Afr.Earth Sci. 4, 21-46.

Vail, J.R., 1983. Pan-African curstal accretion in north-east Africa. J.Afr.Earth Sci. 1, 285-294.

Vail, J.R., 1984. The nature of the Basement Complex of the Nubian Shield in north-east Africa: addendum. J.Afr.Earth Sci. 2, 389-390.

Vail, J.R., 1985a. Relationship between tectonic terrains and favourable metallogenic domains in the central Arabian-Nubian Shield. Trans.Instn. Min. Metall. 94, B1-B6.

Vail, J.R., 1985b. Alkaline ring complexes in Sudan. J.Afr.Earth Sci.3, 51-59.

Vearncombe, J.R. 1983. A dismembered ophiolite from the Mozambique Belt, West Pokot, Kenya. J.Afr.Earth Sci. 1,.133-143.

Warden, A.J., 1981. Correlation and Evolution of the Precambrian of the Horn of Africa and Southwestern Arabia.-Unpublished PhD-Thesis Mining Univ. Leoben, 348 pp, 2 plates, 86 figs., 21 tables, Leoben.

Warden, A.J. and J.L. Daniels, 1984. Evolution of the Precambrian of Northern Somalia. Bull.Fac.Earth Sci. K.A.U. 6, (1983), 145-164.

Geophysics

The following Chapter 12 on geophysical aspects of crustal evolution in Egypt by Makris et al. is necessarily concentrated on the present tectonic situation.

However, crustal thickness and structure in eastern Egypt (apart from limited Cenozoic faulting) are the product of Pan-African orogeny. Tectonic activity after the Pan-African event was rather limited and the Arabian-Nubian Shield remained stable from the deposition of early platform sediments in Cambrian and Ordovician times to the onset of Cenozoic rifting and Red Sea opening.

The most important 'Pan-African' information appears to be the about normal crustal thickness, which may point to the fact that no orogenic root persisted beneath the Pan-African orogen in Egypt. Both generally low metamorphic grade at the present surface and relatively early platform sediments preclude substantial uplift during and after Pan-African orogeny. Consequently, frontal collision and associated crustal thickening may be less likely than lateral accretion without major crustal thickening. Possible mechanisms are discussed, for example, by Church (Chapter 10).

The newly formed Pan-African crust in eastern Egypt was remarkably stable during the Phanerozoic. In contrast, subsidence and basin formation took place in western Egypt, perhaps where the margin of a pre-Pan-African craton could be sought, which acted as a weak zone.

Chapter 12

Some Geophysical Aspects of the Evolution and Structure of the Crust in Egypt

J. Makris[1] / R. Rihm[1] / A. Allam[2]

[1] Institute of Geophysics, University of Hamburg, Bundesstr. 55, D-2000 Hamburg 13, FRG.
[2] Helwan Institute of Astronomy and Geophysics of Egypt, Helwan, Egypt.

Keywords: NE Africa, Egypt, geophysics, seismic section, Bouguer gravity maps, crustal structure

Abstract: The crustal structure of Egypt was established from 10 deep seismic sounding profiles in 1978 and 1981 by the Institute of Geophysics, University of Hamburg, and the Helwan Institute of Astronomy and Geophysics. The profiles are located at the Mediterranean coast, the Siwa Uplift, parts of the Western Desert (the Cairo-Bahariya region), the Red Sea coast, the Gulf of Suez area and the northern Red Sea. The seismic results were used to constrain density models computed along two 2-D sections, one trending N-S from the Western Desert to Sidi Barrani on the Mediterranean coast and the Second in a nearly E-W direction from the Libyan Desert through the Eastern Desert to the Red Sea.

The results obtained along the first profile show an almost N-S stretched continental crust changing in thickness from 34 km below the Western Desert to about 28 km at the Mediterranean coast. The transition from the continental crust to the oceanic crust of the Herodotus Basin is offshore Egypt. The sediments accumulated due to the stretching and subsidence of this passive margin are 6 to 8 km thick at the Mediterranean coast of Egypt, thickening northwards to more than 10 km off the Sidi Barrani Alexandria shoreline. The igneous part of the crust has been reduced to approximately two third of its original thickness over a length of 300 km. The velocity structure is that of a typical continental crust with a normal upper mantle velocity Pn = 8.0 km/s, since the stretching and subsidence episode is very old.

Along the second, E-W cross-section, the crustal thickness changes from approx. 34 km at the Western Desert to approx. 38 km at the Red Sea mountains. The actual increase in thickness of the igneous part of the crust, if we exclude the sediments, is from approx. 28 km below the Western Desert to 38 km below the Red Sea mountains. The crust at the Red Sea coast is only 20 km thick and at a distance of 25 km off the coast of Egypt the crust is oceanic and the thickness of its igneous part is only 8 km. The transition from continent to ocean is therefore very abrupt and the evolution of this passive margin is quite different to what we have observed along the Mediterranean coast of Egypt. This difference may be caused by the fact that a lot of shearing along the Pelusium line has affected the Mediterranean coast, whereas the Red Sea coast is affected mainly by normal faulting.

The upper mantle velocity in the Red Sea and the Red Sea coast of Egypt is 7.5 km/s and therefore comparable to values established in other active rifts. The velocity attains a normal value of 8.0 km/s below the Red Sea mountains and at a very short distance from the coast.

INTRODUCTION

In the following brief account of the geophysical activities of the Institute of Geophysics, Hamburg University, and the Helwan Institute of Astronomy and Geophysics, we intend to present a summary of the crustal structure obtained on two seismic campaigns in 1978 and 1981. The seismic results will also be combined with gravity information published by the General Petroleum Corporation (GPC) of Egypt in 1980, and density models along two lines will be presented. The aim of this brief review is to discuss some aspects of the regional evolution of the area and point to problems which should be considered for future investigations.

LOCATION OF THE SEISMIC LINES AND ASSOCIATED GEOLOGICAL PROBLEMS

The seismic lines were located in different geological provinces, aiming to provide information on the deep structure of the crust and upper mantle and further understanding of the geometry and evolution of the different areas.

One of the profiles is located along the Mediterranean coast and the other perpendicular to it, from Sidi Barrani to El Alamein and from Sidi Barrani to the Oasis of Siwa, respectively. Our main task along these lines was to establish the crustal thickness from the Siwa Oasis to the Mediterranean coast and, in an east-west direction, to observe possible lateral changes of the crust parallel to the Mediterranean Sea. This was necessary, since practically nothing was known about the crustal geometry from Sidi Barrani to the south, apart from some speculative values reported by El Shazly (1977) and based on estimates made by Babaev (1968).

A further profile was observed from Cairo (the Maadi quarries south of Cairo) to the Oasis of Bahariya along a basement high which extends from Cairo up to Gebel Oweinat, limiting the

Qattara Depression to the south (El Shazly, 1977). Along this
NE-SW striking high position of the igneous basement we were
expecting good penetration of the seismic energy from the
"Maadi shots", since lateral changes of the sediments which are
also thin were expected to be minor. Finally, in the Red Sea
area one long profile was observed from Ras Gharib on the
western side of the Gulf of Suez to Ras Banas on the Red Sea
coast (see Fig. 1). This profile, which provided information on
the eastern coastal areas of Egypt, was supplemented by two
more profiles. Both are perpendicular to the Red Sea coast,
extending from Port Safaga to Qena, and from Quseir to Qus in
the Nile Valley. The latter of the two was extended for another
100 km eastwards into the Red Sea (Fig. 1), and provides
information on the transition between the Red Sea crust and the
Eastern Desert Precambrian continental basement which is
exposed between the Nile Valley and the Red Sea. Finally, one
last seismic line was observed parallel to the Sinai coast at
the Gulf of Suez. It provides information on the thickness of
the continental crust along the western Sinai, and supplements
the seismic line observed on the western side of the Gulf of
Suez from Port Safaga to Ras Gharib. Seismic studies did not
exist for the above-mentioned areas, with the exception of the
paper by Ginzburg et al. (1981) on the crustal thickness of the
Dead Sea Rift and the eastern coast of the Sinai.

In the following discussion, some results from these seismic
observations will be presented. They will be combined with
gravity information to produce two-dimensional crustal sections
for the density and velocity distribution along two profiles
characteristic for the crustal structures of Egypt.

THE BOUGUER MAPS OF EGYPT: SOME QUALITATIVE CONSIDERATIONS

In 1980 the General Petroleum Company (GPC), Cairo, Egypt,
published a set of Bouguer maps of Egypt on a 1:500,000 scale
by compiling all available gravity data. In Figure 2 a
simplified and redrawn version of this map is presented. Before

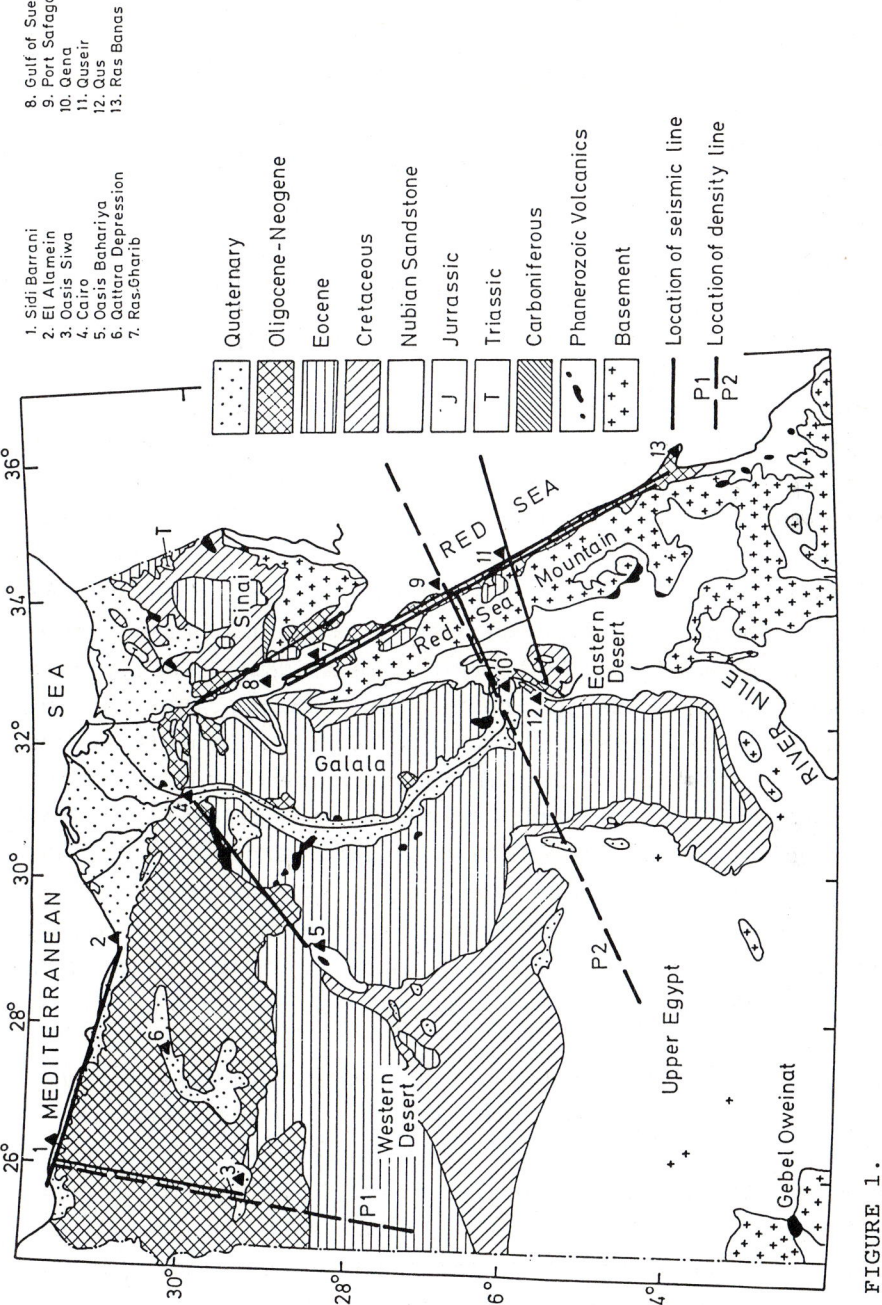

FIGURE 1.

Geological sketch map of Egypt and location of deep seismic lines and density models (modified from El Shazly, 1977).

discussing the anomalies and trends of the field, however, it is necessary to make some comments on the reliability of the gravity anomalies presented in this Bouguer map.

In interpolating the contour lines from the field observations, two conditions must be satisfied (Mundt, 1961):

-   Contour lines should not be drawn unless defined by field-points, and, between two points, preferably not more than one line should be interpolated.
-   The spacing interval of the contour lines should not be smaller than three times the mean error obtained at each gravity station.

These two conditions are only partially fulfilled in the map presented in Figure 2. Northern Egypt, i.e. the area north of $28^{O}N$, has been investigated intensively for hydrocarbons. This region is covered by densely spaced stations, and in most cases contouring of 2.5 mGal isolines is legitimate. The Eastern Desert, and in particular the mountains along the Red Sea coast, are covered by scanty observations limited to a few profiles. In some cases 5 mGal isolines can be interpolated reliably. In upper Egypt south of $26^{O}N$, the situation is very similar to that in the Eastern Desert and the Red Sea mountains. Here too, a 5 mGal spacing of the contoured isolines can only be accepted for some areas. Since only the regional trends of the anomalies will be discussed, the map presented in Figure 2 is drawn in 5 mGal isolines.

The anomalies have been computed according to the International Gravity Formula of 1967 (G.R.S., 1967). They are adjusted to the first order gravity net of Egypt (Kamel and Nahkla, 1978) and reduced with a uniform density of ρ = 2.67 gr/cm^{3} at sea-level. Topographic effects are in most cases very small and have not been considered. This however is an additional problem for the mountainous areas along the Red Sea coast where topographic effects can easily obtain values of 5 mGal.

BOUGUER GRAVITY MAP OF EGYPT

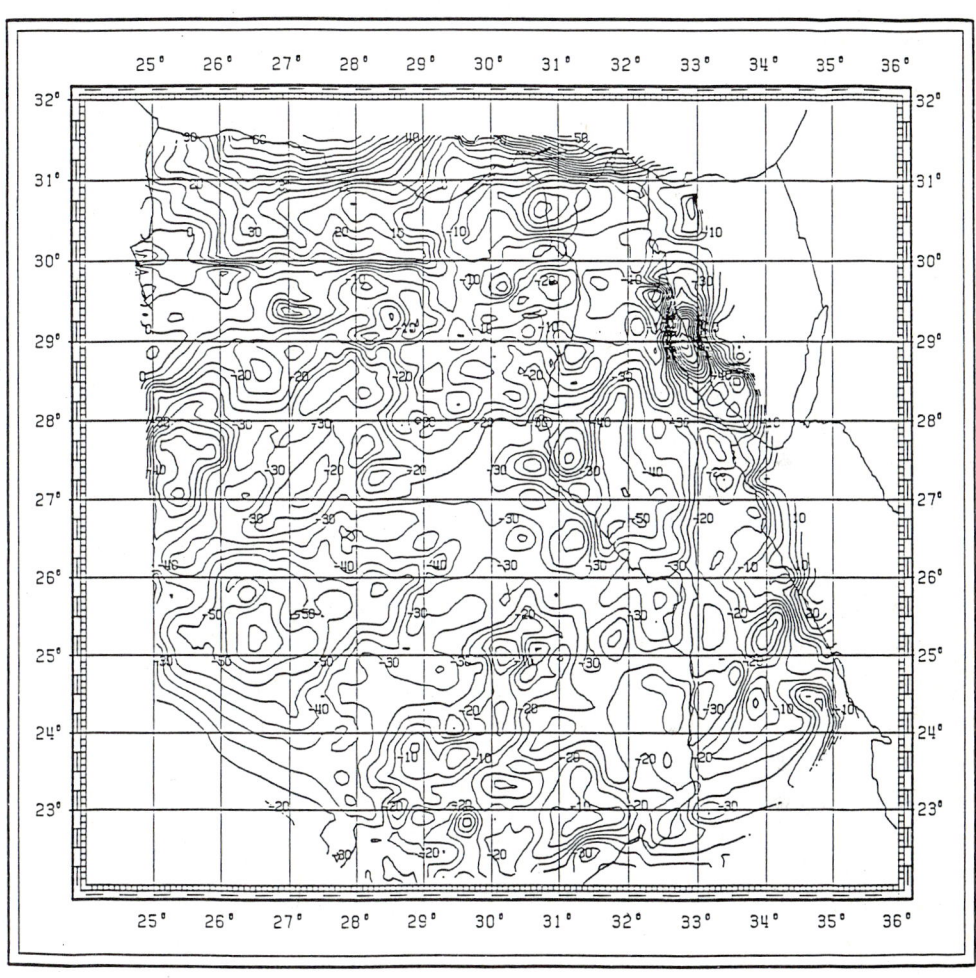

REFERENCE : 27.0 DEG
SCALE 1 : 8 000 000
CONTOUR INTERVAL = 5. MGAL

FIGURE 2.

Bouguer gravity map of Egypt. Gravity is reduced at sea level. Density of reductions is 2,67 g/cm^3 and the International Gravity Formula of 1967 has been used. No topographic reductions have been applied. This map is a simplified version of the 1:500,000 scale maps published by GPC Cairo (1980).

REGIONAL GRAVITY MAP OF EGYPT

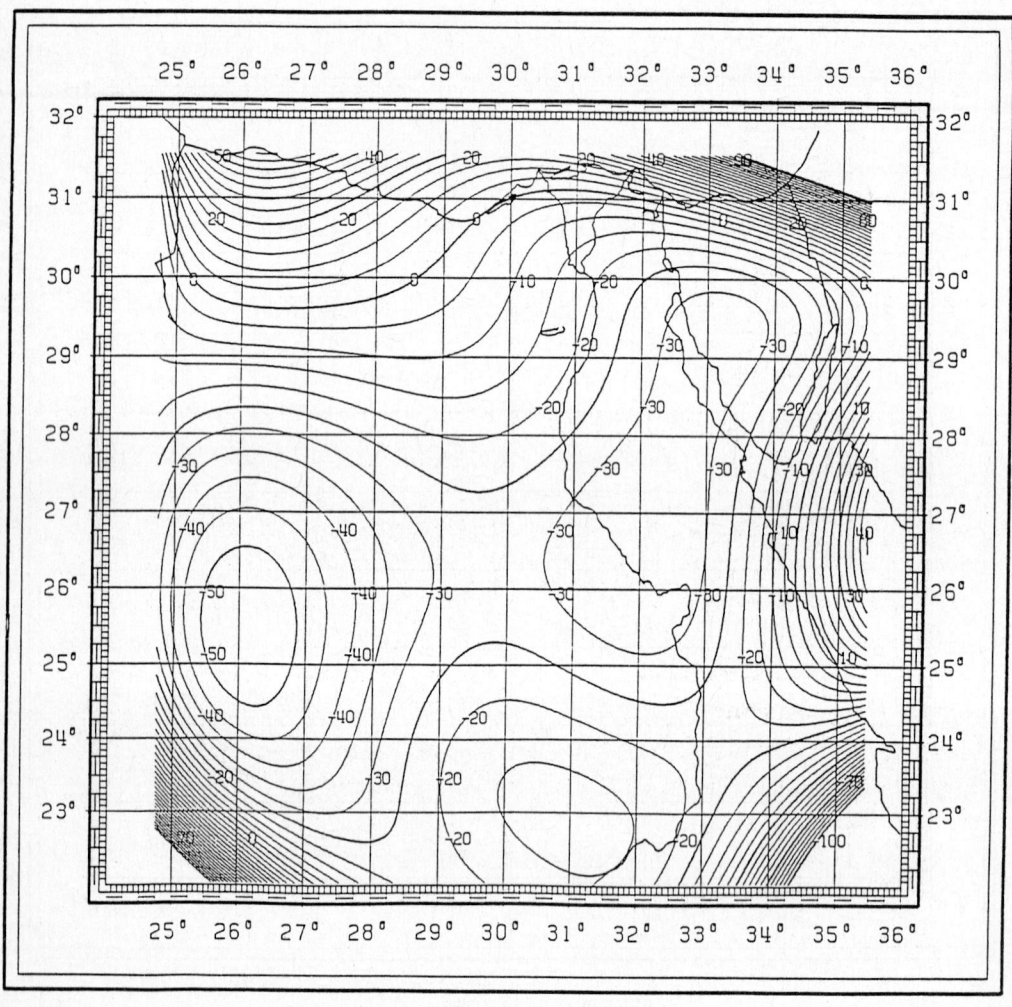

REFERENCE: 27.0 DEG
SCALE 1:8 000 000
CONTOUR INTERVAL 5. MGAL

FIGURE 3.

   Regional gravity anomaly map developed by orthogonal
polynomials of degree 5. See also Marzouk (1987).

The Bouguer anomalies range from +60 mGal in the coastal plains of the Mediterranean Sea to -60 mGal in the Western Desert (Upper Egypt). All other regions have anomalies of smaller amplitudes, and the topographic elevations have no correlation to the level of the gravity field. The highest mountains in the Red Sea ranges for example have values of over 1000 m, and their negative Bouguer anomalies are only about -10 to -30 mGal, aligned parallel to the Red Sea along a NNW trend. To the west of these anomalies from 26° to 28.5^{O}N and 31.5^{O} to 33°E, an elongated gravity minimum with values between -35 and -50 mGal marks the Galala Basin where almost 5 km thick sediments are limited to the east by the outcropping crystalline basement (Makris and Kebeasy, 1982). The elevation of this area is fairly smooth and only 200 to 400 m above sea-level.

The Gulf of Suez is also marked by a strong gravity minimum of approx. -55 mGal (Bayoumi, 1983) caused by the thick Miocene and pre-Miocene sediments which, in some parts of the Gulf, attain a thickness of nearly 6 km. In addition, the minimum in the Western Desert (24.5^{O} to 26°N and 25.5^{O} to 27.5^{O}E) marks an area with a considerable accumulation of sediments (Conoco, personal communication). The thick sediments along the Mediterranean coast (Said, 1962, El Shazly, 1977) are located on a positive gravity level. This indicates that the crustal thickness changes rapidly towards the Mediterranean Sea, and the positive gravity effect of the elevated dense upper mantle overcompensates the negative effect of the thickened low density sediments. This is confirmed by the seismic data from the next section, which will also be used to constrain the density models. Only at the Nile delta do the very thick sediments produce a pronounced gravity low with a slightly negative gravity level.

The trends of the gravity anomalies are aligned in a NE-SW direction, which marks an older tectonic episode of crustal evolution, in the western parts of Egypt and Upper Egypt. In eastern Egypt and the Gulf of Suez, the anomalies have a

NNW-SSE trend which is associated with the Miocene and post-Miocene opening of the Red Sea and the Gulf of Suez.

Although regional and local fields were computed by developing the observed Bouguer anomalies with orthogonal polynomials or filtering methods, we are reluctant to use them in order to delineate trends. Near-surface density contrasts and changes in crustal thickness cause gravity effects of comparable amplitudes and regional extension, and therefore produce anomalies which are not uniquely representative for deep or shallow structures. This can be seen in Figure 3, where the regional field of a 5th degree polynomial development of the Bouguer map is presented. The regional effect of a -30 mGal isoline which crosses the Galala Basin (the basement high of the outcropping Palaeozoic mountains of the Eastern Desert and the Gulf of Suez), contradicts obvious geological features which trend in a NNW-SSE direction and are associated with density and velocity changes in the upper mantle and deeper crust.

SUMMARY OF DEEP SEISMIC STUDIES AND CRUSTAL STRUCTURE

In Figure 1 the locations of all seismic lines shot in 1978 and 1981 are given. The complete results of the seismic evaluation will be presented elsewhere, and here only briefly reviewed. In Figures 4,5,6 and 7 four of the ten seismic lines are shown. Profiles 1 and 2 are from the Mediterranean coast and the Cairo-Bahariya section. Profiles 3 and 4 are from the Red Sea coast and the Red Sea mountains.

The Mediterranean coast is built by a stretched continental crust covered with 6 km thick sediments. The Pn-velocity for the compressional waves travelling along the Moho discontinuity has normal values of approximately 8.0 km/s. The crust attains a thickness of almost 30 km, and has an approximately 24 km thick igneous part. This structure was obviously formed by stretching and subsidence which produced space for the thick

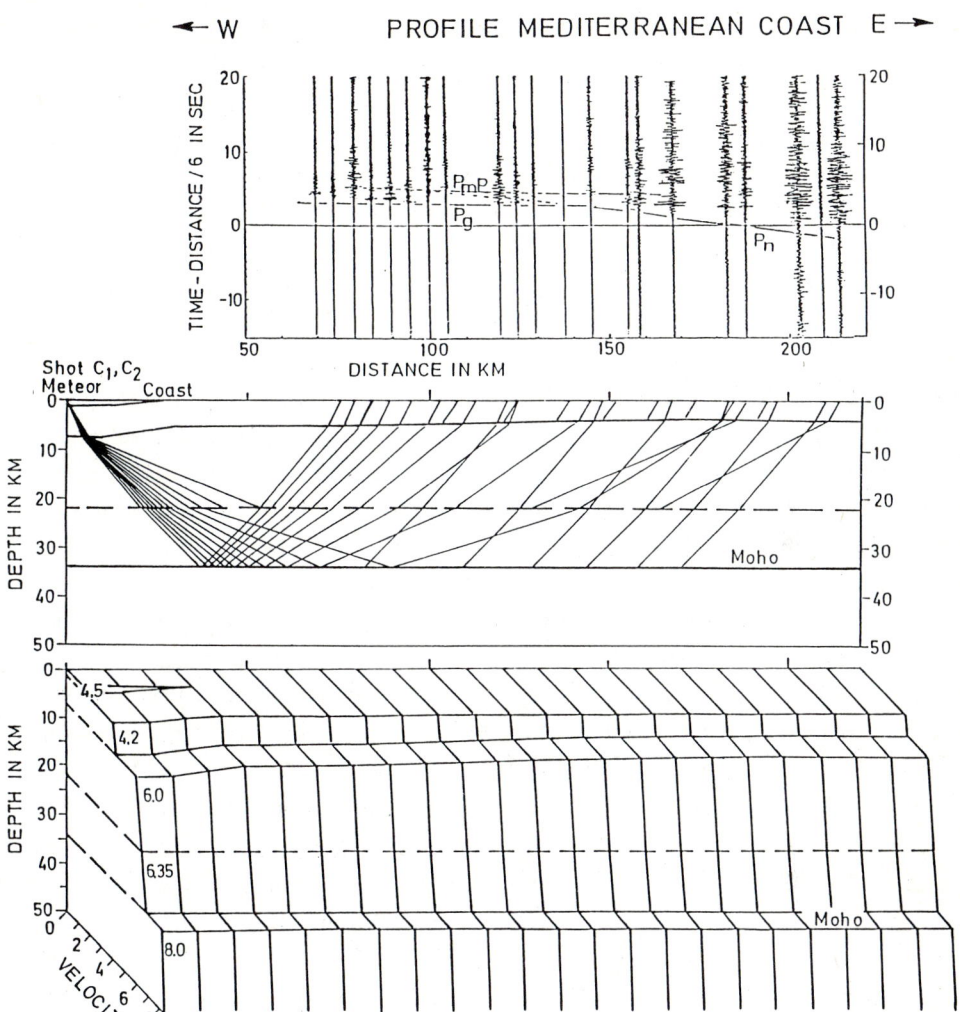

FIGURE 4.

Seismic observations and interpretation along the Mediterranean Sea. The data are reduced with 6 km/s.

sedimentary sequences to accumulate. The Cairo-Bahariya section
on the other hand is positioned on a basement high with a thin
Eocene cover whose thickness does not exceed 1.5 km. Here the
crustal thickness was also found to be about 32 to 34 km. The
Pn-velocity seems to be normal at V_p = 8.0 km/s. The igneous
crust is thicker than along the coastal part of northern Egypt,
since the sediments are thin and the remaining igneous portion
of the crust is approximately 4 to 5 km thicker than along the
Mediterranean coast.

Profile 3 (Fig.6) along the Red Sea coast has a total length of
240 km and provided information on the crustal structure and a
sub-Moho event. The crust is a 20 km thick stretched
continental. The direct wave, Pg-phase, has a velocity of
6.0 km/s, increasing slightly with depth. The Pn-velocity is
7.5 km/s, and thus distinctly lower than that observed in the
other areas of Egypt. This low velocity is identical to that
found in the Red Sea and is typical for active rifting
(Tramontini and Davies, 1969, Berckhemer et al., 1975, Ruegg,
1975, Makris et al., 1980, Makris et al., 1983b). The sub-Moho
event indicated with P^* in Figure 6 has a velocity of 8.3 km/s
and was generated at a first order discontinuity at a depth of
34 to 35 km. The petrological significance of this layer is not
clear. It seems to be identical to a high velocity layer
reported by Mooney et al. (1985) from the Arabian peninsula, or
by Prodehl (1985) from the Arabian peninsula and the Red Sea
north of the Yemen-Saudi Arabian border. Along this Red Sea
coastal profile S-waves were also observed with v_s velocities
of 3.4 km/s from the igneous basement and 3.9 km/s from the
Moho discontinuity. The Poisson ratio σ is therefore 0.26 for
the upper crust, which shows that the latter's composition is
continental and elastic behaviour normal. The upper mantle
however has a σ of approximately 0.3 and is typical for a
"softened", partially melted and mobilised upper mantle in a
tectonically active, rifting zone.

Profile 4 (Fig. 7) was observed perpendicular to the Red Sea
coast. It has a length of 150 km and is located between Quseir

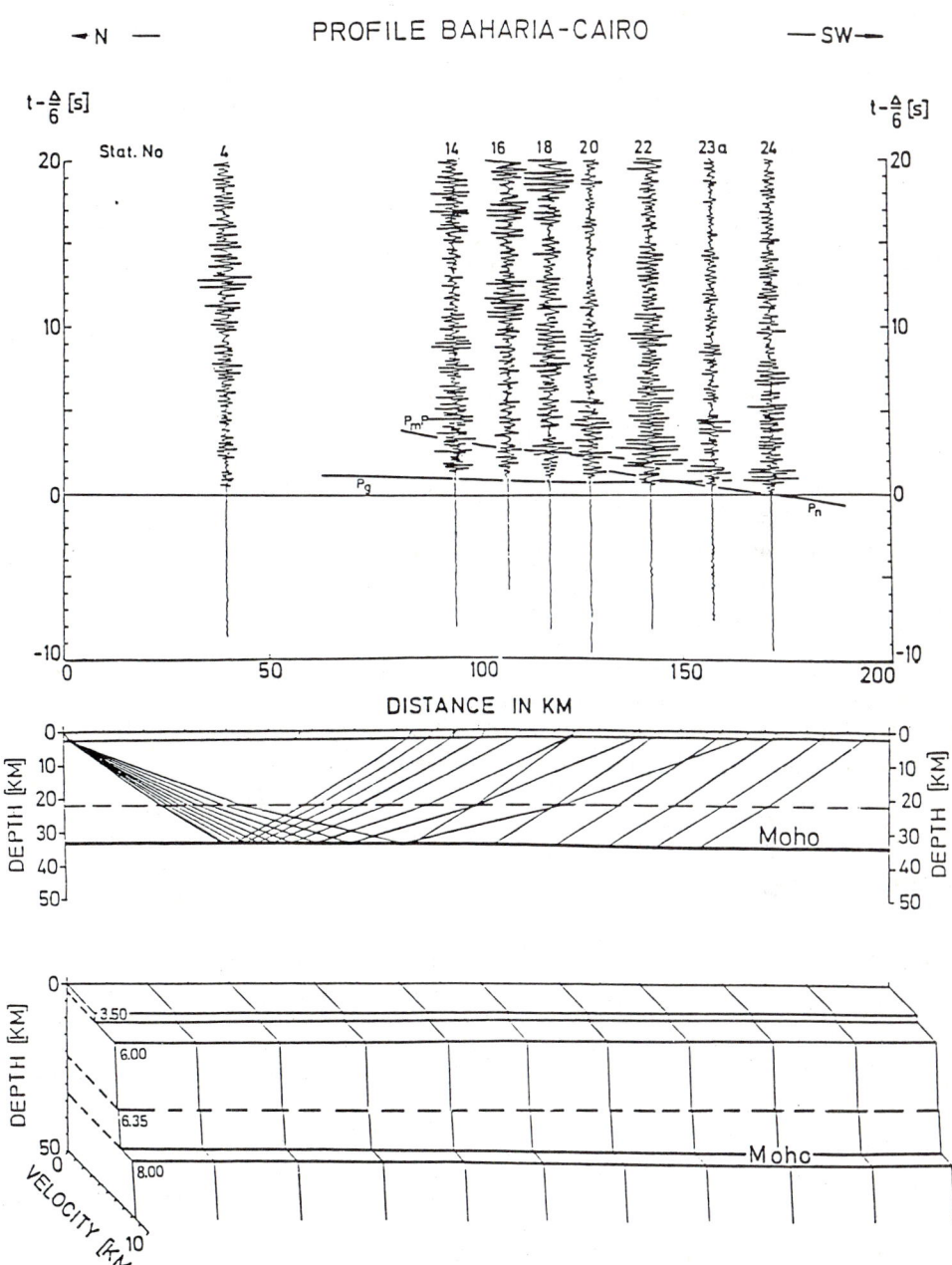

FIGURE 5.

Seismic observations along the Cairo-Bahariya profile, and seismic model.

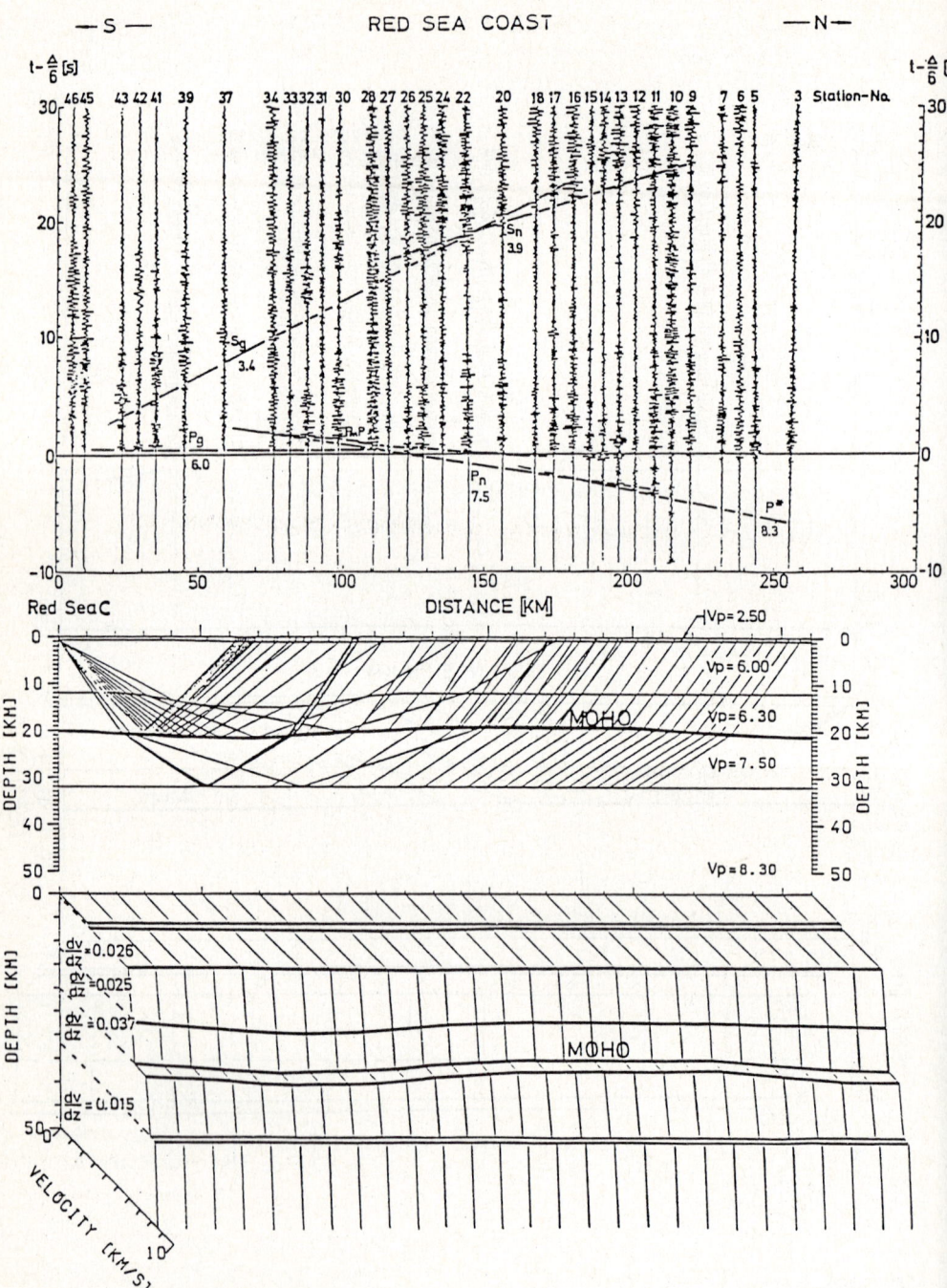

FIGURE 6.

Seismic observations along the Red Sea coast, and seismic model.

on the Red Sea coast and Qus on the River Nile. The seismic
energy generated by shots fired from the RV "Stefan-E" in the
Red Sea propagated very efficiently and produced Pg
(compressional seismic signals critically refracted at the top
of the igneous basement), PiP (wide-angle reflections generated
at the top of the lower crust), PmP (wide-angle reflections
generated at the Moho discontinuity) and Pn (compressional
seismic signals critically refracted at the base of the crust,
Moho discontinuity) seismic phases, as well as S-waves which
will not be discussed here. The evaluation of the section
showed that the continental Precambrian crust below the
elevated Red Sea ranges is 34 km thick with no sediments.
A true Pn-velocity below the high mountains could not be
established since the profile is unreversed. It was not
possible to find a locality along the River Nile suitable for
the firing of a large shot. The Pn-velocity observed however
seems to be between 7.8 and 8.0 km/s. The crust thins from east
to west from 34 km to 30 km. If we also take into consideration
the thickening sediments in the Galala Basin, which were found
by seismic and gravity observations (Makris and Kebeasy, 1982)
to be between 4.5 and 5.5. km thick at the gravity minimum, the
igneous part of the crust is only 26 to 27 km. This area was a
sedimentary basin in Jurassic-Cretaceous times and is now
covered by Jurassic-Triassic and Eocene formations (El-Shazly,
1977). Since the Red Sea mountains have long been exposed to
erosion and have no sediments, the original continental crust
of the Red Sea mountains must have been thicker prior to the
uplift associated with the Red Sea rifting.

TWO-DIMENSIONAL DENSITY MODELS

With the use of the velocity distribution and crustal geometry
from the seismic evaluation, density models for two cross-
sections were computed. The first cross-section is oriented N-S
along the seismic line from Sidi Barrani to Siwa. This line was
extended further south as far as $27^{\circ}N$, and additionally
constrained by drilling and sediment thicknesses obtained by

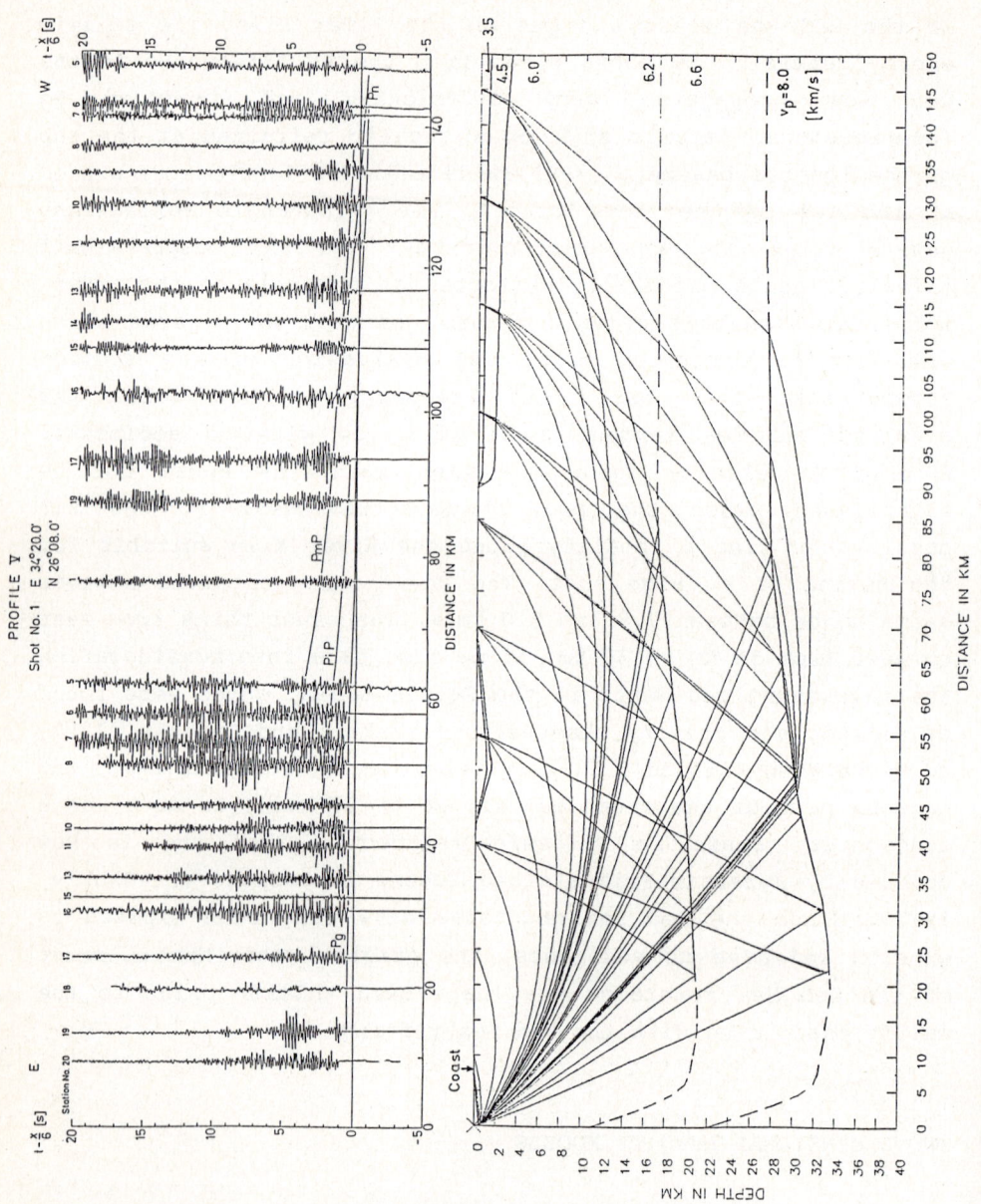

FIGURE 7.

Seismic observations perpendicular to the Red Sea, and seismic model across the Red Sea mountains.

reflection seismic measurements as given by El-Shazly (1977) and Said (1962). In this paper it will be referred to as P-1. The second profile is oriented nearly E-W and crosses from Upper Egypt to the Red Sea. It is constrained by seismic data in the Galala Basin, the Safaga-Quseir seismic line reported in the previous chapter (Makris et al., 1980) and by the seismic results from the Red Sea (Makris and Rihm, in prep.). This profile will be referred to as P-2 in this paper. In Figure 3 the locations of P-1 and P-2 are given.

The density values were constrained by compressional velocities using the Nafe-Drake empirical function for the sediments (Nafe and Drake, 1963) and the Birch relationship (Birch 1961) for the igneous crust and upper mantle.

The values are:

upper crust: v_p = 6.0 km/s ϱ = 2.82 gr/cm^3

lower crust: v_p = 6.4 km/s ϱ = 2.90 gr/cm^3

upper mantle: v_p = 8.0 km/s ϱ = 3.34 gr/cm^3

v_p = 7.5 km/s ϱ = 3.10 gr/cm^3

The densities of the sediments were averaged and range from 2.35 to 2.42 g/cm^3 for the Western Desert and northern Egypt to 2.37 g/cm^3 for the Red Sea. Obviously, these densities are not better than ± 5% of the true values, and their variation within the possible limits can slightly modify the thickness of the crustal models by 1 to 2 km. A variation of this kind is not, however, significant for the results and does not influence the arguments and tectonic implications.

In Figure 8 the density model along P-1 is presented. In the upper part the observed Bouguer values are shown by crosses, while the computed anomalies are indicated by a continuous line. The gravity difference along this 400 km long line is 100 mGal. The crustal thinning towards the Mediterranean coast

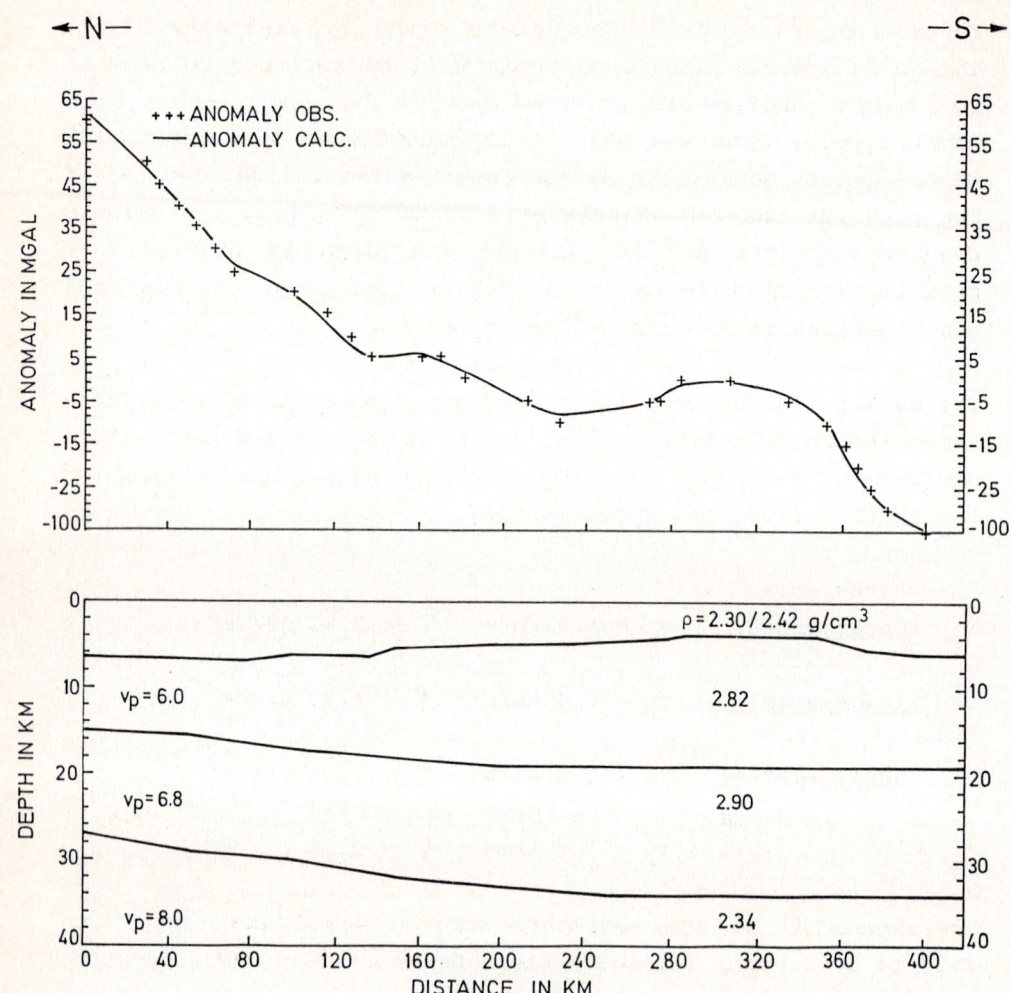

FIGURE 8.

Two-dimensional density model showing the crustal thickness and geometry along the Western Desert. The model is seismically constrained (see also Marzouk, 1987).

is of the order of 6 km. The actual crustal attenuation without the sediments can, however, be estimated if we consider that at km 320 the igneous crust is 30 km thick and thins to approx. 20 km at the coast. This means that the actual crustal stretching which produced this passive margin is considerable, and a complete reconstruction can only be obtained if the offshore continuation of the onshore structures and the limit between continental and oceanic crust can be identified. The distribution of oceanic crust and its limits have been located further to the east off the Sinai coast by Makris and Stobbe (1984) and Makris et al. (1983a). The situation north of the Egyptian coast at Sidi Barrani is not well defined, and the existing seismic data are of too poor quality to permit the identification of the crustal type below 8 to 10 km thick sediments.

A further geological complexity of this passive margin is the Qattara depression. The profile crosses this depression, which has a topographic expression of -120 m, at km 150 from the coast. As can be seen from the model P-1, the crust is not influenced by the depression and nothing particular can be seen in the upper mantle. The hypothesis that major shearing deformation across Africa (Neev et al., 1981) has produced the structures does not contradict the model presented here. This would also be in agreement with the results of Ginzburg et al. (1981) along the Dead Sea-Aqaba shear zone, where the depressions are developed as pull-apart basins due to the transcurrent movement of the Arabian relative to the Sinai blocks. In a similar way small basins also develop along the South Atlas shear zone in Morocco (Makris et al., 1986), having a morphological expression but no effect at the crust-mantle boundary or the upper mantle.

Along profile P-2 (Fig.9) the situation is very different to that described above. The crust thickens from 32 km in the Libyan desert to about 38 km below the Red Sea mountains, where the Precambrian basement is outcropping. The sediments are not very precisely defined since we have measured them only in the

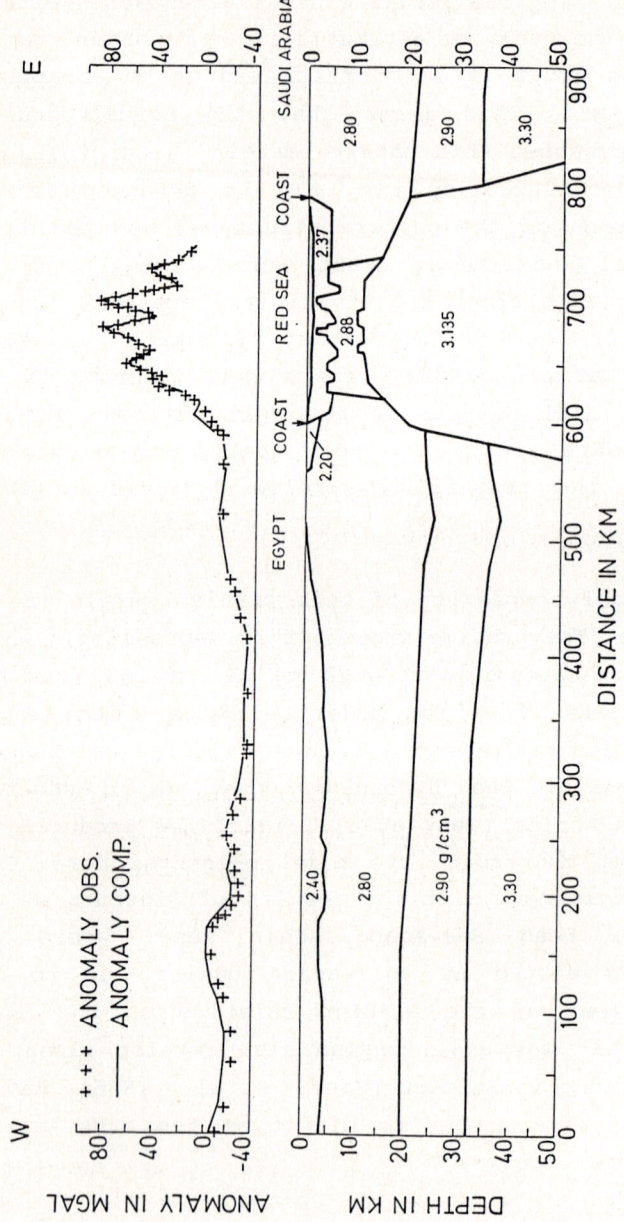

FIGURE 9.

Two-dimensional density model showing the crustal thickness and geometry from the Libyan Desert across the Eastern Desert and the Red Sea Mountains to the Red Sea. The model is seismically constrained.

Galala basin (Makris and Kebeasy, 1982). It seems, however, that in the Libyan desert their thickness is of the order of 4 km, and they are comparable to those in the Galala Basin. Further to the east, on the Red Sea coast, the changes in the crust, sediments and upper mantle are very intense! On the Red Sea coast the crust is still continental, only 20 km thick and, at least in some areas, covered by thick sediments. The density of the upper mantle is very low (ρ = 3.1 gr/cm^3) and in agreement with the low velocities observed along the coast. A few kilometres farther to the east the nature of the crust changes completely. We find an oceanic crust only 8 km thick - without the thick sediments - which seems to cover large parts of the northern Red Sea. The distribution of the oceanic crust is not symmetric to the morphology of the Red Sea basin. The Arabian Shield extends for quite long distances off the Arabian coast. This result, which has been defined by seismic measurements and confirmed by the gravity modelling, is in disagreement with Cochran's concept of a stretched continental crust below the northern part of the Red Sea (Cochran, 1983), but also with the model of Girdler and Styles, (1974) for a symmetrical northern Red Sea produced exclusively by sea-floor spreading. It is therefore obvious that poles of rotation so far published (e.g. Girdler and Styles, 1974, McKenzie et al., 1970) cannot be correct. Their models are too simple and quite clearly contradict the observations. This problem will be more extensively discussed elsewhere (Makris and Rihm, in prep.).

Finally, the lateral density variations in the upper mantle seem to be extremely intense in the coastal zone of the Red Sea - the low density material being confined exclusively to the rifted area. This result is very similar to the findings of Makris and Ginzburg (1987) for the southern Red Sea and Makris et al. (1974) and Berckhemer et al. (1975) in the Afar. This problem will also be discussed elsewhere in Makris and Rihm (in prep.).

SUMMARY AND CONCLUSIONS

The deep seismic profiles measured in Egypt in 1978 and 1981 and their interpretation in connection with the Bouguer maps of GPC have shown that the crust has an average thickness of approximately 30 km, thickening significantly to about 38 km below the Red Sea mountains in the Eastern Desert. The crust is stretched to the north, the igneous part thinning, while the sediments are in the process of thickening. Up to a distance of at least 300 km from the Mediterranean coast, this stretching affects the igneous part of the crust whose thickness changes from approx. 30 km to only 22 km. The development of this passive margin is very old, and has definitely affected the Jurassic sedimentation and eventually even the Palaeozoic.

To the east, along an east-west section from the Libyan Desert to the Red Sea, the sediments are thicker in distinct basins, e.g. in the Libyan Desert and the Eastern Desert-Galala Basin. They are bordered to the east by the Precambrian basement which outcrops along the Red Sea mountains, limiting the Eocene sedimentation to the east. The Red Sea coast and the Red Sea itself are mainly of Miocene age and present a young passive margin affected by fast subsidence and oceanisation. The thickness and nature of the crust change very rapidly at the Red Sea coast where it is seismically defined to be only 20 km thick and continental, followed offshore by a rapid transition to an oceanic type crust which developed at an early stage of rifting of the Red Sea.

The tectonic activity is confined today mainly to the Red Sea, while the Gulf of Suez rift and other areas are affected only by minor deformations.

ACKNOWLEDGEMENTS

The seismic studies were sponsored mainly by the Deutsche Forschungsgemeinschaft und supported by the Helwan Institute of

Astronomy and Geophysics of Egypt. In particular we wish to thank Mr. H. Kamel of the General Petroleum Company, Cairo, for many useful suggestions and encouragement during the experiments, and Prof. El-Shazly for logistic support.

The Federal German Ministry for Research and Technology (BMFT), Bonn, financed the charter of the RV-"Stefan-E".

REFERENCES

Babaev, A.G., 1968. Oil and gas prospects of the U.A.R. Unpublished report, General Petroleum Co., Cairo.
Bayoumi, A.J., 1983. Tectonic origin of the Gulf of Suez, Egypt, as deduced from gravity data. CRC, Handbook of Geophysical Exploration at Sea, eds. R.A. Geyer and J.R. Moore, p. 417-432.
Berckhemer, H., B. Baier, H. Bartelsen, A. Behle, H. Burkhardt, H. Gebrande, J. Makris, H. Menzel, H. Miller and R. Vees, 1975. Deep seismic soundings in the Afar region and on the highland of Ethiopia. In: Pilger, A. and Rössler, A. (edit.). Afar Depression of Ethiopia, Stuttgart (Schweizerbart).
Birch, F., 1961. The Velocity of Compressional Waves in Rocks to 10 Kilobars. J. Geophys. Res., 66, 2199-2224.
Cochran, J.R., 1983. A Model for Development of Red Sea. The American Association of Petroleum Geologists Bulletin, 67, 41-69.
El-Shazly, E.M., 1977. The Geology of the Egyptian Region. In: The Ocean Basins and Margins, vol. 4a, the Eastern Mediterranean, eds. A.E.M. Nairn, W.H. Kanes and F.G. Stehli.
General Petroleum Co., A.S.R.T., 1980. Bouguer gravity map of Egypt. Scale 1:500,000.
Geodetic Reference System, 1967. In: Publication speciale no. 3 of the International Union of Geodesy and Geophysics, Paris 1971.
Ginzburg, A., J. Makris, K. Fuchs and C. Prodehl, 1981. The structure of the crust and upper mantle in the Dead Sea Rift. Tectonophysics, 80, 109-119.
Girdler, R.W. and P. Styles, 1974. Two stage Red Sea floor spreading, Nature, 247, 1-11.
Kamel, H.A. and A.F. Nahkla, 1978. The National Gravity Standardisation Net (N.G.S.B.N. - 1977). General Petroleum Co., Cairo. Internal Report, pp. 85.
Makris, J., H. Menzel, J. Zimmermann and P. Gouin, 1974. Gravity Field and Crustal Structure of North Ethiopia. In: Pilger, A. and Rössler, A. (Edit.), Afar Depression of Ethiopia, Stuttgart (Schweizerbart).

Makris, J., W. Weigel, L. Möller, P. Goldflam, A. Behle, B. Stöfen, A. Allam, M. Maamoun, N. Delibasis, K. Perissoratis, F. Avedik and P. Giese, 1980. Deep seismic soundings in Egypt, parts 1 and 2. Unpublished internal report, Insitute of Geophysics, Hamburg University.

Makris, J. and R. Kebeasy, 1982. South Galala Refraction Seismic Project. Paper presented at the 6th Exploration Seminar of the Egyptian General Petroleum Corporation, Cairo.

Makris, J., Z. Ben-Abraham, A. Behle, A. Ginzburg, P. Giese, L. Steinmetz, R.B. Whitmarsh and S. Eleftheriou, 1983a. Seismic refraction profiles between Cyprus and Israel and their interpretation. Geophys. J.R. Astr. Soc. 75, 575-591.

Makris, J., A. Allam, T. Mokhtar, A. Basahel, G.A. Dehghani and M. Bazari, 1983b. Crustal Structure at the North-western Region of the Saudi-Arabian Peninsula and its transition to the Red Sea. In: Pan-African Crustal Evolution in the Arabian-Nubian Shield, First Symposium I.G.C.P. 164. Bull. Fac. Earth Sci., King Abdulaziz Univ., Jeddah, 6, 435-447.

Makris, J. and C. Stobbe, 1984. Physical properties and state of the crust and upper mantle of the eastern Mediterranean deduced from geophysical data. Marine Geology, 55, 347-363.

Makris, J., S.N. Qureshi and A. Demnati, 1986. Crustal Deformation and Tectonics of Morocco. Paper presented at the 30th Congress and Plenary Assembly of C.I.E.S.M., Palma de Mallorca.

Makris, J. and A. Ginzburg, 1987. The Afar Depression - Transition between Continental Rifting and Seafloor Spreading. In print, Tectonophysics.

Makris, J. and R. Rihm, in prep. Crustal structure of the northern Red Sea from seismic data. To be submitted to Tectonophysics.

Marzouk, I., in prep. A geophysical study of the crustal structure of Egypt based on seismic and gravity data. PhD. thesis, Hamburg University, 1987.

McKenzie, D.P., D. Davies and P. Molnar, 1970. Plate tectonics of the Red Sea and East Africa, Nature, 226, 243-248.

Mooney, W.D., M.E. Gettings, H.R. Blank and J.H. Healy, 1985. Saudi Arabian seismic refraction profile. A traveltime interpretation of crustal and upper mantle structure. Tectonophysics, 111, 173-246.

Mundt, W., 1964. Statistische Bearbeitung und Analyse geomagnetischer Landvermessungen. Deutsche Akademie der Wissenschaften zu Berlin. Geomagnetisches Institut Potsdam, Abhandlung Nr. 31. Akademie Verlag, Berlin.

Nafe, J.E. and C.L. Drake, 1963. Physical properties of marine sediments. In: The Sea, vol. 3, M.N. Hill (ed.). Interscience, pp. 794-814.

Neev, D., J.K. Hall and J.M. Saul, 1981. The Pelusium megashear system across Africa and associated lineament swarms. J. Geophys. Res., 86 (B12), 1015-1030.

Prodehl, C., 1985. Interpretation of a seismic refraction survey across the Arabian Shield in western Saudi Arabia. Tectonophysics 111, 247-282.

Ruegg, J.C., 1975. Main results about the crustal and upper mantle structure of the Djibouti region (T.F.A.I.). - In: Pilger, A. and Rösler, A. (eds.). Afar Depression of Ethiopia, Stuttgart (Schweizerbart).

Said, R., 1962. The Geology of Egypt. Elsevier Publ. Co., Amsterdam and New York.

Tramontini, C. and D. Davies, 1969. A seismic refraction survey in the Red Sea. Geophys. J.R. Astr. Soc., 17, 225-241.

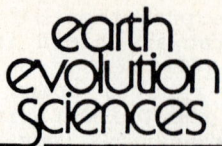

earth
evolution
sciences

International Monograph Series
on Interdisciplinary
Earth Science Research and Applications

Editor
Andreas Vogel, Berlin

Volumes available:

Rodney A. Gayer
**The Tectonic Evolution of the
Caledonide-Appalachian Orogen**

Andreas Vogel
Hubert Miller
Reinhard O. Greiling
The Renish Massif

Jean Pohl
**Research in
Terrestrial Impact Structures**

Chi-Yu King
Roberto Scarpa
Modeling of Volcanic Processes

Samir El-Gaby
Reinhard O. Greiling
**The Pan-African Belt of
Northeast Africa and Adjacent Areas**